VOLUME FIVE HUNDRED AND TWENTY SEVEN

METHODS IN ENZYMOLOGY

Hydrogen Peroxide and Cell Signaling, Part B

METHODS IN ENZYMOLOGY

VOLUME FIVE HUNDRED AND TWENTY SEVEN

METHODS IN ENZYMOLOGY

Hydrogen Peroxide and Cell Signaling, Part B

Edited by

ENRIQUE CADENAS and LESTER PACKER

Pharmacology & Pharmaceutical Sciences
School of Pharmacy
University of Southern California
Los Angeles, CA, USA

AMSTERDAM • BOSTON • HEIDELBERG • LONDON
NEW YORK • OXFORD • PARIS • SAN DIEGO
SAN FRANCISCO • SINGAPORE • SYDNEY • TOKYO
Academic Press is an imprint of Elsevier

Academic Press is an imprint of Elsevier
525 B Street, Suite 1800, San Diego, CA 92101-4495, USA
225 Wyman Street, Waltham, MA 02451, USA
Radarweg 29, PO Box 211, 1000 AE Amsterdam, The Netherlands
The Boulevard, Langford Lane, Kidlington, Oxford, OX5 1GB, UK
32 Jamestown Road, London NW1 7BY, UK

First edition 2013

ISBN: 978-0-12-405882-8
ISSN: 0076-6879

Printed and bound in United States of America
13 14 15 16 11 10 9 8 7 6 5 4 3 2 1

CONTENTS

Contributors xi
Preface xv
Volumes in Series xvii

Section I
H_2O_2 Metabolism: Determination of a Cellular Steady-State

1. The Cellular Steady-State of H_2O_2: Latency Concepts and Gradients **3**

H. Susana Marinho, Luísa Cyrne, Enrique Cadenas, and Fernando Antunes

1. Introduction 4
2. Experimental Components and Considerations When Measuring the H_2O_2 Gradient in *S. cerevisiae* Cells 9
3. Experimental Components and Considerations When Measuring the H_2O_2 Gradient in Mammalian Cell Lines 12
4. Data Handling/Processing 16
5. Summary 17

Acknowledgments 18
References 18

2. Evaluating Peroxiredoxin Sensitivity Toward Inactivation by Peroxide Substrates **21**

Kimberly J. Nelson, Derek Parsonage, P. Andrew Karplus, and Leslie B. Poole

1. Introduction 22
2. Materials 26
3. Measuring Inactivation Sensitivity by Steady-State NADPH-Linked Assays 27
4. Measuring Inactivation Sensitivity by Multiturnover Cycling with ROOH and DTT Followed by Mass Spectrometry Analysis 34
5. Measuring Inactivation Sensitivity of Prx1/AhpC Prxs Under Single Turnover Conditions Followed by Gel Electrophoresis 35
6. Conclusions/Summary 38

Acknowledgment 39
References 39

3. Peroxiredoxins as Preferential Targets in H_2O_2-Induced Signaling **41**

Lía M. Randall, Gerardo Ferrer-Sueta, and Ana Denicola

1. Introduction 42
2. Reaction of H_2O_2 with Cellular Thiols 43
3. H_2O_2 Diffusion Versus Reaction with Cellular Thiols 47
4. Sulfenic Acids as Signal Transduction Intermediates 49
5. Prxs as Preferential Targets for H_2O_2 51
6. Prxs as Primary H_2O_2 Sensors and Transducers 52
7. Prx–Protein Interactions are Needed to Transmit the Signal 52
8. Posttranslational Regulation of Prxs 55
9. Summary 57

Acknowledgments 58
References 58

4. Selenium in the Redox Regulation of the Nrf2 and the Wnt Pathway **65**

Regina Brigelius-Flohé and Anna Patricia Kipp

1. Some Historical Background for Introduction 66
2. Selenium Status and Selenoprotein Synthesis 68
3. Selenium Status and the Keap1/Nrf2 System 69
4. Selenium and the Wnt Pathway 72
5. Common Players and Events in Nrf2 and Wnt Signaling 75
6. How Does Selenium Come into Play? 78

References 80

5. Selenoprotein W as Biomarker for the Efficacy of Selenium Compounds to Act as Source for Selenoprotein Biosynthesis **87**

Anna Patricia Kipp, Janna Frombach, Stefanie Deubel, and Regina Brigelius-Flohé

1. Introduction 88
2. Experimental 90
3. Results 94
4. Discussion 102
5. Conclusions 108

Acknowledgments 109
References 109

6. Peroxiredoxins and Sulfiredoxin at the Crossroads of the NO and H_2O_2 Signaling Pathways **113**

Kahina Abbas, Sylvie Riquier, and Jean-Claude Drapier

1. Introduction 114
2. The Effect of NO on the Level of 2-Cys-Prx Overoxidation 116
3. Detection of Srx 122
4. Comments 123

Acknowledgments 125
References 125

7. Glutathione and γ-Glutamylcysteine in Hydrogen Peroxide Detoxification **129**

Ruben Quintana-Cabrera and Juan P. Bolaños

1. Introduction 130
2. Materials 131
3. Previous Considerations 132
4. Procedure with Purified Enzymes and Substrates 133
5. Analysis in Biological Samples 137
6. Further Applications: H_2O_2 Produced by NOS 140
7. Conclusions 140

Acknowledgments 141
References 141

8. Peroxiredoxin-6 and NADPH Oxidase Activity **145**

Daniel R. Ambruso

1. Introduction 146
2. Experimental Components and Considerations 149
3. Peroxiredoxin Activity of Prdx6 155
4. Effect of Prdx6 on NADPH Oxidase Activity 157
5. Summary 163

References 165

9. Study of the Signaling Function of Sulfiredoxin and Peroxiredoxin III in Isolated Adrenal Gland: Unsuitability of Clonal and Primary Adrenocortical Cells **169**

In Sup Kil, Soo Han Bae, and Sue Goo Rhee

1. Introduction 170
2. Hyperoxidation of PrxIII by H_2O_2 Generated During Corticosterone Synthesis 171

3. Induction of Srx by ACTH 173
4. Unsuitability of Clonal and Primary Adrenocortical Cells for Studies of the Srx–PrxIII Regulatory Pathway 174
5. Adrenal Gland Organ Culture as an *In Vitro* Model for the Srx–PrxIII Regulatory Pathway 176
6. Concluding Remarks 179
Acknowledgment 179
References 179

Section II
H_2O_2 in the Regulation of Cellular Processes in Plants

10. The Use of HyPer to Examine Spatial and Temporal Changes in H_2O_2 in High Light-Exposed Plants 185

Marino Exposito-Rodriguez, Pierre Philippe Laissue, George R. Littlejohn, Nicholas Smirnoff, and Philip M. Mullineaux

1. Introduction 186
2. Experimental Procedures 191
3. Pilot Experiments Using HL Stress 196
4. Conclusions 198
Acknowledgments 198
References 198

11. A Simple and Powerful Approach for Isolation of *Arabidopsis* Mutants with Increased Tolerance to H_2O_2-Induced Cell Death 203

Tsanko Gechev, Nikolay Mehterov, Iliya Denev, and Jacques Hille

1. Introduction 204
2. Generation and Isolation of Mutants More Tolerant to H_2O_2-Induced Oxidative Stress 206
3. Identification of Mutations in the Genome 209
4. Analysis of the Mutants with Enhanced Tolerance to H_2O_2-Induced Oxidative Stress 211
5. Conclusion 217
Acknowledgments 218
References 218

12. Analysis of Environmental Stress in Plants with the Aid of Marker Genes for H_2O_2 Responses **221**

Ayaka Hieno, Hushna Ara Naznin, Katsunobu Sawaki, Hiroyuki Koyama, Yusaku Sakai, Haruka Ishino, Mitsuro Hyakumachi, and Yoshiharu Y. Yamamoto

1. Introduction 222
2. Experimental Materials and Procedures 224
3. Example of Analysis 231

Appendix: Recipes of Stock Solutions, Buffers, and Media 235

References 236

13. The Role of Plant Bax Inhibitor-1 in Suppressing H_2O_2-Induced Cell Death **239**

Toshiki Ishikawa, Hirofumi Uchimiya, and Maki Kawai-Yamada

1. Introduction 240
2. Morphological Changes of Mitochondria Under ROS Stress 242
3. Assay for Inhibitory Effect of BI-1 on ROS Stress-Induced Cell Death Using Heterologous Expression System in Suspension Cultured Cells 246
4. Summary 253

References 253

14. Comparative Analysis of Cyanobacterial and Plant Peroxiredoxins and Their Electron Donors: Peroxidase Activity and Susceptibility to Overoxidation **257**

Marika Lindahl and Francisco Javier Cejudo

1. Introduction 258
2. Expression and Purification of Recombinant Prxs and Thioredoxins 259
3. Prx Activity Assays *In Vitro* 260
4. Peroxide Decomposition in Cyanobacteria *In Vivo* 264
5. Overoxidation of Plant and Cyanobacterial 2-Cys Prx 267
6. Concluding Remarks 272

References 272

15. Using Hyper as a Molecular Probe to Visualize Hydrogen Peroxide in Living Plant Cells: A Method with Virtually Unlimited Potential in Plant Biology **275**

Alejandra Hernández-Barrera, Carmen Quinto, Eric A. Johnson, Hen-Ming Wu, Alice Y. Cheung, and Luis Cárdenas

1. Introduction 276
2. NADPH Oxidase in Plant Cells 277
3. Plant Cells Respond to External and Internal Stimuli 279
4. Visualizing Hydrogen Peroxide in Living Plant Cells 280
5. Hyper as a New Genetically Encoded Probe 281
6. Vector Description and Plant Transformation 282
7. Preparation and Sterilization of Modified Petri Dishes for Growing *Arabidopsis* Plants for Microscopy Analysis 284
8. Seeds Sterilization and Stratification 284
9. Growth Conditions 285
10. Image Acquisition and Processing 286

Acknowledgments 288
References 288

Author Index *291*
Subject Index *313*

CONTRIBUTORS

Kahina Abbas
Institut de Chimie des Substances Naturelles, Centre National de la Recherche Scientifique, Gif-sur-Yvette, France

Daniel R. Ambruso
Department of Pediatrics, University of Colorado Denver, Anschutz Medical Campus, and Center for Cancer and Blood Disorders, Children's Hospital Colorado, Aurora, Colorado, USA

Fernando Antunes
Departamento de Química e Bioquímica and Centro de Química e Bioquímica, Faculdade de Ciências, Universidade de Lisboa, Lisboa, Portugal

Soo Han Bae
Yonsei Biomedical Research Institute, Yonsei University College of Medicine, Seodaemun-gu, Seoul, South Korea

Juan P. Bolaños
Institute of Functional Biology and Genomics (IBFG), Department of Biochemistry and Molecular Biology, University of Salamanca-CSIC, Salamanca, Spain

Regina Brigelius-Flohé
Department Biochemistry of Micronutrients, German Institute of Human Nutrition Potsdam-Rehbruecke, Nuthetal, Germany

Enrique Cadenas
Pharmacology & Pharmaceutical Sciences, School of Pharmacy, University of Southern California, Los Angeles, California, USA

Luis Cárdenas
Instituto de Biotecnología, Universidad Nacional Autónoma de México, Morelos, Mexico

Francisco Javier Cejudo
Instituto de Bioquímica Vegetal y Fotosíntesis, Universidad de Sevilla, CSIC IBVF(CSIC/US), Seville, Spain

Alice Y. Cheung
Department of Biochemistry and Molecular Biology; Molecular Cell Biology Program, and Plant Biology Program, University of Massachusetts, Amherst, Massachusetts, USA

Luísa Cyrne
Departamento de Química e Bioquímica and Centro de Química e Bioquímica, Faculdade de Ciências, Universidade de Lisboa, Lisboa, Portugal

Iliya Denev
Department of Plant Physiology and Plant Molecular Biology, University of Plovdiv, Plovdiv, Bulgaria

Ana Denicola
Laboratorio de Fisicoquímica Biológica, Instituto de Química Biológica, Facultad de Ciencias, and Center for Free Radical and Biomedical Research, Facultad de Medicina, Universidad de la República, Montevideo, Uruguay

Stefanie Deubel
Department Biochemistry of Micronutrients, German Institute of Human Nutrition Potsdam-Rehbruecke, Nuthetal, Germany

Jean-Claude Drapier
Institut de Chimie des Substances Naturelles, Centre National de la Recherche Scientifique, Gif-sur-Yvette, France

Marino Exposito-Rodriguez
School of Biological Sciences, University of Essex, Colchester, United Kingdom

Gerardo Ferrer-Sueta
Laboratorio de Fisicoquímica Biológica, Instituto de Química Biológica, Facultad de Ciencias, and Center for Free Radical and Biomedical Research, Facultad de Medicina, Universidad de la República, Montevideo, Uruguay

Janna Frombach
Department Biochemistry of Micronutrients, German Institute of Human Nutrition Potsdam-Rehbruecke, Nuthetal, Germany

Tsanko Gechev
Department of Plant Physiology and Plant Molecular Biology, University of Plovdiv, and Institute of Molecular Biology and Biotechnologies, Plovdiv, Bulgaria

Alejandra Hernández-Barrera
Instituto de Biotecnología, Universidad Nacional Autónoma de México, Morelos, Mexico

Ayaka Hieno
The United Graduate School of Agricultural Sciences, Gifu University, Gifu, Japan

Jacques Hille
Department Molecular Biology of Plants, University of Groningen, Groningen, The Netherlands

Mitsuro Hyakumachi
The United Graduate School of Agricultural Sciences, and Faculty of Applied Biological Sciences, Gifu University, Gifu, Japan

Toshiki Ishikawa
Graduate School of Science and Engineering, Saitama University, Sakura-ku, Saitama City, Saitama, Japan

Haruka Ishino
Faculty of Applied Biological Sciences, Gifu University, Gifu, Japan

Eric A. Johnson
Department of Biochemistry and Molecular Biology, and Molecular Cell Biology Program, University of Massachusetts, Amherst, Massachusetts, USA

P. Andrew Karplus
Department of Biochemistry and Biophysics, Oregon State University, Corvallis, Oregon, USA

Maki Kawai-Yamada
Graduate School of Science and Engineering, and Institute for Environmental Science and Technology, Saitama University, Sakura-ku, Saitama City, Saitama, Japan

In Sup Kil
Yonsei Biomedical Research Institute, Yonsei University College of Medicine, Seodaemun-gu, Seoul, South Korea

Anna Patricia Kipp
Department Biochemistry of Micronutrients, German Institute of Human Nutrition Potsdam-Rehbruecke, Nuthetal, Germany

Hiroyuki Koyama
The United Graduate School of Agricultural Sciences, and Faculty of Applied Biological Sciences, Gifu University, Gifu, Japan

Pierre Philippe Laissue
School of Biological Sciences, University of Essex, Colchester, United Kingdom

Marika Lindahl
Instituto de Bioquímica Vegetal y Fotosíntesis, Universidad de Sevilla, CSIC IBVF(CSIC/US), Seville, Spain

George R. Littlejohn
Biosciences, College of Life and Environmental Sciences, University of Exeter, Exeter, United Kingdom

H. Susana Marinho
Departamento de Química e Bioquímica and Centro de Química e Bioquímica, Faculdade de Ciências, Universidade de Lisboa, Lisboa, Portugal

Nikolay Mehterov
Department of Plant Physiology and Plant Molecular Biology, University of Plovdiv, and Institute of Molecular Biology and Biotechnologies, Plovdiv, Bulgaria

Philip M. Mullineaux
School of Biological Sciences, University of Essex, Colchester, United Kingdom

Hushna Ara Naznin
The United Graduate School of Agricultural Sciences, Gifu University, Gifu, Japan

Kimberly J. Nelson
Department of Biochemistry, Wake Forest School of Medicine, Winston-Salem, North Carolina, USA

Derek Parsonage
Department of Biochemistry, Wake Forest School of Medicine, Winston-Salem, North Carolina, USA

Leslie B. Poole
Department of Biochemistry, Wake Forest School of Medicine, Winston-Salem, North Carolina, USA

Ruben Quintana-Cabrera
Institute of Functional Biology and Genomics (IBFG), Department of Biochemistry and Molecular Biology, University of Salamanca-CSIC, Salamanca, Spain

Carmen Quinto
Instituto de Biotecnología, Universidad Nacional Autónoma de México, Morelos, Mexico

Lía M. Randall
Laboratorio de Fisicoquímica Biológica, Instituto de Química Biológica, Facultad de Ciencias, and Center for Free Radical and Biomedical Research, Facultad de Medicina, Universidad de la República, Montevideo, Uruguay

Sue Goo Rhee
Yonsei Biomedical Research Institute, Yonsei University College of Medicine, Seodaemun-gu, Seoul, South Korea

Sylvie Riquier
Institut de Chimie des Substances Naturelles, Centre National de la Recherche Scientifique, Gif-sur-Yvette, France

Yusaku Sakai
Faculty of Applied Biological Sciences, Gifu University, Gifu, Japan

Katsunobu Sawaki
The United Graduate School of Agricultural Sciences, Gifu University, Gifu, Japan

Nicholas Smirnoff
Biosciences, College of Life and Environmental Sciences, University of Exeter, Exeter, United Kingdom

Hirofumi Uchimiya
Institute for Environmental Science and Technology, Saitama University, Sakura-ku, Saitama City, Saitama, Japan

Hen-Ming Wu
Department of Biochemistry and Molecular Biology, and Molecular Cell Biology Program, University of Massachusetts, Amherst, Massachusetts, USA

Yoshiharu Y. Yamamoto
The United Graduate School of Agricultural Sciences, and Faculty of Applied Biological Sciences, Gifu University, Gifu, Japan

PREFACE

The identification of hydrogen peroxide in regulation of cell signaling and gene expression was a significant breakthrough in oxygen biology. Hydrogen peroxide is probably the most important redox signaling molecule that, among others, can activate NFκB, Nrf2, and other universal transcription factors and is involved in the regulation of insulin- and MAPK signaling. These pleiotropic effects of hydrogen peroxide are largely accounted for by changes in the thiol/disulfide status of the cell, an important determinant of the cell's redox status. Moreover, disruption of redox signaling and control recognizes the occurrence of compartmentalized cell redox circuits.

Hydrogen peroxide signaling has been of central importance in cell research for some time and some previous volumes of *Methods in Enzymology* have covered in part some aspects of the physiological roles of hydrogen peroxide. However, there have been new developments and techniques that warrant these three volumes of *Methods in Enzymology*, which were designed to be the premier place for a compendium of hydrogen peroxide detection and delivery methods, microdomain imaging, and determinants of hydrogen peroxide steady-state levels; in addition, the role of hydrogen peroxide in cellular processes entailing redox regulation of cell signaling and transcription was covered by experts in mammalian and plant biochemistry and physiology.

In bringing this volume to fruition credit must be given to the experts in various aspects of hydrogen peroxide signaling research, whose thorough and innovative work is the basis of these three *Methods in Enzymology* volumes. Special thanks to the Advisory Board Members—Christopher J. Chang, So Goo Rhee, and Balyanaraman Kalyanaram—who provided guidance in the selection of topics and contributors. We hope that these volumes would be of help to both new and established investigators in this field.

Enrique Cadenas
Lester Packer
May 2013

METHODS IN ENZYMOLOGY

Volume I. Preparation and Assay of Enzymes
Edited by Sidney P. Colowick and Nathan O. Kaplan

Volume II. Preparation and Assay of Enzymes
Edited by Sidney P. Colowick and Nathan O. Kaplan

Volume III. Preparation and Assay of Substrates
Edited by Sidney P. Colowick and Nathan O. Kaplan

Volume IV. Special Techniques for the Enzymologist
Edited by Sidney P. Colowick and Nathan O. Kaplan

Volume V. Preparation and Assay of Enzymes
Edited by Sidney P. Colowick and Nathan O. Kaplan

Volume VI. Preparation and Assay of Enzymes (*Continued*)
Preparation and Assay of Substrates
Special Techniques
Edited by Sidney P. Colowick and Nathan O. Kaplan

Volume VII. Cumulative Subject Index
Edited by Sidney P. Colowick and Nathan O. Kaplan

Volume VIII. Complex Carbohydrates
Edited by Elizabeth F. Neufeld and Victor Ginsburg

Volume IX. Carbohydrate Metabolism
Edited by Willis A. Wood

Volume X. Oxidation and Phosphorylation
Edited by Ronald W. Estabrook and Maynard E. Pullman

Volume XI. Enzyme Structure
Edited by C. H. W. Hirs

Volume XII. Nucleic Acids (Parts A and B)
Edited by Lawrence Grossman and Kivie Moldave

Volume XIII. Citric Acid Cycle
Edited by J. M. Lowenstein

Volume XIV. Lipids
Edited by J. M. Lowenstein

VOLUME XV. Steroids and Terpenoids
Edited by RAYMOND B. CLAYTON

VOLUME XVI. Fast Reactions
Edited by KENNETH KUSTIN

VOLUME XVII. Metabolism of Amino Acids and Amines (Parts A and B)
Edited by HERBERT TABOR AND CELIA WHITE TABOR

VOLUME XVIII. Vitamins and Coenzymes (Parts A, B, and C)
Edited by DONALD B. MCCORMICK AND LEMUEL D. WRIGHT

VOLUME XIX. Proteolytic Enzymes
Edited by GERTRUDE E. PERLMANN AND LASZLO LORAND

VOLUME XX. Nucleic Acids and Protein Synthesis (Part C)
Edited by KIVIE MOLDAVE AND LAWRENCE GROSSMAN

VOLUME XXI. Nucleic Acids (Part D)
Edited by LAWRENCE GROSSMAN AND KIVIE MOLDAVE

VOLUME XXII. Enzyme Purification and Related Techniques
Edited by WILLIAM B. JAKOBY

VOLUME XXIII. Photosynthesis (Part A)
Edited by ANTHONY SAN PIETRO

VOLUME XXIV. Photosynthesis and Nitrogen Fixation (Part B)
Edited by ANTHONY SAN PIETRO

VOLUME XXV. Enzyme Structure (Part B)
Edited by C. H. W. HIRS AND SERGE N. TIMASHEFF

VOLUME XXVI. Enzyme Structure (Part C)
Edited by C. H. W. HIRS AND SERGE N. TIMASHEFF

VOLUME XXVII. Enzyme Structure (Part D)
Edited by C. H. W. HIRS AND SERGE N. TIMASHEFF

VOLUME XXVIII. Complex Carbohydrates (Part B)
Edited by VICTOR GINSBURG

VOLUME XXIX. Nucleic Acids and Protein Synthesis (Part E)
Edited by LAWRENCE GROSSMAN AND KIVIE MOLDAVE

VOLUME XXX. Nucleic Acids and Protein Synthesis (Part F)
Edited by KIVIE MOLDAVE AND LAWRENCE GROSSMAN

VOLUME XXXI. Biomembranes (Part A)
Edited by SIDNEY FLEISCHER AND LESTER PACKER

VOLUME XXXII. Biomembranes (Part B)
Edited by SIDNEY FLEISCHER AND LESTER PACKER

VOLUME XXXIII. Cumulative Subject Index Volumes I-XXX
Edited by MARTHA G. DENNIS AND EDWARD A. DENNIS

VOLUME XXXIV. Affinity Techniques (Enzyme Purification: Part B)
Edited by WILLIAM B. JAKOBY AND MEIR WILCHEK

VOLUME XXXV. Lipids (Part B)
Edited by JOHN M. LOWENSTEIN

VOLUME XXXVI. Hormone Action (Part A: Steroid Hormones)
Edited by BERT W. O'MALLEY AND JOEL G. HARDMAN

VOLUME XXXVII. Hormone Action (Part B: Peptide Hormones)
Edited by BERT W. O'MALLEY AND JOEL G. HARDMAN

VOLUME XXXVIII. Hormone Action (Part C: Cyclic Nucleotides)
Edited by JOEL G. HARDMAN AND BERT W. O'MALLEY

VOLUME XXXIX. Hormone Action (Part D: Isolated Cells, Tissues, and Organ Systems)
Edited by JOEL G. HARDMAN AND BERT W. O'MALLEY

VOLUME XL. Hormone Action (Part E: Nuclear Structure and Function)
Edited by BERT W. O'MALLEY AND JOEL G. HARDMAN

VOLUME XLI. Carbohydrate Metabolism (Part B)
Edited by W. A. WOOD

VOLUME XLII. Carbohydrate Metabolism (Part C)
Edited by W. A. WOOD

VOLUME XLIII. Antibiotics
Edited by JOHN H. HASH

VOLUME XLIV. Immobilized Enzymes
Edited by KLAUS MOSBACH

VOLUME XLV. Proteolytic Enzymes (Part B)
Edited by LASZLO LORAND

VOLUME XLVI. Affinity Labeling
Edited by WILLIAM B. JAKOBY AND MEIR WILCHEK

VOLUME XLVII. Enzyme Structure (Part E)
Edited by C. H. W. HIRS AND SERGE N. TIMASHEFF

Volume XLVIII. Enzyme Structure (Part F)
Edited by C. H. W. Hirs and Serge N. Timasheff

Volume XLIX. Enzyme Structure (Part G)
Edited by C. H. W. Hirs and Serge N. Timasheff

Volume L. Complex Carbohydrates (Part C)
Edited by Victor Ginsburg

Volume LI. Purine and Pyrimidine Nucleotide Metabolism
Edited by Patricia A. Hoffee and Mary Ellen Jones

Volume LII. Biomembranes (Part C: Biological Oxidations)
Edited by Sidney Fleischer and Lester Packer

Volume LIII. Biomembranes (Part D: Biological Oxidations)
Edited by Sidney Fleischer and Lester Packer

Volume LIV. Biomembranes (Part E: Biological Oxidations)
Edited by Sidney Fleischer and Lester Packer

Volume LV. Biomembranes (Part F: Bioenergetics)
Edited by Sidney Fleischer and Lester Packer

Volume LVI. Biomembranes (Part G: Bioenergetics)
Edited by Sidney Fleischer and Lester Packer

Volume LVII. Bioluminescence and Chemiluminescence
Edited by Marlene A. DeLuca

Volume LVIII. Cell Culture
Edited by William B. Jakoby and Ira Pastan

Volume LIX. Nucleic Acids and Protein Synthesis (Part G)
Edited by Kivie Moldave and Lawrence Grossman

Volume LX. Nucleic Acids and Protein Synthesis (Part H)
Edited by Kivie Moldave and Lawrence Grossman

Volume 61. Enzyme Structure (Part H)
Edited by C. H. W. Hirs and Serge N. Timasheff

Volume 62. Vitamins and Coenzymes (Part D)
Edited by Donald B. McCormick and Lemuel D. Wright

Volume 63. Enzyme Kinetics and Mechanism (Part A: Initial Rate and Inhibitor Methods)
Edited by Daniel L. Purich

VOLUME 64. Enzyme Kinetics and Mechanism (Part B: Isotopic Probes and Complex Enzyme Systems)
Edited by DANIEL L. PURICH

VOLUME 65. Nucleic Acids (Part I)
Edited by LAWRENCE GROSSMAN AND KIVIE MOLDAVE

VOLUME 66. Vitamins and Coenzymes (Part E)
Edited by DONALD B. MCCORMICK AND LEMUEL D. WRIGHT

VOLUME 67. Vitamins and Coenzymes (Part F)
Edited by DONALD B. MCCORMICK AND LEMUEL D. WRIGHT

VOLUME 68. Recombinant DNA
Edited by RAY WU

VOLUME 69. Photosynthesis and Nitrogen Fixation (Part C)
Edited by ANTHONY SAN PIETRO

VOLUME 70. Immunochemical Techniques (Part A)
Edited by HELEN VAN VUNAKIS AND JOHN J. LANGONE

VOLUME 71. Lipids (Part C)
Edited by JOHN M. LOWENSTEIN

VOLUME 72. Lipids (Part D)
Edited by JOHN M. LOWENSTEIN

VOLUME 73. Immunochemical Techniques (Part B)
Edited by JOHN J. LANGONE AND HELEN VAN VUNAKIS

VOLUME 74. Immunochemical Techniques (Part C)
Edited by JOHN J. LANGONE AND HELEN VAN VUNAKIS

VOLUME 75. Cumulative Subject Index Volumes XXXI, XXXII, XXXIV–LX
Edited by EDWARD A. DENNIS AND MARTHA G. DENNIS

VOLUME 76. Hemoglobins
Edited by ERALDO ANTONINI, LUIGI ROSSI-BERNARDI, AND EMILIA CHIANCONE

VOLUME 77. Detoxication and Drug Metabolism
Edited by WILLIAM B. JAKOBY

VOLUME 78. Interferons (Part A)
Edited by SIDNEY PESTKA

VOLUME 79. Interferons (Part B)
Edited by SIDNEY PESTKA

VOLUME 80. Proteolytic Enzymes (Part C)
Edited by LASZLO LORAND

VOLUME 81. Biomembranes (Part H: Visual Pigments and Purple Membranes, I)
Edited by LESTER PACKER

VOLUME 82. Structural and Contractile Proteins (Part A: Extracellular Matrix)
Edited by LEON W. CUNNINGHAM AND DIXIE W. FREDERIKSEN

VOLUME 83. Complex Carbohydrates (Part D)
Edited by VICTOR GINSBURG

VOLUME 84. Immunochemical Techniques (Part D: Selected Immunoassays)
Edited by JOHN J. LANGONE AND HELEN VAN VUNAKIS

VOLUME 85. Structural and Contractile Proteins (Part B: The Contractile Apparatus and the Cytoskeleton)
Edited by DIXIE W. FREDERIKSEN AND LEON W. CUNNINGHAM

VOLUME 86. Prostaglandins and Arachidonate Metabolites
Edited by WILLIAM E. M. LANDS AND WILLIAM L. SMITH

VOLUME 87. Enzyme Kinetics and Mechanism (Part C: Intermediates, Stereo-chemistry, and Rate Studies)
Edited by DANIEL L. PURICH

VOLUME 88. Biomembranes (Part I: Visual Pigments and Purple Membranes, II)
Edited by LESTER PACKER

VOLUME 89. Carbohydrate Metabolism (Part D)
Edited by WILLIS A. WOOD

VOLUME 90. Carbohydrate Metabolism (Part E)
Edited by WILLIS A. WOOD

VOLUME 91. Enzyme Structure (Part I)
Edited by C. H. W. HIRS AND SERGE N. TIMASHEFF

VOLUME 92. Immunochemical Techniques (Part E: Monoclonal Antibodies and General Immunoassay Methods)
Edited by JOHN J. LANGONE AND HELEN VAN VUNAKIS

VOLUME 93. Immunochemical Techniques (Part F: Conventional Antibodies, Fc Receptors, and Cytotoxicity)
Edited by JOHN J. LANGONE AND HELEN VAN VUNAKIS

VOLUME 94. Polyamines
Edited by HERBERT TABOR AND CELIA WHITE TABOR

VOLUME 95. Cumulative Subject Index Volumes 61–74, 76–80
Edited by EDWARD A. DENNIS AND MARTHA G. DENNIS

VOLUME 96. Biomembranes [Part J: Membrane Biogenesis: Assembly and Targeting (General Methods; Eukaryotes)]
Edited by SIDNEY FLEISCHER AND BECCA FLEISCHER

VOLUME 97. Biomembranes [Part K: Membrane Biogenesis: Assembly and Targeting (Prokaryotes, Mitochondria, and Chloroplasts)]
Edited by SIDNEY FLEISCHER AND BECCA FLEISCHER

VOLUME 98. Biomembranes (Part L: Membrane Biogenesis: Processing and Recycling)
Edited by SIDNEY FLEISCHER AND BECCA FLEISCHER

VOLUME 99. Hormone Action (Part F: Protein Kinases)
Edited by JACKIE D. CORBIN AND JOEL G. HARDMAN

VOLUME 100. Recombinant DNA (Part B)
Edited by RAY WU, LAWRENCE GROSSMAN, AND KIVIE MOLDAVE

VOLUME 101. Recombinant DNA (Part C)
Edited by RAY WU, LAWRENCE GROSSMAN, AND KIVIE MOLDAVE

VOLUME 102. Hormone Action (Part G: Calmodulin and Calcium-Binding Proteins)
Edited by ANTHONY R. MEANS AND BERT W. O'MALLEY

VOLUME 103. Hormone Action (Part H: Neuroendocrine Peptides)
Edited by P. MICHAEL CONN

VOLUME 104. Enzyme Purification and Related Techniques (Part C)
Edited by WILLIAM B. JAKOBY

VOLUME 105. Oxygen Radicals in Biological Systems
Edited by LESTER PACKER

VOLUME 106. Posttranslational Modifications (Part A)
Edited by FINN WOLD AND KIVIE MOLDAVE

Volume 107. Posttranslational Modifications (Part B)
Edited by Finn Wold and Kivie Moldave

Volume 108. Immunochemical Techniques (Part G: Separation and Characterization of Lymphoid Cells)
Edited by Giovanni Di Sabato, John J. Langone, and Helen Van Vunakis

Volume 109. Hormone Action (Part I: Peptide Hormones)
Edited by Lutz Birnbaumer and Bert W. O'Malley

Volume 110. Steroids and Isoprenoids (Part A)
Edited by John H. Law and Hans C. Rilling

Volume 111. Steroids and Isoprenoids (Part B)
Edited by John H. Law and Hans C. Rilling

Volume 112. Drug and Enzyme Targeting (Part A)
Edited by Kenneth J. Widder and Ralph Green

Volume 113. Glutamate, Glutamine, Glutathione, and Related Compounds
Edited by Alton Meister

Volume 114. Diffraction Methods for Biological Macromolecules (Part A)
Edited by Harold W. Wyckoff, C. H. W. Hirs, and Serge N. Timasheff

Volume 115. Diffraction Methods for Biological Macromolecules (Part B)
Edited by Harold W. Wyckoff, C. H. W. Hirs, and Serge N. Timasheff

Volume 116. Immunochemical Techniques
(Part H: Effectors and Mediators of Lymphoid Cell Functions)
Edited by Giovanni Di Sabato, John J. Langone, and Helen Van Vunakis

Volume 117. Enzyme Structure (Part J)
Edited by C. H. W. Hirs and Serge N. Timasheff

Volume 118. Plant Molecular Biology
Edited by Arthur Weissbach and Herbert Weissbach

Volume 119. Interferons (Part C)
Edited by Sidney Pestka

Volume 120. Cumulative Subject Index Volumes 81–94, 96–101

Volume 121. Immunochemical Techniques (Part I: Hybridoma Technology and Monoclonal Antibodies)
Edited by John J. Langone and Helen Van Vunakis

Volume 122. Vitamins and Coenzymes (Part G)
Edited by Frank Chytil and Donald B. McCormick

Volume 123. Vitamins and Coenzymes (Part H)
Edited by Frank Chytil and Donald B. McCormick

Volume 124. Hormone Action (Part J: Neuroendocrine Peptides)
Edited by P. Michael Conn

Volume 125. Biomembranes (Part M: Transport in Bacteria, Mitochondria, and Chloroplasts: General Approaches and Transport Systems)
Edited by Sidney Fleischer and Becca Fleischer

Volume 126. Biomembranes (Part N: Transport in Bacteria, Mitochondria, and Chloroplasts: Protonmotive Force)
Edited by Sidney Fleischer and Becca Fleischer

Volume 127. Biomembranes (Part O: Protons and Water: Structure and Translocation)
Edited by Lester Packer

Volume 128. Plasma Lipoproteins (Part A: Preparation, Structure, and Molecular Biology)
Edited by Jere P. Segrest and John J. Albers

Volume 129. Plasma Lipoproteins (Part B: Characterization, Cell Biology, and Metabolism)
Edited by John J. Albers and Jere P. Segrest

Volume 130. Enzyme Structure (Part K)
Edited by C. H. W. Hirs and Serge N. Timasheff

Volume 131. Enzyme Structure (Part L)
Edited by C. H. W. Hirs and Serge N. Timasheff

Volume 132. Immunochemical Techniques (Part J: Phagocytosis and Cell-Mediated Cytotoxicity)
Edited by Giovanni Di Sabato and Johannes Everse

Volume 133. Bioluminescence and Chemiluminescence (Part B)
Edited by Marlene DeLuca and William D. McElroy

Volume 134. Structural and Contractile Proteins (Part C: The Contractile Apparatus and the Cytoskeleton)
Edited by Richard B. Vallee

Volume 135. Immobilized Enzymes and Cells (Part B)
Edited by Klaus Mosbach

VOLUME 136. Immobilized Enzymes and Cells (Part C)
Edited by KLAUS MOSBACH

VOLUME 137. Immobilized Enzymes and Cells (Part D)
Edited by KLAUS MOSBACH

VOLUME 138. Complex Carbohydrates (Part E)
Edited by VICTOR GINSBURG

VOLUME 139. Cellular Regulators (Part A: Calcium- and Calmodulin-Binding Proteins)
Edited by ANTHONY R. MEANS AND P. MICHAEL CONN

VOLUME 140. Cumulative Subject Index Volumes 102–119, 121–134

VOLUME 141. Cellular Regulators (Part B: Calcium and Lipids)
Edited by P. MICHAEL CONN AND ANTHONY R. MEANS

VOLUME 142. Metabolism of Aromatic Amino Acids and Amines
Edited by SEYMOUR KAUFMAN

VOLUME 143. Sulfur and Sulfur Amino Acids
Edited by WILLIAM B. JAKOBY AND OWEN GRIFFITH

VOLUME 144. Structural and Contractile Proteins (Part D: Extracellular Matrix)
Edited by LEON W. CUNNINGHAM

VOLUME 145. Structural and Contractile Proteins (Part E: Extracellular Matrix)
Edited by LEON W. CUNNINGHAM

VOLUME 146. Peptide Growth Factors (Part A)
Edited by DAVID BARNES AND DAVID A. SIRBASKU

VOLUME 147. Peptide Growth Factors (Part B)
Edited by DAVID BARNES AND DAVID A. SIRBASKU

VOLUME 148. Plant Cell Membranes
Edited by LESTER PACKER AND ROLAND DOUCE

VOLUME 149. Drug and Enzyme Targeting (Part B)
Edited by RALPH GREEN AND KENNETH J. WIDDER

VOLUME 150. Immunochemical Techniques (Part K: *In Vitro* Models of B and T Cell Functions and Lymphoid Cell Receptors)
Edited by GIOVANNI DI SABATO

VOLUME 151. Molecular Genetics of Mammalian Cells
Edited by MICHAEL M. GOTTESMAN

VOLUME 152. Guide to Molecular Cloning Techniques
Edited by SHELBY L. BERGER AND ALAN R. KIMMEL

VOLUME 153. Recombinant DNA (Part D)
Edited by RAY WU AND LAWRENCE GROSSMAN

VOLUME 154. Recombinant DNA (Part E)
Edited by RAY WU AND LAWRENCE GROSSMAN

VOLUME 155. Recombinant DNA (Part F)
Edited by RAY WU

VOLUME 156. Biomembranes (Part P: ATP-Driven Pumps and Related Transport: The Na, K-Pump)
Edited by SIDNEY FLEISCHER AND BECCA FLEISCHER

VOLUME 157. Biomembranes (Part Q: ATP-Driven Pumps and Related Transport: Calcium, Proton, and Potassium Pumps)
Edited by SIDNEY FLEISCHER AND BECCA FLEISCHER

VOLUME 158. Metalloproteins (Part A)
Edited by JAMES F. RIORDAN AND BERT L. VALLEE

VOLUME 159. Initiation and Termination of Cyclic Nucleotide Action
Edited by JACKIE D. CORBIN AND ROGER A. JOHNSON

VOLUME 160. Biomass (Part A: Cellulose and Hemicellulose)
Edited by WILLIS A. WOOD AND SCOTT T. KELLOGG

VOLUME 161. Biomass (Part B: Lignin, Pectin, and Chitin)
Edited by WILLIS A. WOOD AND SCOTT T. KELLOGG

VOLUME 162. Immunochemical Techniques (Part L: Chemotaxis and Inflammation)
Edited by GIOVANNI DI SABATO

VOLUME 163. Immunochemical Techniques (Part M: Chemotaxis and Inflammation)
Edited by GIOVANNI DI SABATO

VOLUME 164. Ribosomes
Edited by HARRY F. NOLLER, JR., AND KIVIE MOLDAVE

VOLUME 165. Microbial Toxins: Tools for Enzymology
Edited by SIDNEY HARSHMAN

Volume 166. Branched-Chain Amino Acids
Edited by Robert Harris and John R. Sokatch

Volume 167. Cyanobacteria
Edited by Lester Packer and Alexander N. Glazer

Volume 168. Hormone Action (Part K: Neuroendocrine Peptides)
Edited by P. Michael Conn

Volume 169. Platelets: Receptors, Adhesion, Secretion (Part A)
Edited by Jacek Hawiger

Volume 170. Nucleosomes
Edited by Paul M. Wassarman and Roger D. Kornberg

Volume 171. Biomembranes (Part R: Transport Theory: Cells and Model Membranes)
Edited by Sidney Fleischer and Becca Fleischer

Volume 172. Biomembranes (Part S: Transport: Membrane Isolation and Characterization)
Edited by Sidney Fleischer and Becca Fleischer

Volume 173. Biomembranes [Part T: Cellular and Subcellular Transport: Eukaryotic (Nonepithelial) Cells]
Edited by Sidney Fleischer and Becca Fleischer

Volume 174. Biomembranes [Part U: Cellular and Subcellular Transport: Eukaryotic (Nonepithelial) Cells]
Edited by Sidney Fleischer and Becca Fleischer

Volume 175. Cumulative Subject Index Volumes 135–139, 141–167

Volume 176. Nuclear Magnetic Resonance (Part A: Spectral Techniques and Dynamics)
Edited by Norman J. Oppenheimer and Thomas L. James

Volume 177. Nuclear Magnetic Resonance (Part B: Structure and Mechanism)
Edited by Norman J. Oppenheimer and Thomas L. James

Volume 178. Antibodies, Antigens, and Molecular Mimicry
Edited by John J. Langone

Volume 179. Complex Carbohydrates (Part F)
Edited by Victor Ginsburg

VOLUME 180. RNA Processing (Part A: General Methods)
Edited by JAMES E. DAHLBERG AND JOHN N. ABELSON

VOLUME 181. RNA Processing (Part B: Specific Methods)
Edited by JAMES E. DAHLBERG AND JOHN N. ABELSON

VOLUME 182. Guide to Protein Purification
Edited by MURRAY P. DEUTSCHER

VOLUME 183. Molecular Evolution: Computer Analysis of Protein and Nucleic Acid Sequences
Edited by RUSSELL F. DOOLITTLE

VOLUME 184. Avidin-Biotin Technology
Edited by MEIR WILCHEK AND EDWARD A. BAYER

VOLUME 185. Gene Expression Technology
Edited by DAVID V. GOEDDEL

VOLUME 186. Oxygen Radicals in Biological Systems (Part B: Oxygen Radicals and Antioxidants)
Edited by LESTER PACKER AND ALEXANDER N. GLAZER

VOLUME 187. Arachidonate Related Lipid Mediators
Edited by ROBERT C. MURPHY AND FRANK A. FITZPATRICK

VOLUME 188. Hydrocarbons and Methylotrophy
Edited by MARY E. LIDSTROM

VOLUME 189. Retinoids (Part A: Molecular and Metabolic Aspects)
Edited by LESTER PACKER

VOLUME 190. Retinoids (Part B: Cell Differentiation and Clinical Applications)
Edited by LESTER PACKER

VOLUME 191. Biomembranes (Part V: Cellular and Subcellular Transport: Epithelial Cells)
Edited by SIDNEY FLEISCHER AND BECCA FLEISCHER

VOLUME 192. Biomembranes (Part W: Cellular and Subcellular Transport: Epithelial Cells)
Edited by SIDNEY FLEISCHER AND BECCA FLEISCHER

VOLUME 193. Mass Spectrometry
Edited by JAMES A. MCCLOSKEY

VOLUME 194. Guide to Yeast Genetics and Molecular Biology
Edited by CHRISTINE GUTHRIE AND GERALD R. FINK

VOLUME 195. Adenylyl Cyclase, G Proteins, and Guanylyl Cyclase
Edited by ROGER A. JOHNSON AND JACKIE D. CORBIN

VOLUME 196. Molecular Motors and the Cytoskeleton
Edited by RICHARD B. VALLEE

VOLUME 197. Phospholipases
Edited by EDWARD A. DENNIS

VOLUME 198. Peptide Growth Factors (Part C)
Edited by DAVID BARNES, J. P. MATHER, AND GORDON H. SATO

VOLUME 199. Cumulative Subject Index Volumes 168–174, 176–194

VOLUME 200. Protein Phosphorylation (Part A: Protein Kinases: Assays, Purification, Antibodies, Functional Analysis, Cloning, and Expression)
Edited by TONY HUNTER AND BARTHOLOMEW M. SEFTON

VOLUME 201. Protein Phosphorylation (Part B: Analysis of Protein Phosphorylation, Protein Kinase Inhibitors, and Protein Phosphatases)
Edited by TONY HUNTER AND BARTHOLOMEW M. SEFTON

VOLUME 202. Molecular Design and Modeling: Concepts and Applications (Part A: Proteins, Peptides, and Enzymes)
Edited by JOHN J. LANGONE

VOLUME 203. Molecular Design and Modeling: Concepts and Applications (Part B: Antibodies and Antigens, Nucleic Acids, Polysaccharides, and Drugs)
Edited by JOHN J. LANGONE

VOLUME 204. Bacterial Genetic Systems
Edited by JEFFREY H. MILLER

VOLUME 205. Metallobiochemistry (Part B: Metallothionein and Related Molecules)
Edited by JAMES F. RIORDAN AND BERT L. VALLEE

VOLUME 206. Cytochrome P450
Edited by MICHAEL R. WATERMAN AND ERIC F. JOHNSON

VOLUME 207. Ion Channels
Edited by BERNARDO RUDY AND LINDA E. IVERSON

VOLUME 208. Protein–DNA Interactions
Edited by ROBERT T. SAUER

VOLUME 209. Phospholipid Biosynthesis
Edited by EDWARD A. DENNIS AND DENNIS E. VANCE

VOLUME 210. Numerical Computer Methods
Edited by LUDWIG BRAND AND MICHAEL L. JOHNSON

VOLUME 211. DNA Structures (Part A: Synthesis and Physical Analysis of DNA)
Edited by DAVID M. J. LILLEY AND JAMES E. DAHLBERG

VOLUME 212. DNA Structures (Part B: Chemical and Electrophoretic Analysis of DNA)
Edited by DAVID M. J. LILLEY AND JAMES E. DAHLBERG

VOLUME 213. Carotenoids (Part A: Chemistry, Separation, Quantitation, and Antioxidation)
Edited by LESTER PACKER

VOLUME 214. Carotenoids (Part B: Metabolism, Genetics, and Biosynthesis)
Edited by LESTER PACKER

VOLUME 215. Platelets: Receptors, Adhesion, Secretion (Part B)
Edited by JACEK J. HAWIGER

VOLUME 216. Recombinant DNA (Part G)
Edited by RAY WU

VOLUME 217. Recombinant DNA (Part H)
Edited by RAY WU

VOLUME 218. Recombinant DNA (Part I)
Edited by RAY WU

VOLUME 219. Reconstitution of Intracellular Transport
Edited by JAMES E. ROTHMAN

VOLUME 220. Membrane Fusion Techniques (Part A)
Edited by NEJAT DÜZGÜNEŞ

VOLUME 221. Membrane Fusion Techniques (Part B)
Edited by NEJAT DÜZGÜNEŞ

VOLUME 222. Proteolytic Enzymes in Coagulation, Fibrinolysis, and Complement Activation (Part A: Mammalian Blood Coagulation

Factors and Inhibitors)
Edited by Laszlo Lorand and Kenneth G. Mann

Volume 223. Proteolytic Enzymes in Coagulation, Fibrinolysis, and Complement Activation (Part B: Complement Activation, Fibrinolysis, and Nonmammalian Blood Coagulation Factors)
Edited by Laszlo Lorand and Kenneth G. Mann

Volume 224. Molecular Evolution: Producing the Biochemical Data
Edited by Elizabeth Anne Zimmer, Thomas J. White, Rebecca L. Cann, and Allan C. Wilson

Volume 225. Guide to Techniques in Mouse Development
Edited by Paul M. Wassarman and Melvin L. DePamphilis

Volume 226. Metallobiochemistry (Part C: Spectroscopic and Physical Methods for Probing Metal Ion Environments in Metalloenzymes and Metalloproteins)
Edited by James F. Riordan and Bert L. Vallee

Volume 227. Metallobiochemistry (Part D: Physical and Spectroscopic Methods for Probing Metal Ion Environments in Metalloproteins)
Edited by James F. Riordan and Bert L. Vallee

Volume 228. Aqueous Two-Phase Systems
Edited by Harry Walter and Göte Johansson

Volume 229. Cumulative Subject Index Volumes 195–198, 200–227

Volume 230. Guide to Techniques in Glycobiology
Edited by William J. Lennarz and Gerald W. Hart

Volume 231. Hemoglobins (Part B: Biochemical and Analytical Methods)
Edited by Johannes Everse, Kim D. Vandegriff, and Robert M. Winslow

Volume 232. Hemoglobins (Part C: Biophysical Methods)
Edited by Johannes Everse, Kim D. Vandegriff, and Robert M. Winslow

Volume 233. Oxygen Radicals in Biological Systems (Part C)
Edited by Lester Packer

Volume 234. Oxygen Radicals in Biological Systems (Part D)
Edited by Lester Packer

Volume 235. Bacterial Pathogenesis (Part A: Identification and Regulation of Virulence Factors)
Edited by Virginia L. Clark and Patrik M. Bavoil

VOLUME 236. Bacterial Pathogenesis (Part B: Integration of Pathogenic Bacteria with Host Cells)
Edited by VIRGINIA L. CLARK AND PATRIK M. BAVOIL

VOLUME 237. Heterotrimeric G Proteins
Edited by RAVI IYENGAR

VOLUME 238. Heterotrimeric G-Protein Effectors
Edited by RAVI IYENGAR

VOLUME 239. Nuclear Magnetic Resonance (Part C)
Edited by THOMAS L. JAMES AND NORMAN J. OPPENHEIMER

VOLUME 240. Numerical Computer Methods (Part B)
Edited by MICHAEL L. JOHNSON AND LUDWIG BRAND

VOLUME 241. Retroviral Proteases
Edited by LAWRENCE C. KUO AND JULES A. SHAFER

VOLUME 242. Neoglycoconjugates (Part A)
Edited by Y. C. LEE AND REIKO T. LEE

VOLUME 243. Inorganic Microbial Sulfur Metabolism
Edited by HARRY D. PECK, JR., AND JEAN LEGALL

VOLUME 244. Proteolytic Enzymes: Serine and Cysteine Peptidases
Edited by ALAN J. BARRETT

VOLUME 245. Extracellular Matrix Components
Edited by E. RUOSLAHTI AND E. ENGVALL

VOLUME 246. Biochemical Spectroscopy
Edited by KENNETH SAUER

VOLUME 247. Neoglycoconjugates (Part B: Biomedical Applications)
Edited by Y. C. LEE AND REIKO T. LEE

VOLUME 248. Proteolytic Enzymes: Aspartic and Metallo Peptidases
Edited by ALAN J. BARRETT

VOLUME 249. Enzyme Kinetics and Mechanism (Part D: Developments in Enzyme Dynamics)
Edited by DANIEL L. PURICH

VOLUME 250. Lipid Modifications of Proteins
Edited by PATRICK J. CASEY AND JANICE E. BUSS

VOLUME 251. Biothiols (Part A: Monothiols and Dithiols, Protein Thiols, and Thiyl Radicals)
Edited by LESTER PACKER

VOLUME 252. Biothiols (Part B: Glutathione and Thioredoxin; Thiols in Signal Transduction and Gene Regulation)
Edited by LESTER PACKER

VOLUME 253. Adhesion of Microbial Pathogens
Edited by RON J. DOYLE AND ITZHAK OFEK

VOLUME 254. Oncogene Techniques
Edited by PETER K. VOGT AND INDER M. VERMA

VOLUME 255. Small GTPases and Their Regulators (Part A: Ras Family)
Edited by W. E. BALCH, CHANNING J. DER, AND ALAN HALL

VOLUME 256. Small GTPases and Their Regulators (Part B: Rho Family)
Edited by W. E. BALCH, CHANNING J. DER, AND ALAN HALL

VOLUME 257. Small GTPases and Their Regulators (Part C: Proteins Involved in Transport)
Edited by W. E. BALCH, CHANNING J. DER, AND ALAN HALL

VOLUME 258. Redox-Active Amino Acids in Biology
Edited by JUDITH P. KLINMAN

VOLUME 259. Energetics of Biological Macromolecules
Edited by MICHAEL L. JOHNSON AND GARY K. ACKERS

VOLUME 260. Mitochondrial Biogenesis and Genetics (Part A)
Edited by GIUSEPPE M. ATTARDI AND ANNE CHOMYN

VOLUME 261. Nuclear Magnetic Resonance and Nucleic Acids
Edited by THOMAS L. JAMES

VOLUME 262. DNA Replication
Edited by JUDITH L. CAMPBELL

VOLUME 263. Plasma Lipoproteins (Part C: Quantitation)
Edited by WILLIAM A. BRADLEY, SANDRA H. GIANTURCO, AND JERE P. SEGREST

VOLUME 264. Mitochondrial Biogenesis and Genetics (Part B)
Edited by GIUSEPPE M. ATTARDI AND ANNE CHOMYN

VOLUME 265. Cumulative Subject Index Volumes 228, 230–262

VOLUME 266. Computer Methods for Macromolecular Sequence Analysis
Edited by RUSSELL F. DOOLITTLE

VOLUME 267. Combinatorial Chemistry
Edited by JOHN N. ABELSON

VOLUME 268. Nitric Oxide (Part A: Sources and Detection of NO; NO Synthase)
Edited by LESTER PACKER

VOLUME 269. Nitric Oxide (Part B: Physiological and Pathological Processes)
Edited by LESTER PACKER

VOLUME 270. High Resolution Separation and Analysis of Biological Macromolecules (Part A: Fundamentals)
Edited by BARRY L. KARGER AND WILLIAM S. HANCOCK

VOLUME 271. High Resolution Separation and Analysis of Biological Macromolecules (Part B: Applications)
Edited by BARRY L. KARGER AND WILLIAM S. HANCOCK

VOLUME 272. Cytochrome P450 (Part B)
Edited by ERIC F. JOHNSON AND MICHAEL R. WATERMAN

VOLUME 273. RNA Polymerase and Associated Factors (Part A)
Edited by SANKAR ADHYA

VOLUME 274. RNA Polymerase and Associated Factors (Part B)
Edited by SANKAR ADHYA

VOLUME 275. Viral Polymerases and Related Proteins
Edited by LAWRENCE C. KUO, DAVID B. OLSEN, AND STEVEN S. CARROLL

VOLUME 276. Macromolecular Crystallography (Part A)
Edited by CHARLES W. CARTER, JR., AND ROBERT M. SWEET

VOLUME 277. Macromolecular Crystallography (Part B)
Edited by CHARLES W. CARTER, JR., AND ROBERT M. SWEET

VOLUME 278. Fluorescence Spectroscopy
Edited by LUDWIG BRAND AND MICHAEL L. JOHNSON

VOLUME 279. Vitamins and Coenzymes (Part I)
Edited by DONALD B. MCCORMICK, JOHN W. SUTTIE, AND CONRAD WAGNER

VOLUME 280. Vitamins and Coenzymes (Part J)
Edited by DONALD B. MCCORMICK, JOHN W. SUTTIE, AND CONRAD WAGNER

VOLUME 281. Vitamins and Coenzymes (Part K)
Edited by DONALD B. MCCORMICK, JOHN W. SUTTIE, AND CONRAD WAGNER

VOLUME 282. Vitamins and Coenzymes (Part L)
Edited by DONALD B. MCCORMICK, JOHN W. SUTTIE, AND CONRAD WAGNER

VOLUME 283. Cell Cycle Control
Edited by WILLIAM G. DUNPHY

VOLUME 284. Lipases (Part A: Biotechnology)
Edited by BYRON RUBIN AND EDWARD A. DENNIS

VOLUME 285. Cumulative Subject Index Volumes 263, 264, 266–284, 286–289

VOLUME 286. Lipases (Part B: Enzyme Characterization and Utilization)
Edited by BYRON RUBIN AND EDWARD A. DENNIS

VOLUME 287. Chemokines
Edited by RICHARD HORUK

VOLUME 288. Chemokine Receptors
Edited by RICHARD HORUK

VOLUME 289. Solid Phase Peptide Synthesis
Edited by GREGG B. FIELDS

VOLUME 290. Molecular Chaperones
Edited by GEORGE H. LORIMER AND THOMAS BALDWIN

VOLUME 291. Caged Compounds
Edited by GERARD MARRIOTT

VOLUME 292. ABC Transporters: Biochemical, Cellular, and Molecular Aspects
Edited by SURESH V. AMBUDKAR AND MICHAEL M. GOTTESMAN

VOLUME 293. Ion Channels (Part B)
Edited by P. MICHAEL CONN

VOLUME 294. Ion Channels (Part C)
Edited by P. MICHAEL CONN

VOLUME 295. Energetics of Biological Macromolecules (Part B)
Edited by GARY K. ACKERS AND MICHAEL L. JOHNSON

VOLUME 296. Neurotransmitter Transporters
Edited by SUSAN G. AMARA

VOLUME 297. Photosynthesis: Molecular Biology of Energy Capture
Edited by LEE MCINTOSH

VOLUME 298. Molecular Motors and the Cytoskeleton (Part B)
Edited by RICHARD B. VALLEE

VOLUME 299. Oxidants and Antioxidants (Part A)
Edited by LESTER PACKER

VOLUME 300. Oxidants and Antioxidants (Part B)
Edited by LESTER PACKER

VOLUME 301. Nitric Oxide: Biological and Antioxidant Activities (Part C)
Edited by LESTER PACKER

VOLUME 302. Green Fluorescent Protein
Edited by P. MICHAEL CONN

VOLUME 303. cDNA Preparation and Display
Edited by SHERMAN M. WEISSMAN

VOLUME 304. Chromatin
Edited by PAUL M. WASSARMAN AND ALAN P. WOLFFE

VOLUME 305. Bioluminescence and Chemiluminescence (Part C)
Edited by THOMAS O. BALDWIN AND MIRIAM M. ZIEGLER

VOLUME 306. Expression of Recombinant Genes in Eukaryotic Systems
Edited by JOSEPH C. GLORIOSO AND MARTIN C. SCHMIDT

VOLUME 307. Confocal Microscopy
Edited by P. MICHAEL CONN

VOLUME 308. Enzyme Kinetics and Mechanism (Part E: Energetics of Enzyme Catalysis)
Edited by DANIEL L. PURICH AND VERN L. SCHRAMM

VOLUME 309. Amyloid, Prions, and Other Protein Aggregates
Edited by RONALD WETZEL

VOLUME 310. Biofilms
Edited by RON J. DOYLE

VOLUME 311. Sphingolipid Metabolism and Cell Signaling (Part A)
Edited by ALFRED H. MERRILL, JR., AND YUSUF A. HANNUN

VOLUME 312. Sphingolipid Metabolism and Cell Signaling (Part B)
Edited by ALFRED H. MERRILL, JR., AND YUSUF A. HANNUN

Volume 313. Antisense Technology
(Part A: General Methods, Methods of Delivery, and RNA Studies)
Edited by M. Ian Phillips

Volume 314. Antisense Technology (Part B: Applications)
Edited by M. Ian Phillips

Volume 315. Vertebrate Phototransduction and the Visual Cycle (Part A)
Edited by Krzysztof Palczewski

Volume 316. Vertebrate Phototransduction and the Visual Cycle (Part B)
Edited by Krzysztof Palczewski

Volume 317. RNA–Ligand Interactions (Part A: Structural Biology Methods)
Edited by Daniel W. Celander and John N. Abelson

Volume 318. RNA–Ligand Interactions (Part B: Molecular Biology Methods)
Edited by Daniel W. Celander and John N. Abelson

Volume 319. Singlet Oxygen, UV-A, and Ozone
Edited by Lester Packer and Helmut Sies

Volume 320. Cumulative Subject Index Volumes 290–319

Volume 321. Numerical Computer Methods (Part C)
Edited by Michael L. Johnson and Ludwig Brand

Volume 322. Apoptosis
Edited by John C. Reed

Volume 323. Energetics of Biological Macromolecules (Part C)
Edited by Michael L. Johnson and Gary K. Ackers

Volume 324. Branched-Chain Amino Acids (Part B)
Edited by Robert A. Harris and John R. Sokatch

Volume 325. Regulators and Effectors of Small GTPases
(Part D: Rho Family)
Edited by W. E. Balch, Channing J. Der, and Alan Hall

Volume 326. Applications of Chimeric Genes and Hybrid Proteins
(Part A: Gene Expression and Protein Purification)
Edited by Jeremy Thorner, Scott D. Emr, and John N. Abelson

VOLUME 327. Applications of Chimeric Genes and Hybrid Proteins (Part B: Cell Biology and Physiology)
Edited by JEREMY THORNER, SCOTT D. EMR, AND JOHN N. ABELSON

VOLUME 328. Applications of Chimeric Genes and Hybrid Proteins (Part C: Protein–Protein Interactions and Genomics)
Edited by JEREMY THORNER, SCOTT D. EMR, AND JOHN N. ABELSON

VOLUME 329. Regulators and Effectors of Small GTPases (Part E: GTPases Involved in Vesicular Traffic)
Edited by W. E. BALCH, CHANNING J. DER, AND ALAN HALL

VOLUME 330. Hyperthermophilic Enzymes (Part A)
Edited by MICHAEL W. W. ADAMS AND ROBERT M. KELLY

VOLUME 331. Hyperthermophilic Enzymes (Part B)
Edited by MICHAEL W. W. ADAMS AND ROBERT M. KELLY

VOLUME 332. Regulators and Effectors of Small GTPases (Part F: Ras Family I)
Edited by W. E. BALCH, CHANNING J. DER, AND ALAN HALL

VOLUME 333. Regulators and Effectors of Small GTPases (Part G: Ras Family II)
Edited by W. E. BALCH, CHANNING J. DER, AND ALAN HALL

VOLUME 334. Hyperthermophilic Enzymes (Part C)
Edited by MICHAEL W. W. ADAMS AND ROBERT M. KELLY

VOLUME 335. Flavonoids and Other Polyphenols
Edited by LESTER PACKER

VOLUME 336. Microbial Growth in Biofilms (Part A: Developmental and Molecular Biological Aspects)
Edited by RON J. DOYLE

VOLUME 337. Microbial Growth in Biofilms (Part B: Special Environments and Physicochemical Aspects)
Edited by RON J. DOYLE

VOLUME 338. Nuclear Magnetic Resonance of Biological Macromolecules (Part A)
Edited by THOMAS L. JAMES, VOLKER DÖTSCH, AND ULI SCHMITZ

VOLUME 339. Nuclear Magnetic Resonance of Biological Macromolecules (Part B)
Edited by THOMAS L. JAMES, VOLKER DÖTSCH, AND ULI SCHMITZ

VOLUME 340. Drug–Nucleic Acid Interactions
Edited by JONATHAN B. CHAIRES AND MICHAEL J. WARING

VOLUME 341. Ribonucleases (Part A)
Edited by ALLEN W. NICHOLSON

VOLUME 342. Ribonucleases (Part B)
Edited by ALLEN W. NICHOLSON

VOLUME 343. G Protein Pathways (Part A: Receptors)
Edited by RAVI IYENGAR AND JOHN D. HILDEBRANDT

VOLUME 344. G Protein Pathways (Part B: G Proteins and Their Regulators)
Edited by RAVI IYENGAR AND JOHN D. HILDEBRANDT

VOLUME 345. G Protein Pathways (Part C: Effector Mechanisms)
Edited by RAVI IYENGAR AND JOHN D. HILDEBRANDT

VOLUME 346. Gene Therapy Methods
Edited by M. IAN PHILLIPS

VOLUME 347. Protein Sensors and Reactive Oxygen Species (Part A: Selenoproteins and Thioredoxin)
Edited by HELMUT SIES AND LESTER PACKER

VOLUME 348. Protein Sensors and Reactive Oxygen Species (Part B: Thiol Enzymes and Proteins)
Edited by HELMUT SIES AND LESTER PACKER

VOLUME 349. Superoxide Dismutase
Edited by LESTER PACKER

VOLUME 350. Guide to Yeast Genetics and Molecular and Cell Biology (Part B)
Edited by CHRISTINE GUTHRIE AND GERALD R. FINK

VOLUME 351. Guide to Yeast Genetics and Molecular and Cell Biology (Part C)
Edited by CHRISTINE GUTHRIE AND GERALD R. FINK

VOLUME 352. Redox Cell Biology and Genetics (Part A)
Edited by CHANDAN K. SEN AND LESTER PACKER

VOLUME 353. Redox Cell Biology and Genetics (Part B)
Edited by CHANDAN K. SEN AND LESTER PACKER

Volume 354. Enzyme Kinetics and Mechanisms (Part F: Detection and Characterization of Enzyme Reaction Intermediates)
Edited by Daniel L. Purich

Volume 355. Cumulative Subject Index Volumes 321–354

Volume 356. Laser Capture Microscopy and Microdissection
Edited by P. Michael Conn

Volume 357. Cytochrome P450, Part C
Edited by Eric F. Johnson and Michael R. Waterman

Volume 358. Bacterial Pathogenesis (Part C: Identification, Regulation, and Function of Virulence Factors)
Edited by Virginia L. Clark and Patrik M. Bavoil

Volume 359. Nitric Oxide (Part D)
Edited by Enrique Cadenas and Lester Packer

Volume 360. Biophotonics (Part A)
Edited by Gerard Marriott and Ian Parker

Volume 361. Biophotonics (Part B)
Edited by Gerard Marriott and Ian Parker

Volume 362. Recognition of Carbohydrates in Biological Systems (Part A)
Edited by Yuan C. Lee and Reiko T. Lee

Volume 363. Recognition of Carbohydrates in Biological Systems (Part B)
Edited by Yuan C. Lee and Reiko T. Lee

Volume 364. Nuclear Receptors
Edited by David W. Russell and David J. Mangelsdorf

Volume 365. Differentiation of Embryonic Stem Cells
Edited by Paul M. Wassauman and Gordon M. Keller

Volume 366. Protein Phosphatases
Edited by Susanne Klumpp and Josef Krieglstein

Volume 367. Liposomes (Part A)
Edited by Nejat Düzgüneş

Volume 368. Macromolecular Crystallography (Part C)
Edited by Charles W. Carter, Jr., and Robert M. Sweet

Volume 369. Combinational Chemistry (Part B)
Edited by Guillermo A. Morales and Barry A. Bunin

VOLUME 370. RNA Polymerases and Associated Factors (Part C)
Edited by SANKAR L. ADHYA AND SUSAN GARGES

VOLUME 371. RNA Polymerases and Associated Factors (Part D)
Edited by SANKAR L. ADHYA AND SUSAN GARGES

VOLUME 372. Liposomes (Part B)
Edited by NEJAT DÜZGÜNEŞ

VOLUME 373. Liposomes (Part C)
Edited by NEJAT DÜZGÜNEŞ

VOLUME 374. Macromolecular Crystallography (Part D)
Edited by CHARLES W. CARTER, JR., AND ROBERT W. SWEET

VOLUME 375. Chromatin and Chromatin Remodeling Enzymes (Part A)
Edited by C. DAVID ALLIS AND CARL WU

VOLUME 376. Chromatin and Chromatin Remodeling Enzymes (Part B)
Edited by C. DAVID ALLIS AND CARL WU

VOLUME 377. Chromatin and Chromatin Remodeling Enzymes (Part C)
Edited by C. DAVID ALLIS AND CARL WU

VOLUME 378. Quinones and Quinone Enzymes (Part A)
Edited by HELMUT SIES AND LESTER PACKER

VOLUME 379. Energetics of Biological Macromolecules (Part D)
Edited by JO M. HOLT, MICHAEL L. JOHNSON, AND GARY K. ACKERS

VOLUME 380. Energetics of Biological Macromolecules (Part E)
Edited by JO M. HOLT, MICHAEL L. JOHNSON, AND GARY K. ACKERS

VOLUME 381. Oxygen Sensing
Edited by CHANDAN K. SEN AND GREGG L. SEMENZA

VOLUME 382. Quinones and Quinone Enzymes (Part B)
Edited by HELMUT SIES AND LESTER PACKER

VOLUME 383. Numerical Computer Methods (Part D)
Edited by LUDWIG BRAND AND MICHAEL L. JOHNSON

VOLUME 384. Numerical Computer Methods (Part E)
Edited by LUDWIG BRAND AND MICHAEL L. JOHNSON

VOLUME 385. Imaging in Biological Research (Part A)
Edited by P. MICHAEL CONN

VOLUME 386. Imaging in Biological Research (Part B)
Edited by P. MICHAEL CONN

VOLUME 387. Liposomes (Part D)
Edited by NEJAT DÜZGÜNEŞ

VOLUME 388. Protein Engineering
Edited by DAN E. ROBERTSON AND JOSEPH P. NOEL

VOLUME 389. Regulators of G-Protein Signaling (Part A)
Edited by DAVID P. SIDEROVSKI

VOLUME 390. Regulators of G-Protein Signaling (Part B)
Edited by DAVID P. SIDEROVSKI

VOLUME 391. Liposomes (Part E)
Edited by NEJAT DÜZGÜNEŞ

VOLUME 392. RNA Interference
Edited by ENGELKE ROSSI

VOLUME 393. Circadian Rhythms
Edited by MICHAEL W. YOUNG

VOLUME 394. Nuclear Magnetic Resonance of Biological Macromolecules (Part C)
Edited by THOMAS L. JAMES

VOLUME 395. Producing the Biochemical Data (Part B)
Edited by ELIZABETH A. ZIMMER AND ERIC H. ROALSON

VOLUME 396. Nitric Oxide (Part E)
Edited by LESTER PACKER AND ENRIQUE CADENAS

VOLUME 397. Environmental Microbiology
Edited by JARED R. LEADBETTER

VOLUME 398. Ubiquitin and Protein Degradation (Part A)
Edited by RAYMOND J. DESHAIES

VOLUME 399. Ubiquitin and Protein Degradation (Part B)
Edited by RAYMOND J. DESHAIES

VOLUME 400. Phase II Conjugation Enzymes and Transport Systems
Edited by HELMUT SIES AND LESTER PACKER

VOLUME 401. Glutathione Transferases and Gamma Glutamyl Transpeptidases
Edited by HELMUT SIES AND LESTER PACKER

VOLUME 402. Biological Mass Spectrometry
Edited by A. L. BURLINGAME

Volume 403. GTPases Regulating Membrane Targeting and Fusion
Edited by William E. Balch, Channing J. Der, and Alan Hall

Volume 404. GTPases Regulating Membrane Dynamics
Edited by William E. Balch, Channing J. Der, and Alan Hall

Volume 405. Mass Spectrometry: Modified Proteins and Glycoconjugates
Edited by A. L. Burlingame

Volume 406. Regulators and Effectors of Small GTPases: Rho Family
Edited by William E. Balch, Channing J. Der, and Alan Hall

Volume 407. Regulators and Effectors of Small GTPases: Ras Family
Edited by William E. Balch, Channing J. Der, and Alan Hall

Volume 408. DNA Repair (Part A)
Edited by Judith L. Campbell and Paul Modrich

Volume 409. DNA Repair (Part B)
Edited by Judith L. Campbell and Paul Modrich

Volume 410. DNA Microarrays (Part A: Array Platforms and Web-Bench Protocols)
Edited by Alan Kimmel and Brian Oliver

Volume 411. DNA Microarrays (Part B: Databases and Statistics)
Edited by Alan Kimmel and Brian Oliver

Volume 412. Amyloid, Prions, and Other Protein Aggregates (Part B)
Edited by Indu Kheterpal and Ronald Wetzel

Volume 413. Amyloid, Prions, and Other Protein Aggregates (Part C)
Edited by Indu Kheterpal and Ronald Wetzel

Volume 414. Measuring Biological Responses with Automated Microscopy
Edited by James Inglese

Volume 415. Glycobiology
Edited by Minoru Fukuda

Volume 416. Glycomics
Edited by Minoru Fukuda

Volume 417. Functional Glycomics
Edited by Minoru Fukuda

Volume 418. Embryonic Stem Cells
Edited by Irina Klimanskaya and Robert Lanza

VOLUME 419. Adult Stem Cells
Edited by IRINA KLIMANSKAYA AND ROBERT LANZA

VOLUME 420. Stem Cell Tools and Other Experimental Protocols
Edited by IRINA KLIMANSKAYA AND ROBERT LANZA

VOLUME 421. Advanced Bacterial Genetics: Use of Transposons and Phage for Genomic Engineering
Edited by KELLY T. HUGHES

VOLUME 422. Two-Componcnt Signaling Systems, Part A
Edited by MELVIN I. SIMON, BRIAN R. CRANE, AND ALEXANDRINE CRANE

VOLUME 423. Two-Component Signaling Systems, Part B
Edited by MELVIN I. SIMON, BRIAN R. CRANE, AND ALEXANDRINE CRANE

VOLUME 424. RNA Editing
Edited by JONATHA M. GOTT

VOLUME 425. RNA Modification
Edited by JONATHA M. GOTT

VOLUME 426. Integrins
Edited by DAVID CHERESH

VOLUME 427. MicroRNA Methods
Edited by JOHN J. ROSSI

VOLUME 428. Osmosensing and Osmosignaling
Edited by HELMUT SIES AND DIETER HAUSSINGER

VOLUME 429. Translation Initiation: Extract Systems and Molecular Genetics
Edited by JON LORSCH

VOLUME 430. Translation Initiation: Reconstituted Systems and Biophysical Methods
Edited by JON LORSCH

VOLUME 431. Translation Initiation: Cell Biology, High-Throughput and Chemical-Based Approaches
Edited by JON LORSCH

VOLUME 432. Lipidomics and Bioactive Lipids: Mass-Spectrometry–Based Lipid Analysis
Edited by H. ALEX BROWN

VOLUME 433. Lipidomics and Bioactive Lipids: Specialized Analytical Methods and Lipids in Disease
Edited by H. ALEX BROWN

VOLUME 434. Lipidomics and Bioactive Lipids: Lipids and Cell Signaling
Edited by H. ALEX BROWN

VOLUME 435. Oxygen Biology and Hypoxia
Edited by HELMUT SIES AND BERNHARD BRÜNE

VOLUME 436. Globins and Other Nitric Oxide-Reactive Protiens (Part A)
Edited by ROBERT K. POOLE

VOLUME 437. Globins and Other Nitric Oxide-Reactive Protiens (Part B)
Edited by ROBERT K. POOLE

VOLUME 438. Small GTPases in Disease (Part A)
Edited by WILLIAM E. BALCH, CHANNING J. DER, AND ALAN HALL

VOLUME 439. Small GTPases in Disease (Part B)
Edited by WILLIAM E. BALCH, CHANNING J. DER, AND ALAN HALL

VOLUME 440. Nitric Oxide, Part F Oxidative and Nitrosative Stress in Redox Regulation of Cell Signaling
Edited by ENRIQUE CADENAS AND LESTER PACKER

VOLUME 441. Nitric Oxide, Part G Oxidative and Nitrosative Stress in Redox Regulation of Cell Signaling
Edited by ENRIQUE CADENAS AND LESTER PACKER

VOLUME 442. Programmed Cell Death, General Principles for Studying Cell Death (Part A)
Edited by ROYA KHOSRAVI-FAR, ZAHRA ZAKERI, RICHARD A. LOCKSHIN, AND MAURO PIACENTINI

VOLUME 443. Angiogenesis: *In Vitro* Systems
Edited by DAVID A. CHERESH

VOLUME 444. Angiogenesis: *In Vivo* Systems (Part A)
Edited by DAVID A. CHERESH

VOLUME 445. Angiogenesis: *In Vivo* Systems (Part B)
Edited by DAVID A. CHERESH

VOLUME 446. Programmed Cell Death, The Biology and Therapeutic Implications of Cell Death (Part B)
Edited by ROYA KHOSRAVI-FAR, ZAHRA ZAKERI, RICHARD A. LOCKSHIN, AND MAURO PIACENTINI

VOLUME 447. RNA Turnover in Bacteria, Archaea and Organelles
Edited by LYNNE E. MAQUAT AND CECILIA M. ARRAIANO

VOLUME 448. RNA Turnover in Eukaryotes: Nucleases, Pathways and Analysis of mRNA Decay
Edited by LYNNE E. MAQUAT AND MEGERDITCH KILEDJIAN

VOLUME 449. RNA Turnover in Eukaryotes: Analysis of Specialized and Quality Control RNA Decay Pathways
Edited by LYNNE E. MAQUAT AND MEGERDITCH KILEDJIAN

VOLUME 450. Fluorescence Spectroscopy
Edited by LUDWIG BRAND AND MICHAEL L. JOHNSON

VOLUME 451. Autophagy: Lower Eukaryotes and Non-Mammalian Systems (Part A)
Edited by DANIEL J. KLIONSKY

VOLUME 452. Autophagy in Mammalian Systems (Part B)
Edited by DANIEL J. KLIONSKY

VOLUME 453. Autophagy in Disease and Clinical Applications (Part C)
Edited by DANIEL J. KLIONSKY

VOLUME 454. Computer Methods (Part A)
Edited by MICHAEL L. JOHNSON AND LUDWIG BRAND

VOLUME 455. Biothermodynamics (Part A)
Edited by MICHAEL L. JOHNSON, JO M. HOLT, AND GARY K. ACKERS (RETIRED)

VOLUME 456. Mitochondrial Function, Part A: Mitochondrial Electron Transport Complexes and Reactive Oxygen Species
Edited by WILLIAM S. ALLISON AND IMMO E. SCHEFFLER

VOLUME 457. Mitochondrial Function, Part B: Mitochondrial Protein Kinases, Protein Phosphatases and Mitochondrial Diseases
Edited by WILLIAM S. ALLISON AND ANNE N. MURPHY

VOLUME 458. Complex Enzymes in Microbial Natural Product Biosynthesis, Part A: Overview Articles and Peptides
Edited by DAVID A. HOPWOOD

VOLUME 459. Complex Enzymes in Microbial Natural Product Biosynthesis, Part B: Polyketides, Aminocoumarins and Carbohydrates
Edited by DAVID A. HOPWOOD

Volume 460. Chemokines, Part A
Edited by Tracy M. Handel and Damon J. Hamel

Volume 461. Chemokines, Part B
Edited by Tracy M. Handel and Damon J. Hamel

Volume 462. Non-Natural Amino Acids
Edited by Tom W. Muir and John N. Abelson

Volume 463. Guide to Protein Purification, 2nd Edition
Edited by Richard R. Burgess and Murray P. Deutscher

Volume 464. Liposomes, Part F
Edited by Nejat Düzgüneş

Volume 465. Liposomes, Part G
Edited by Nejat Düzgüneş

Volume 466. Biothermodynamics, Part B
Edited by Michael L. Johnson, Gary K. Ackers, and Jo M. Holt

Volume 467. Computer Methods Part B
Edited by Michael L. Johnson and Ludwig Brand

Volume 468. Biophysical, Chemical, and Functional Probes of RNA Structure, Interactions and Folding: Part A
Edited by Daniel Herschlag

Volume 469. Biophysical, Chemical, and Functional Probes of RNA Structure, Interactions and Folding: Part B
Edited by Daniel Herschlag

Volume 470. Guide to Yeast Genetics: Functional Genomics, Proteomics, and Other Systems Analysis, 2nd Edition
Edited by Gerald Fink, Jonathan Weissman, and Christine Guthrie

Volume 471. Two-Component Signaling Systems, Part C
Edited by Melvin I. Simon, Brian R. Crane, and Alexandrine Crane

Volume 472. Single Molecule Tools, Part A: Fluorescence Based Approaches
Edited by Nils G. Walter

Volume 473. Thiol Redox Transitions in Cell Signaling, Part A Chemistry and Biochemistry of Low Molecular Weight and Protein Thiols
Edited by Enrique Cadenas and Lester Packer

VOLUME 474. Thiol Redox Transitions in Cell Signaling, Part B Cellular Localization and Signaling
Edited by ENRIQUE CADENAS AND LESTER PACKER

VOLUME 475. Single Molecule Tools, Part B: Super-Resolution, Particle Tracking, Multiparameter, and Force Based Methods
Edited by NILS G. WALTER

VOLUME 476. Guide to Techniques in Mouse Development, Part A Mice, Embryos, and Cells, 2nd Edition
Edited by PAUL M. WASSARMAN AND PHILIPPE M. SORIANO

VOLUME 477. Guide to Techniques in Mouse Development, Part B Mouse Molecular Genetics, 2nd Edition
Edited by PAUL M. WASSARMAN AND PHILIPPE M. SORIANO

VOLUME 478. Glycomics
Edited by MINORU FUKUDA

VOLUME 479. Functional Glycomics
Edited by MINORU FUKUDA

VOLUME 480. Glycobiology
Edited by MINORU FUKUDA

VOLUME 481. Cryo-EM, Part A: Sample Preparation and Data Collection
Edited by GRANT J. JENSEN

VOLUME 482. Cryo-EM, Part B: 3-D Reconstruction
Edited by GRANT J. JENSEN

VOLUME 483. Cryo-EM, Part C: Analyses, Interpretation, and Case Studies
Edited by GRANT J. JENSEN

VOLUME 484. Constitutive Activity in Receptors and Other Proteins, Part A
Edited by P. MICHAEL CONN

VOLUME 485. Constitutive Activity in Receptors and Other Proteins, Part B
Edited by P. MICHAEL CONN

VOLUME 486. Research on Nitrification and Related Processes, Part A
Edited by MARTIN G. KLOTZ

VOLUME 487. Computer Methods, Part C
Edited by MICHAEL L. JOHNSON AND LUDWIG BRAND

VOLUME 488. Biothermodynamics, Part C
Edited by MICHAEL L. JOHNSON, JO M. HOLT, AND GARY K. ACKERS

VOLUME 489. The Unfolded Protein Response and Cellular Stress, Part A
Edited by P. MICHAEL CONN

VOLUME 490. The Unfolded Protein Response and Cellular Stress, Part B
Edited by P. MICHAEL CONN

VOLUME 491. The Unfolded Protein Response and Cellular Stress, Part C
Edited by P. MICHAEL CONN

VOLUME 492. Biothermodynamics, Part D
Edited by MICHAEL L. JOHNSON, JO M. HOLT, AND GARY K. ACKERS

VOLUME 493. Fragment-Based Drug Design Tools,
Practical Approaches, and Examples
Edited by LAWRENCE C. KUO

VOLUME 494. Methods in Methane Metabolism, Part A
Methanogenesis
Edited by AMY C. ROSENZWEIG AND STEPHEN W. RAGSDALE

VOLUME 495. Methods in Methane Metabolism, Part B
Methanotrophy
Edited by AMY C. ROSENZWEIG AND STEPHEN W. RAGSDALE

VOLUME 496. Research on Nitrification and Related Processes, Part B
Edited by MARTIN G. KLOTZ AND LISA Y. STEIN

VOLUME 497. Synthetic Biology, Part A
Methods for Part/Device Characterization and Chassis Engineering
Edited by CHRISTOPHER VOIGT

VOLUME 498. Synthetic Biology, Part B
Computer Aided Design and DNA Assembly
Edited by CHRISTOPHER VOIGT

VOLUME 499. Biology of Serpins
Edited by JAMES C. WHISSTOCK AND PHILLIP I. BIRD

VOLUME 500. Methods in Systems Biology
Edited by DANIEL JAMESON, MALKHEY VERMA, AND HANS V. WESTERHOFF

VOLUME 501. Serpin Structure and Evolution
Edited by JAMES C. WHISSTOCK AND PHILLIP I. BIRD

VOLUME 502. Protein Engineering for Therapeutics, Part A
Edited by K. DANE WITTRUP AND GREGORY L. VERDINE

VOLUME 503. Protein Engineering for Therapeutics, Part B
Edited by K. DANE WITTRUP AND GREGORY L. VERDINE

VOLUME 504. Imaging and Spectroscopic Analysis of Living Cells
Optical and Spectroscopic Techniques
Edited by P. MICHAEL CONN

VOLUME 505. Imaging and Spectroscopic Analysis of Living Cells
Live Cell Imaging of Cellular Elements and Functions
Edited by P. MICHAEL CONN

VOLUME 506. Imaging and Spectroscopic Analysis of Living Cells
Imaging Live Cells in Health and Disease
Edited by P. MICHAEL CONN

VOLUME 507. Gene Transfer Vectors for Clinical Application
Edited by THEODORE FRIEDMANN

VOLUME 508. Nanomedicine
Cancer, Diabetes, and Cardiovascular, Central Nervous System, Pulmonary and Inflammatory Diseases
Edited by NEJAT DÜZGÜNEŞ

VOLUME 509. Nanomedicine
Infectious Diseases, Immunotherapy, Diagnostics, Antifibrotics, Toxicology and Gene Medicine
Edited by NEJAT DÜZGÜNEŞ

VOLUME 510. Cellulases
Edited by HARRY J. GILBERT

VOLUME 511. RNA Helicases
Edited by ECKHARD JANKOWSKY

VOLUME 512. Nucleosomes, Histones & Chromatin, Part A
Edited by CARL WU AND C. DAVID ALLIS

VOLUME 513. Nucleosomes, Histones & Chromatin, Part B
Edited by CARL WU AND C. DAVID ALLIS

VOLUME 514. Ghrelin
Edited by MASAYASU KOJIMA AND KENJI KANGAWA

VOLUME 515. Natural Product Biosynthesis by Microorganisms and Plants, Part A
Edited by DAVID A. HOPWOOD

Volume 516. Natural Product Biosynthesis by Microorganisms and Plants, Part B
Edited by David A. Hopwood

Volume 517. Natural Product Biosynthesis by Microorganisms and Plants, Part C
Edited by David A. Hopwood

Volume 518. Fluorescence Fluctuation Spectroscopy (FFS), Part A
Edited by Sergey Y. Tetin

Volume 519. Fluorescence Fluctuation Spectroscopy (FFS), Part B
Edited by Sergey Y. Tetin

Volume 520. G Protein Couple Receptors
Structure
Edited by P. Michael Conn

Volume 521. G Protein Couple Receptors
Trafficking and Oligomerization
Edited by P. Michael Conn

Volume 522. G Protein Couple Receptors
Modeling, Activation, Interactions and Virtual Screening
Edited by P. Michael Conn

Volume 523. Methods in Protein Design
Edited by Amy E. Keating

Volume 524. Cilia, Part A
Edited by Wallace F. Marshall

Volume 525. Cilia, Part B
Edited by Wallace F. Marshall

Volume 526. Hydrogen Peroxide and Cell Signaling, Part A
Edited by Enrique Cadenas and Lester Packer

Volume 527. Hydrogen Peroxide and Cell Signaling, Part B
Edited by Enrique Cadenas and Lester Packer

SECTION I

H_2O_2 Metabolism: Determination of a Cellular Steady-State

CHAPTER ONE

The Cellular Steady-State of H_2O_2: Latency Concepts and Gradients

H. Susana Marinho*, Luísa Cyrne*, Enrique Cadenas†, Fernando Antunes*,[1]

*Departamento de Química e Bioquímica and Centro de Química e Bioquímica, Faculdade de Ciências, Universidade de Lisboa, Lisboa, Portugal
†Pharmacology & Pharmaceutical Sciences, School of Pharmacy, University of Southern California, Los Angeles, California, USA
[1]Corresponding author: e-mail address: fantunes@fc.ul.pt

Contents

1. Introduction 4
   1.1 Measuring H_2O_2 gradients across the plasma membrane 6
2. Experimental Components and Considerations When Measuring the H_2O_2 Gradient in *S. cerevisiae* Cells 9
   2.1 H_2O_2 calibration curve 10
   2.2 Determination of H_2O_2 consumption by intact and permeabilized yeast cells 11
   2.3 Additional points 12
3. Experimental Components and Considerations When Measuring the H_2O_2 Gradient in Mammalian Cell Lines 12
   3.1 Determination of H_2O_2 consumption by intact cells 13
   3.2 Determination of H_2O_2 consumption by disrupted cells 14
4. Data Handling/Processing 16
   4.1 Yeast cells 16
   4.2 Cell lines 16
5. Summary 17
Acknowledgments 18
References 18

Abstract

Hydrogen peroxide (H_2O_2) is able to diffuse across biomembranes but, when cells are exposed to external H_2O_2, the fast consumption of H_2O_2 inside the cells due to H_2O_2-removing enzymes provides the driving force for setting up a H_2O_2 gradient across the plasma membrane. Knowing this gradient is fundamental to standardize studies with H_2O_2 as for the same extracellular H_2O_2 concentration cells with different H_2O_2 gradients may be exposed to different intracellular H_2O_2 concentrations. Here, we present the kinetic background behind the establishment of the H_2O_2 gradient and show how the gradient can be determined experimentally using the principle of enzyme latency. Furthermore, we discuss some of the caveats that may arise when determining the H_2O_2

Methods in Enzymology, Volume 527
ISSN 0076-6879
http://dx.doi.org/10.1016/B978-0-12-405882-8.00001-5

gradient. Finally, we describe detailed protocols for the experimental determination of the H_2O_2 gradient across the plasma membrane in *Saccharomyces cerevisiae* cells and in mammalian cell lines.

1. INTRODUCTION

In the 1950s, de Duve and coworkers (De Duve, 1965) found that active enzymes entrapped within a compartment, such as a membrane, may display either lower or lack of activity, the so-called latency of the enzyme, when compared with enzymes free in solution. This latency is not due to inactivation or inhibition of the enzyme within the compartment, but it is caused by the impermeability of the membrane toward the substrate. There is another possible cause for the latency, as shown by de Duve in his studies on peroxisomal catalase latency. Catalase latency is due to the fact that the enzyme is present at a very high concentration and displays first-order kinetics and so diffusion of the substrate through the membrane becomes a rate-limiting step. Therefore, although H_2O_2 is able to diffuse across biomembranes, when a membrane separates the sites where H_2O_2 is produced and consumed, a H_2O_2 gradient is formed across the membrane. When cells are exposed to external H_2O_2, the fast consumption of H_2O_2 inside the cells due to the presence of H_2O_2-removing enzymes provides the driving force for setting up a gradient across the plasma membrane (Fig. 1.1).

In most studies involving H_2O_2, it is assumed that the rate of H_2O_2 diffusion through biomembranes (v_1) is much faster than the rate of H_2O_2 intracellular consumption (v_3), which has an outcome that the H_2O_2 concentration would be the same extracellularly and intracellularly:

$$k_1 = k_2 \gg k_3 \Rightarrow \frac{[H_2O_2]_{out}}{[H_2O_2]_{in}} = \frac{k_2}{k_2} = 1 \tag{1.1}$$

However, if we consider that the rate of intracellular H_2O_2 removal is not negligible, we have:

$$k_3 \geq k_1 = k_2 \Rightarrow \frac{[H_2O_2]_{out}}{[H_2O_2]_{in}} = \frac{k_2 + k_3}{k_2} \tag{1.2}$$

Therefore, the magnitude of the gradient will depend on the overall cellular capability to remove H_2O_2 (v_3) and on the membrane permeability to H_2O_2 (v_1, v_2), which depends on the properties of the membrane and on the ratio of cellular area/volume. In mammalian cells, cellular H_2O_2 consumption is

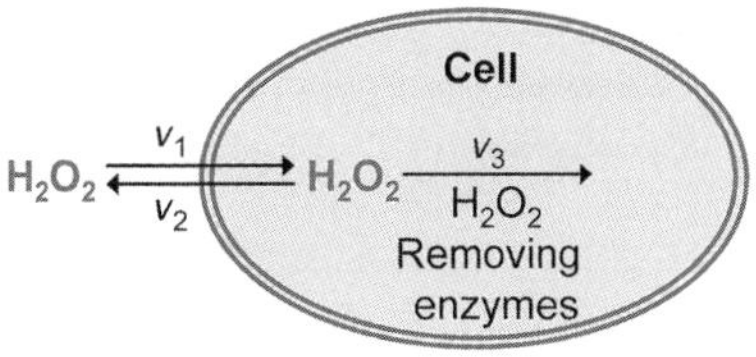

$$v_1 = k_1[H_2O_2]_{out}$$
$$v_2 = k_2[H_2O_2]_{in}$$
$$v_3 = k_3[H_2O_2]_{in} = k_{catabolism}[H_2O_2]_{in}$$
$$k_1 = k_2 = P_s \frac{A}{V_{in}}$$

At steady-state conditions:

$$v_1 = v_2 + v_3$$
$$k_2[H_2O_2]_{out} = k_2[H_2O_2]_{in} + k_3[H_2O_2]_{in}$$
$$[H_2O_2]_{out} = \frac{(k_2 + k_3)}{k_2}[H_2O_2]_{in}$$

Figure 1.1 Kinetic reasons for the establishment of a H_2O_2 gradient across the cell membrane. At steady-state conditions, and if the intracellular consumption of H_2O_2 is not negligible, the H_2O_2 concentration inside the cell ($[H_2O_2]_{in}$) will be lower than the H_2O_2 extracellular concentration ($[H_2O_2]_{out}$). The magnitude of the gradient is dependent on the relative rate values for the intracellular H_2O_2 consumption (v_3) and the diffusion through the plasma membrane (v_2) and is given by the ratio $(k_2+k_3)/k_2$. P_s is the permeability coefficient for H_2O_2 and A/V_{in} is the ratio between the area and volume of the cell. *Adapted from Antunes and Cadenas (2000).* (For color version of this figure, the reader is referred to the online version of this chapter.)

largely due to catalysis by catalase and GSH peroxidase (GPx), with a lesser role for peroxiredoxins. Catalase (Chance, Sies, & Boveris, 1979) and GPx (Flohé, 1979) both display first-order kinetics and not saturation kinetics under the conditions prevalent *in vivo*. So, the gradient between the intracellular (cytosolic) H_2O_2 concentration ($[H_2O_2]_{in}$) and the extracellular concentration ($[H_2O_2]_{out}$) is independent of the H_2O_2 concentration and is given by Eq. (1.3), where R is the ratio of activity of the first-order process of H_2O_2 removal between disrupted and intact cells (Nicholls, 1965).

$$\frac{[H_2O_2]_{out}}{[H_2O_2]_{in}} = R \qquad (1.3)$$

The higher the gradient value obtained, the higher the difference between the extracellular and intracellular H_2O_2 concentrations. Since cells have different H_2O_2 gradients (Table 1.1), this means that for the same extracellular H_2O_2 concentration two different cell types, or the same cells under different phases of growth or when adapted to H_2O_2, may be exposed to different intracellular H_2O_2 concentrations (Sousa-Lopes et al., 2004).

Table 1.1 The H_2O_2 gradient across the plasma membrane

Cells	$[H_2O_2]_{extracellular}$/ $[H_2O_2]_{cytosol}$	References
Jurkat T	6.7	Antunes and Cadenas (2000)
MCF-7	1.9±0.6	Oliveira-Marques, Cyrne, Marinho, and Antunes (2007)
E. coli	8.3	Seaver and Imlay (2001)
S. cerevisiae (haploid)		
Exponential phase		
wt (BY4741)	1.6±0.2	Branco, Marinho, Cyrne, and Antunes (2004)
wt adapted to H_2O_2 (15 min)	2.0±0.2	Folmer et al. (2008)
wt adapted to H_2O_2 (90 min)	3.0±0.5	Branco et al. (2004)
*erg3*Δ	0.9±0.2	Branco et al. (2004)
*erg6*Δ	1.0±0.2	Branco et al. (2004)
Stationary phase		
wt (BY4741)	19.8±6.4	Sousa-Lopes, Antunes, Cyrne, and Marinho (2004)
*ccp1*Δ	18.9±6.0	Sousa-Lopes et al. (2004)
S. cerevisiae (diploid) in exponential phase		
wt (BY4743)	3.0±0.7	Matias et al. (2007)
*fas1*Δ	2.3±0.1	Matias et al. (2007)

In the case of yeast cells, the gradient reported is only due to catalase.

1.1. Measuring H_2O_2 gradients across the plasma membrane

1.1.1 Mammalian cell lines

The H_2O_2 gradient across the plasma membrane may be obtained using the principle of enzyme latency from which Eq. (1.4) may be derived, by measuring the consumption of H_2O_2 by intact cells ($k_{intact\ cells}$) and disrupted cells ($k_{catabolism}$). In mammalian cells, the H_2O_2 gradient imposed by both catalase and GPx, which have first-order kinetics, may be obtained from determinations of k_{GPx}, the rate constant for the reaction of H_2O_2 with glutathione peroxidase (GPx) and $k_{catalase}$, the rate constant for the reaction of H_2O_2 with catalase.

$$R = \frac{k_{\text{catabolism}}}{k_{\text{intact cells}}} = \frac{k_{\text{GPx}} + k_{\text{catalase}}}{k_{\text{intact cells}}} \tag{1.4}$$

The gradient obtained by applying Eq. (1.4) will be a lower limit, since if other H_2O_2-consuming activities would be considered, the gradient would be steeper. Determination of the contribution of catalase for the gradient can be easily done in disrupted cells by determining the enzyme activity *in situ*, using digitonin to disrupt the plasma membrane but keeping peroxisomes intact. Determination of the contribution of GPx to cellular H_2O_2 removal is more complex than that due to catalase. This is due to the GPx reaction mechanism, which involves an oxidation–reduction cycle of the Se-cysteine moiety at the active center, using GSH as the reducing agent (Flohé, 1979):

$$H_2O_2 + GPx_{red} + H^+ \xrightarrow{k_{GPxred}} GPx_{ox} + H_2O \tag{1.5}$$

$$GPx_{ox} + GSH \xrightarrow{k_6} GS - GPx + H_2O \tag{1.6}$$

$$GS - GPx + GSH \xrightarrow{k_7} GPx_{red} + GSSG + H^+ \tag{1.7}$$

Flohé (1979) showed that the kinetics of glutathione peroxidase follows the equation:

$$\frac{[GPx]_{Total}}{v} = \frac{\phi_1}{[H_2O_2]} + \frac{\phi_2}{[GSH]} \tag{1.8}$$

where $[GPx]_{Total}$ is the total concentration of the enzyme, v is the reaction rate, $\phi_1 = 1/k_{GPx\ red}$ and $\phi_2 = 1/k_6 + 1/k_7$. GPx activity is usually assayed by following GSH oxidation using relatively high H_2O_2 concentrations. Under these conditions, catalysis will be mainly limited by the reduction reactions (Eqs. 1.6 and 1.7 with their respective rate constants, k_6 and k_7). However, for most *in vivo* conditions and also in experiments where H_2O_2 is added to intact cells, catalysis will be mainly limited by the oxidation step (Eq. 1.5 with a rate constant k_{GPxred}). So, common GPx activity assays are not adequate for an accurate determination of the contribution of GPx for H_2O_2 consumption. Therefore, it is necessary to use a method (Forstrom, Stults, & Tappel, 1979) which, by using the integrated rate equation of GPx kinetics, allows the determination of changes in the initial rate of GPx-catalyzed reaction with low concentrations of H_2O_2. The integrated rate equation of GPx is given by Eq. (1.9), where t is time.

$$\frac{[\text{GPx}]_{\text{Total}} \times t}{[\text{H}_2\text{O}_2]_0 - [\text{H}_2\text{O}_2]_t} = \frac{\phi_2}{[\text{GSH}]} + \phi_1 \frac{\ln\left([\text{H}_2\text{O}_2]_0/[\text{H}_2\text{O}_2]_t\right)}{[\text{H}_2\text{O}_2]_0 - [\text{H}_2\text{O}_2]_t} \tag{1.9}$$

GPx requires GSH to complete its catalytic cycle. Titration of GPx activity with digitonin is not as useful as for catalase because a) GPx latency will be determined by both H_2O_2 and GSH, and b) GSH is not transported into cells (Anderson, Powrie, Puri, & Meister, 1985). While GPx is present in the cytosol, in mitochondria (Stults, Forstrom, Chiu, & Tappel, 1977), and in the nucleus (Soboll et al., 1995), most GPx activity is present in the cytosol and, therefore, an approximation of the cytosolic activity can be inferred by considering the value found in cell lysates.

1.1.2 Yeast cells

The H_2O_2 gradient imposed by catalase can be determined by measuring H_2O_2 consumption in intact and permeabilized *Saccharomyces cerevisiae* cells. In *S. cerevisiae* cells, the main H_2O_2-removing enzymes are cytosolic catalase (Ctt1p) and mitochondrial cytochrome *c* peroxidase (Ccp1p) (Minard & McAlister-Henn, 2001), although yeast cells also contain a peroxisomal catalase and several thiol-dependent peroxidases, including three non-selenium-dependent glutathione peroxidases (Avery & Avery, 2001; Wong, Siu, & Jin, 2004). A key issue to apply Eq. (1.4) is that H_2O_2-removing enzymes operating in intact cells and in disrupted cells must be the same, otherwise the gradient obtained is not only due to cellular permeabilization. All the H_2O_2 consumption in yeast cells permeabilized with digitonin is due to cytosolic catalase, under our experimental conditions. This is due to the fact that the other H_2O_2-removing enzyme activities, namely, cytochrome *c* peroxidase activity, are limited in permeabilized cells by the availability of their reduced substrates. In intact cells, this does not happen. Therefore, to determine the H_2O_2 gradient imposed by catalase, it is important to make sure when measuring H_2O_2 consumption by intact cells that H_2O_2 is removed only via catalase. This can be accomplished by measuring H_2O_2 consumption by intact cells only after oxidation of the reducing substrates and by incubating cells in buffer so that the oxidized substrates cannot be reduced (Fig. 1.2). We confirmed by using *ctt1Δ* cells that there was only a minor H_2O_2 consumption under these experimental conditions thus validating the assumption that we measure a catalase-imposed H_2O_2 gradient.

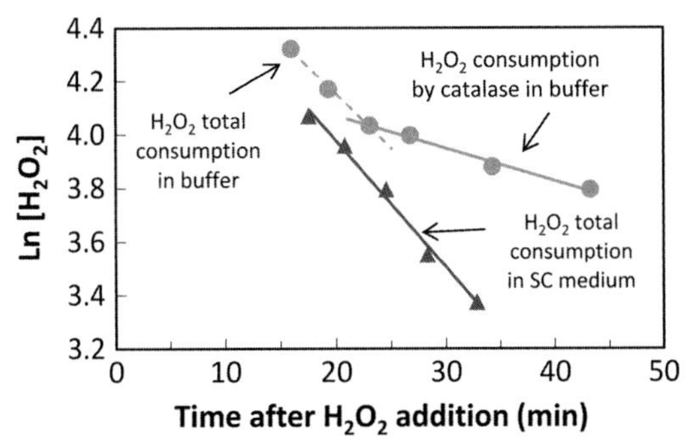

Figure 1.2 H_2O_2 consumption in wild-type intact *S. cerevisiae* cells. 120 μ*M* H_2O_2 was added to wild-type cells (BY4741, EUROSCARF, Germany) incubated at 30 °C either in 0.1 *M* potassium phosphate buffer 6.5 (circles) or in synthetic complete (SC) medium (triangles), and aliquots were taken for measurement of H_2O_2 consumption using an oxygen electrode at the indicated times. When cells are suspended in buffer, the addition of H_2O_2 leads to an oxidation of the reducing substrates of H_2O_2-metabolizing enzymes and the remaining rate of H_2O_2 consumption is only due to catalase. On the contrary, cells in SC medium are able to reduce the substrates oxidized by H_2O_2 and, unlike cells in buffer, the rate does not decrease along the assay. (For color version of this figure, the reader is referred to the online version of this chapter.)

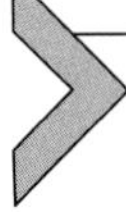

2. EXPERIMENTAL COMPONENTS AND CONSIDERATIONS WHEN MEASURING THE H_2O_2 GRADIENT IN *S. CEREVISIAE* CELLS

Reagents

1. H_2O_2—Make fresh every day the solution using concentrated H_2O_2 [Perhydrol 30% (m/m), density 1.11 g/mL, MW = 34.02, 9.79 *M*]. To obtain the stock solution of H_2O_2 (approximately 9–10 m*M*), dilute 1/1000 the concentrated H_2O_2 solution. For that, make an initial dilution of 1/100 in water (10 μL + 990 μL water or, alternatively, two sequential 1/10 dilutions) in an *eppendorf* tube. Then perform a 1/10 dilution using 500 μL of H_2O_2 diluted 1/100 and adding 4.5 mL of water. Confirm the concentration by reading the absorbance of the stock solution at 240 nm ($\varepsilon = 43.4\ M^{-1}\text{cm}^{-1}$). Keep on ice.
2. Catalase (bovine liver, Sigma C-1345, 2000–5000 U/mg protein) 1 mg/mL (in water).
3. Digitonin—Either make a solution of 10 mg/mL purified digitonin in water (make the solution just before using it) or make a 10-fold dilution from a dimethyl sulfoxide 100 mg/mL stock solution of nonpurified

digitonin. Digitonin when not purified is not soluble in water, and purification is done by recrystallization according to Kun, Kirsten, and Piper (1979).

4. 0.1 *M* potassium phosphate buffer (pH 6.5).

2.1. H_2O_2 calibration curve

H_2O_2 is assayed by the formation of O_2 after the addition of catalase (Eq. 1.10) using an oxygen electrode.

$$2H_2O_2 \xrightarrow{\text{catalase}} O_2 + 2H_2O \tag{1.10}$$

We use a chamber oxygen electrode (Oxygraph system from Hansatech Instruments Ltd., Norfolk, UK) with a magnetic stirrer and temperature control. All measurements are performed either at room temperature or at 30 °C and with a final volume of 800 μL. The electrode should be giving a stable baseline. For that, it is recommended to add 800 μL of distilled water and to connect the stirring a few hours before the measurements.

A H_2O_2 calibration curve within the range 10–90 μ*M*, where the O_2 electrode has a linear response, should be made (see Fig. 1.3). For that:

1. Pipette 10 μL up to 100 μL of the H_2O_2 stock solution to different test tubes and add distilled water to a final volume of 5 mL. Keep the test tubes either at room temperature or at 30 °C.
2. Add 400 μL of H_2O_2 from one of the test tubes, starting with the lowest concentration, to the 400 μL of 0.1 *M* potassium phosphate buffer (pH 6.5) already in the electrode chamber. Readings can also be done without using the buffer, but we found that with the buffer the oxygen electrode has a more stable output.
3. Start recording and, when a baseline is established, rapidly add 15 μL of catalase using a Hamilton syringe (being careful not to add air bubbles since they interfere in the measurement). This addition should cause an increase in the reading since O_2 is formed from the H_2O_2 present by the action of catalase. After a new baseline is established, stop the recording. The value of the difference between the two baselines should be used to make a plot versus H_2O_2 concentration.
4. Remove the content of the oxygen electrode chamber and clean thoroughly with distilled water (fill the chamber up until the middle at least four times and up until the top four times also) in order to be sure to remove all the catalase before the next H_2O_2 assay.

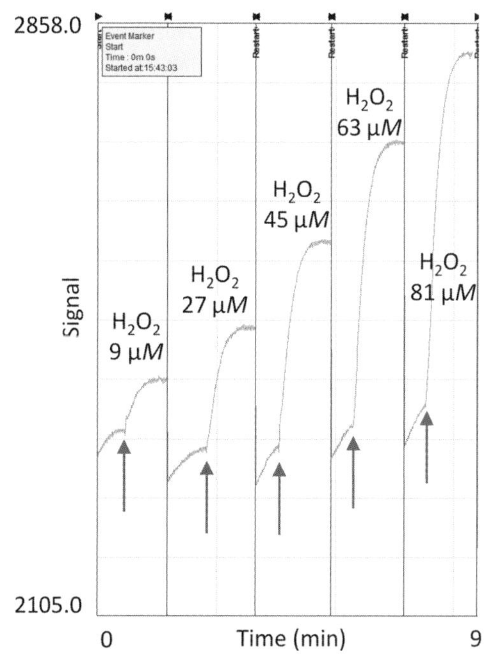

Figure 1.3 Typical recording of a H_2O_2 calibration curve made with the O_2 electrode. The indicated amounts of H_2O_2 are in the electrode chamber and, where indicated by the arrows, catalase is added. Due to the O_2 formed, there is an increase in the signal. In between each of the assays, the electrode chamber should be washed thoroughly in order that all catalase is removed. The difference in the signal between the initial and final baselines for each assay should be plotted versus the H_2O_2 concentration. (For color version of this figure, the reader is referred to the online version of this chapter.)

2.2. Determination of H_2O_2 consumption by intact and permeabilized yeast cells

This procedure is optimized for BY4741 (haploid) and BY4743 (diploid) wild-type cells obtained from EUROSCARF (Germany). A 60-mL culture of yeast cells at 0.075 OD_{600} (1 $OD_{600} \sim 2 \times 10^7$ cells/mL) in SC medium should be made and let grow at 30 °C until it reaches 0.3 OD.

1. Take 50 mL of the culture, put them in Falcon tubes prewarmed at 30 °C, and centrifuge at 5000 × *g* for 2 min. Remove the supernatant and wash the pellet with potassium phosphate buffer (pH 6.5) at 30 °C. Centrifuge again at 5000 × *g* for 2 min.
2. Remove the supernatant and resuspend the pellet with 30 mL of 0.1 *M* potassium phosphate buffer (pH 6.5) at 30 °C to obtain a cell suspension

of 0.5 OD_{600}. Make a reading of absorbance at 600 nm to confirm the OD_{600}.

3. Take aliquots (10 mL) in two conical flasks preheated at 30 °C. One is to be used for intact cells (flask-I) and the other for permeabilized cells (flask-P). Put both conical flasks in the incubator at 30 °C by shaking at 160 rpm where they should stay for the rest of the experiment. Wait 5 min.
4. Add to flask-I 200 μ*M* H_2O_2 (diploid cells) or 150 μ*M* H_2O_2 (haploid cells).
5. Add to flask-P 100 μL of 10 mg/mL digitonin. Wait 10 min and then add 100 μ*M* H_2O_2.
6. Start measuring H_2O_2 consumption in permeabilized cells (flask-P) at the oxygen electrode. The final volume in the electrode chamber should be 800 μL for all H_2O_2 assays. Take aliquots (400–800 μL) from flask-P at different times, add them to electrode chamber, and measure O_2 formation after adding 15 μL of catalase. Aliquots with the higher volume should be used when most H_2O_2 has been consumed by cells.
7. After 10 min (20 min after the addition of H_2O_2), start measuring H_2O_2 consumption in intact cells (flask-I) by removing aliquots (400–800 μL) every 10 min up until 90 min. H_2O_2 consumption should be followed only after this 20 min lag period in order to oxidize cofactors of other H_2O_2-removing enzymes.

2.3. Additional points

This protocol is optimized for the BY4741 and BY4743 wild-type strains (EUROSCARF, Germany). For other strains, it should be checked, when first making the assay, whether an incubation of 10 min with digitonin is enough for full permeabilization of the cells. Also, for other strains, it should be checked, when measuring H_2O_2 consumption in intact cells, whether this consumption is due only to catalase. For that, it may be necessary to use *ctt1*Δ strains. A critical point is to perform the whole procedure at 30 °C, as changes in temperature may lead to induction of catalase.

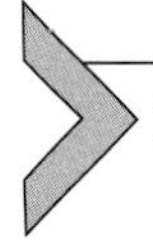

3. EXPERIMENTAL COMPONENTS AND CONSIDERATIONS WHEN MEASURING THE H_2O_2 GRADIENT IN MAMMALIAN CELL LINES

In mammalian cell lines, a procedure similar to that described above for yeast is often not possible because a catalase knockout cell line is not available. On the other hand, it is possible to estimate a gradient driven by catalase and glutathione peroxidase taking in account the kinetics of these two enzymes, using Eq. (1.4). Three measurements are needed: (1) consumption of H_2O_2 in intact cells ($k_{\text{intact cells}}$); (2) determination of catalase activity *in situ*, using digitonin to disrupt the plasma membrane but keeping peroxisomes intact (k_{catalase}); and (3) determination of GPx activity in a lysate generated in the cuvette (k_{GPx}). Usually, a large amount of cells are needed to measure GPx activity, and it is possible that the light scattering caused by the cell debris generated from the lysate in the cuvette impairs the assay. If this occurs, GPx activity is determined in a postnuclear protein extract ($k_{\text{GPxPostNuclear}}$). $k_{\text{GPxPostNuclear}}$ cannot be compared directly with $k_{\text{intact cells}}$ or k_{catalase} because the yield of the extraction of the postnuclear cell fraction is not 100%. So, an additional measurement is needed, namely, the determination of catalase activity in the postnuclear protein extract used for determination of GPx activity ($k_{\text{catalasePostNuclear}}$), and instead of Eq. (1.4), Eq. (1.11) is used.

$$R = \frac{k_{\text{catabolism}}}{k_{\text{intact cells}}} = \frac{k_{\text{GPxPostNuclear}}\left(k_{\text{catalase}}/k_{\text{catalasePostNuclear}}\right) + k_{\text{catalase}}}{k_{\text{intact cells}}} \quad (1.11)$$

The protocol described here was optimized for MCF-7 cells. For other cell types, adjustments in the number of cells used in the assays may be needed.

3.1. Determination of H_2O_2 consumption by intact cells

Reagents

1. Phosphate-buffered saline (PBS) pH 7.2 (1.5 m*M* KH_2PO_4, 155 m*M* NaCl, and 2.7 m*M* Na_2HPO_4).
2. H_2O_2 9 m*M* (see Section 2).
3. Catalase (see Section 2).
4. 0.05% Trypsin–EDTA (Gibco, Invitrogen Life Technologies, Grand Island, NY, USA)

   Cells (15×10^4 cells/cm^2 at, approximately, 30% of confluence) should be plated onto 100-mm dishes 48 h before the experiment, in order to get 60% confluence at the day of the experiment, and culture medium should be replaced by fresh medium 1 h before starting the experiment. H_2O_2 consumption by intact cells is determined by adding 90 μ*M* H_2O_2 (see Section 2) and following the decrease of H_2O_2 concentration along time by removing aliquots (800 μL) and using catalase and an oxygen electrode. A calibration curve of H_2O_2 as described in Section 2.1 should be done. H_2O_2 consumption is reported as a first-order rate constant.

3.2. Determination of H_2O_2 consumption by disrupted cells

3.2.1 Preparation of a postnuclear protein extract

A total of 18×10^6 cells should be used for the preparation of the lysate and fresh medium should be added to cells 1 h before starting the experiment. After trypsinization, cells are transferred to a single centrifuge tube. Centrifuge at $200 \times g$ for 5 min. Resuspend in 2 mL of PBS. Add 40 μL of Triton X-100 10% (v/v). Vortex and incubate on ice for 5 min. Centrifuge at $5000 \times g$ for 5 min. Remove the supernatant and measure the volume. Use part of the supernatant to measure protein concentration (Peterson, 1977). The rest should be used to measure both GPx and catalase activities.

3.2.2 Assay of glutathione peroxidase activity

Reagents

1. Potassium phosphate buffer (0.05 *M*, pH 7.0, with DTPA 1 m*M*; pH should be corrected after (diethylene triamine pentaacetic acid) DTPA addition).
2. Glutathione reductase 140 U/ml.
3. NaN_3 10 m*M* (stored at room temperature).
4. GSH 33 m*M* in phosphate buffer (can be stored at −20 °C).
5. NADPH 10 m*M* (stock solution can be kept in −20 °C in alkaline solution 0.5% $NaHCO_3$).
6. H_2O_2 0.35 m*M* (in water prepared daily).
7. Triton X-100 10% (v/v).

Postnuclear protein extract

GPx activity is measured as previously described (Antunes & Cadenas, 2000; Forstrom et al., 1979). The assay mixture (total volume 1 mL) contains 0.05 *M* phosphate buffer (pH 7.0) and successively (final concentrations) 50 μ*M* NaN_3, 1.5 U/mL glutathione reductase, 1–3.5 m*M* of reduced glutathione, 0.1 m*M* NADPH, and the postnuclear protein extract (corresponding to 0.8×10^6 MCF 7 cells). Preincubate at 37 °C for 10 min and then add 35 μ*M* H_2O_2 to the *cuvette* to start the reaction. NADPH consumption is followed at 340 nm ($\varepsilon = 6220\ M^{-1}\ cm^{-1}$). The absorbance should be recorded at short time intervals (0.1 or 0.2 s) until well after all peroxide is consumed (the rate of reaction decreases rapidly when all peroxide is consumed). It is important to continue to record the absorbance after this because the rate of

NADPH oxidation in the absence of peroxide is needed to analyse the data. In the analysis of the data, use the part of the curve corresponding to concentrations of H_2O_2 between 16 and 1.6 μ*M*.

Cuvette cell lysate

For the determination with a cuvette cell lysate, the protocol is identical to that of the postnuclear extract except that the total volume in the cuvette is 3 mL and, instead of the extract, cells (10×10^6 MCF-7 cells) and Triton X-100 are added to the assay mixture. Stirring and a spectrophotometer that deals well with light scattering is of essence.

3.2.3 Assay of catalase activity

Reagents

1. Potassium phosphate buffer 0.05 *M*, pH 7.0, 1 m*M* DTPA.
2. Digitonin—make a stock solution of 100 mg/mL digitonin either in DMSO or water (see Section 2). From that stock solution, make digitonin solutions of 0.1, 1, and 10 mg/mL.
3. 100 m*M* H_2O_2 (see Section 2).

Catalase activity can be measured by following H_2O_2 consumption at 240 nm at 25 °C as previously described (Aebi, 1978). The assay can be done either *in situ,* using approximately 2.5×10^5 cells (MCF-7) in the presence of 1 μg/mL digitonin to permeabilize the plasma membrane, but not the peroxisomal membrane, or using cell lysates.

In situ

For the *in situ* determination, the total volume of the assay medium is 3 mL and the assay should be done using stirring. Add to a quartz cuvette sequentially:

1. 2.59 mL of potassium phosphate buffer 0.05 *M*, pH 7.0, 1 m*M* DTPA.
2. 30 μL of digitonin 0.1 mg/mL.
3. 80 μL of cell suspension (2.5×10^5 MCF-7 cells).
4. Mix, autozero absorbance at 240 nm. Wait 2 min for cells to be disrupted. Mix and autozero (important to be zero at this point).
5. To start reaction, add 300 μL of 100 m*M* H_2O_2.
6. Measure absorbance at 240 nm for 2 min registering absorbance every second.

The assay should also be done with higher concentrations of digitonin in order to confirm that the peroxisomal membrane stays intact. Results are to be analyzed in a semilogarithmic plot using $\varepsilon_{240nm} = 43.4\ M^{-1}\ cm^{-1}$.

Activity is the slope of the plot.

Cells cause light scattering, and so the decrease in the absorbance recorded is going to be not only due to the consumption of H_2O_2 but also due to the sedimentation of cells. Controls have to be performed to make sure that most of the absorbance measured is due to H_2O_2 consumption and not cell sedimentation. A spectrophotometer with stirring is advisable.

Cell lysates

For the determination of catalase using cell lysates, the total assay volume is 1 mL. The assay mixture contains potassium phosphate buffer 0.05 *M*, pH 7.0, 1 m*M* DTPA, a volume of cell lysate corresponding to 2×10^5 cells and the reaction starts with 10 μL of 100 m*M* H_2O_2.

Results are to be analyzed in a semilogarithmic plot using $\varepsilon_{240nm} = 43.4\ M^{-1}\ cm^{-1}$. Activity is the slope of the plot.

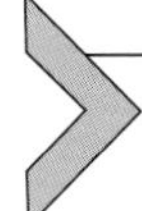

4. DATA HANDLING/PROCESSING

4.1. Yeast cells

Make a calibration curve for H_2O_2 to establish the relation between the variation in O_2 concentration when adding catalase and H_2O_2. Then, make a plot of ln $[H_2O_2]$ versus time (min) for intact cells and permeabilized cells. The two first-order rate constants $k_{\text{intact cells}}$ and $k_{catabolism}$ in min^{-1} can be obtained from the slopes (Fig. 1.4) and then should be divided by the OD_{600} to account for the number of cells present. The gradient can be calculated from Eq. (1.4).

4.2. Cell lines

The rate of H_2O_2 consumption by intact cells ($k_{\text{intact cells}}$) can be obtained using a plot of ln $[H_2O_2]$ versus time for cells growing in suspension. The rate constant is the slope of the line obtained and should be divided by the number of cells in the assay times the volume of the assay. For adherent cells, this plot cannot be done, because after taking each aliquot to measure H_2O_2 the reaction volume covering the cells decreases, and H_2O_2 will be consumed faster as the number of the cells in the assay remains constant. So, before plotting the data, data points should be corrected according to the procedure described in Marinho, Cyrne, Cadenas, and Antunes (2013). Catalase activity (k_{catalase}) is obtained by doing semilogarithmic plot of the results using $\varepsilon_{240nm} = 43.4\ M^{-1}\ cm^{-1}$. Activity is the slope of the plot divided by

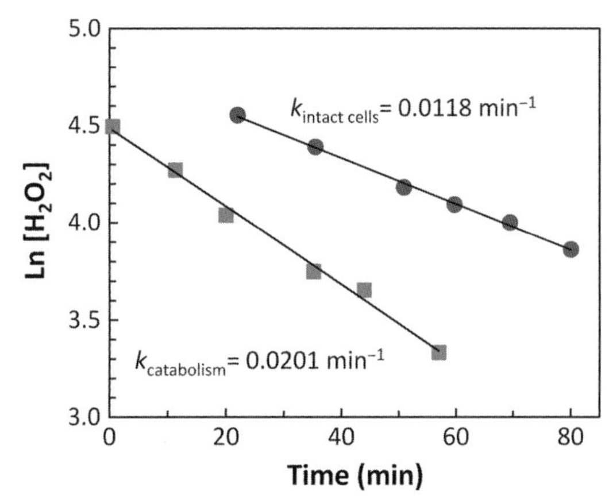

Figure 1.4 Determination of the H_2O_2 gradient in wild-type *Saccharomyces cerevisiae* cells. 150 μM H_2O_2 was added to wild-type cells (BY4741, EUROSCARF, Germany) at 0.4 OD_{600}, and H_2O_2 consumption was measured in intact cells (circles) and cells permeabilized with digitonin (squares) using an oxygen electrode. The rate constants are the slopes. A gradient of 0.0201/0.0118 = 1.7 was obtained. (For color version of this figure, the reader is referred to the online version of this chapter.)

the number of cells in the assay times the volume of the assay. To obtain the rate constant of H_2O_2 reaction with GPx (k_{GPx}), a plot of

$$\frac{t}{[H_2O_2]_0 - [H_2O_2]_t} \text{ vs. } \frac{\ln([H_2O_2]_0/[H_2O_2]_t)}{[H_2O_2]_0 - [H_2O_2]_t}$$

should be done to obtain $\phi_1/[GPx]_{Total}$. At low H_2O_2 concentrations, and since most of the enzyme is in its reduced form, the apparent rate constant for H_2O_2 consumption via GPx will be given by $[GPx]_{Total}/\phi_1 = k_{GPxred} \times [GPx]_{Total}$. The value should be divided by the number of cells in the assay times the volume of the assay. The gradient is calculated from Eq. (1.4).

If postnuclear lysates are being used for the assay of GPx and catalase, $k_{GPxPostNuclear}$ and $k_{catalasePostNuclear}$ are calculated in a similar way to k_{GPx} and $k_{catalase}$, but the slopes obtained in the experiment are divided by the amount of protein. The gradient is calculated from Eq. (1.11).

5. SUMMARY

This chapter presented the kinetics leading to the establishment of a H_2O_2 gradient across the plasma membrane when cells are treated with

H_2O_2. Moreover, a general method for the determination of this gradient in both *S. cerevisiae* cells and mammalian cell lines is also presented. This determination of the H_2O_2 gradient is important to standardize the exposure of different types of cells to extracellular H_2O_2, as it allows establishing the intracellular H_2O_2 concentration that those cells are subject to.

ACKNOWLEDGMENTS

Supported by Fundação para a Ciência e a Tecnologia (FCT), Portugal (Grants PTDC/QUI/69466/2006, PTDC/QUI-BIQ/104311/2008, and PEst-OE/QUI/UI0612/2013), and by NIH Grants RO1AG016718 (to E. C.) and PO1AG026572 (to R.D.B.).

REFERENCES

Aebi, H. E. (1978). In H.-U. Bergmeyer (Ed.), *Methods in enzymatic analysis* (pp. 273–286). Weinheim: Verlag Chemie.

Anderson, M. E., Powrie, F., Puri, R. N., & Meister, A. (1985). Glutathione monoethyl ester: Preparation, uptake by tissues, and conversion to glutathione. *Archives of Biochemistry and Biophysics*, *239*, 538–548.

Antunes, F., & Cadenas, E. (2000). Estimation of H_2O_2 gradients across biomembranes. *FEBS Letters*, *475*, 121–126.

Avery, A. M., & Avery, S. V. (2001). *Saccharomyces cerevisiae* expresses three phospholipid hydroperoxide glutathione peroxidases. *The Journal of Biological Chemistry*, *276*, 33730–33735.

Branco, M. R., Marinho, H. S., Cyrne, L., & Antunes, F. (2004). Decrease of H2O2 plasma membrane permeability during adaptation to H2O2 in *Saccharomyces cerevisiae*. *The Journal of Biological Chemistry*, *279*, 6501–6506.

Chance, B., Sies, H., & Boveris, A. (1979). Hydroperoxide metabolism in mammalian organs. *Physiological Reviews*, *59*, 527–605.

De Duve, C. (1965). The separation and characterization of subcellular particles. *Harvey Lectures*, *59*, 49–87.

Flohé, L. (1979). Glutathione peroxidase: Fact and fiction. In D. W. Fitzsimmons (Ed.), *Oxygen free radicals and tissue damage* (pp. 95–122). Amsterdam: Excerpta Medica.

Folmer, V., Pedroso, N., Matias, A. C., Lopes, S. C. D. N., Antunes, F., Cyrne, L., et al. (2008). H_2O_2 induces rapid biophysical and permeability changes in the plasma membrane of *Saccharomyces cerevisiae*. *Biochimica et Biophysica Acta*, *1778*, 1141–1147.

Forstrom, J. W., Stults, F. H., & Tappel, A. L. (1979). Rat liver cytosolic glutathione peroxidase: Reactivity with linoleic acid hydroperoxide and cumene hydroperoxide. *Archives of Biochemistry and Biophysics*, *193*, 51–55.

Kun, E., Kirsten, E., & Piper, W. N. (1979). Stabilization of mitochondrial functions with digitonin. *Methods in Enzymology*, *55*, 115–118.

Marinho, H. S., Cyrne, L., Cadenas, E., & Antunes, F. (2013). H_2O_2 delivery to cells: Steady-state versus bolus addition. *Methods in Enzymology*, *526*, 159–173.

Matias, A. C., Pedroso, N., Teodoro, N., Marinho, H. S., Antunes, F., Nogueira, J. M., et al. (2007). Down-regulation of fatty acid synthase increases the resistance of *Saccharomyces cerevisiae* cells to H_2O_2. *Free Radical Biology and Medicine*, *43*, 1458–1465.

Minard, K. I., & McAlister-Henn, L. (2001). Antioxidant function of cytosolic sources of NADPH in yeast. *Free Radical Biology and Medicine*, *31*, 832–843.

Nicholls, P. (1965). Activity of catalase in the red cell. *Biochimica et Biophysica Acta*, *99*, 286–297.

Oliveira-Marques, V., Cyrne, L., Marinho, H. S., & Antunes, F. (2007). A quantitative study of NF-kB activation by H_2O_2: Relevance in inflammation and synergy with TNF-alpha. *The Journal of Immunology*, *178*, 3893–3902.

Peterson, G. L. (1977). A simplification of the protein assay method of Lowry et al. which is more generally applicable. *Analytical Biochemistry*, *83*, 346–356.

Seaver, L. C., & Imlay, J. A. (2001). Hydrogen peroxide fluxes and compartmentalization inside growing Escherichia coli. *Journal of Bacteriology*, *183*, 7182–7189.

Soboll, S., Grundel, S., Harris, J., Kolb-Bachofen, V., Ketterer, B., & Sies, H. (1995). The content of glutathione and glutathione S-transferases and the glutathione peroxidase activity in rat liver nuclei determined by a non-aqueous technique of cell fractionation. *The Biochemical Journal*, *311*, 889–894.

Sousa-Lopes, A., Antunes, F., Cyrne, L., & Marinho, H. S. (2004). Decreased cellular permeability to H2O2 protects *Saccharomyces cerevisiae* cells in stationary phase against oxidative stress. *FEBS Letters*, *578*, 152–156.

Stults, F. H., Forstrom, J. W., Chiu, D. T. Y., & Tappel, A. L. (1977). Rat liver glutathione peroxidase: Purification and study of multiple forms. *Archives of Biochemistry and Biophysics*, *183*, 490–497.

Wong, C. M., Siu, K. L., & Jin, D. Y. (2004). Peroxiredoxin-null yeast cells are hypersensitive to oxidative stress and are genomically unstable. *The Journal of Biological Chemistry*, *279*, 23207–23213.

CHAPTER TWO

Evaluating Peroxiredoxin Sensitivity Toward Inactivation by Peroxide Substrates

Kimberly J. Nelson*, Derek Parsonage*, P. Andrew Karplus†,[1], Leslie B. Poole*,[1]

*Department of Biochemistry, Wake Forest School of Medicine, Winston-Salem, North Carolina, USA
†Department of Biochemistry and Biophysics, Oregon State University, Corvallis, Oregon, USA
[1]Corresponding authors: e-mail address: karplusp@science.oregonstate.edu; lbpoole@wakehealth.edu

Contents

1. Introduction 22
   1.1 Background 22
   1.2 Theory of the peroxide dependence of sensitivity 24
2. Materials 26
   2.1 Solutions 26
   2.2 Proteins 26
3. Measuring Inactivation Sensitivity by Steady-State NADPH-Linked Assays 27
   3.1 Evaluation of peroxide sensitivity of *E. coli* thiol peroxidase with thioredoxin and thioredoxin reductase 27
   3.2 Evaluation of substrate specificity for inactivation 31
4. Measuring Inactivation Sensitivity by Multiturnover Cycling with ROOH and DTT Followed by Mass Spectrometry Analysis 34
5. Measuring Inactivation Sensitivity of Prx1/AhpC Prxs Under Single Turnover Conditions Followed by Gel Electrophoresis 35
6. Conclusions/Summary 38
Acknowledgment 39
References 39

Abstract

Peroxiredoxins (Prxs) not only are very effective peroxide-reducing enzymes but also are susceptible to being oxidatively inactivated by their own substrates. The level of sensitivity to such hyperoxidation varies depending on both the enzyme involved and the type of peroxide substrate. For some Prxs, the hyperoxidation has physiological relevance, so it is important to define approaches that can be used to quantify sensitivity. Here, we describe three distinct approaches that can be used to obtain quantitative or semiquantitative estimates of Prx sensitivity and define $C_{hyp1\%}$ as a simple way of quantifying sensitivity so that values can easily be compared.

Methods in Enzymology, Volume 527
ISSN 0076-6879
http://dx.doi.org/10.1016/B978-0-12-405882-8.00002-7

1. INTRODUCTION

1.1. Background

Peroxiredoxins (Prxs) are a widespread family of cysteine (Cys)-dependent peroxidases that appear to be the dominant peroxidases in many organisms (reviewed in Hall, Karplus, & Poole, 2009; Winterbourn, 2008). The Prx catalytic cycle centers around an active site Cys residue (C_p for peroxidatic Cys) that is oxidized to Cys-sulfenic acid (C_p-SOH) and then reacts with another (resolving) thiol to make a disulfide, that is, in turn, reduced to regenerate the substrate-ready form of the enzyme (Fig. 2.1). An additional key point is that, in this enzyme family, the C_p residue in the fully folded active site is protected and a local unfolding event must occur prior to disulfide formation. It is now well established that certain Prxs are readily

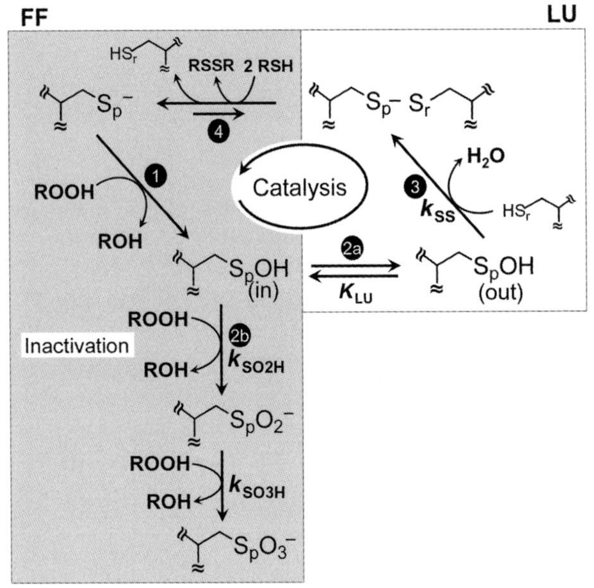

Figure 2.1 The catalytic cycle and inactivation pathways of 2-Cys peroxiredoxins. The peroxidatic cysteine is depicted as the thiolate (S_p^-), sulfenic acid (S_pOH), sulfinic acid ($S_pO_2^-$) or sulfonic acid ($S_pO_3^-$), or in a disulfide with the resolving cysteine (S_r). Reaction of the thiolate with hydroperoxide substrates (step 1) in the fully folded (FF) active site results in a sulfenic acid which either flips out of the active site due to localized unfolding (LU) to generate the disulfide (steps 2a and 3) or reacts with an additional peroxide (step 2b) to become inactivated. Reductive recycling of the disulfide species (step 4) is catalyzed by a reducing system like thioredoxin and thioredoxin reductase.

inactivated by hydrogen peroxide at submillimolar concentrations (Chae, Chung, & Rhee, 1994; Yang et al., 2002), whereas others are much less sensitive (Wood, Poole, & Karplus, 2003). This oxidative inactivation, or "hyperoxidation," occurs when a second molecule of peroxide reacts with the C_p-SOH to make Cys-sulfinic acid (or sulfinate, C_p-SO_2^-), which also can even be further oxidized by a third peroxide to Cys-sulfonic acid (or sulfonate, C_p-SO_3^-) (Fig. 2.1).

Introducing the terms "sensitive" and "robust" to characterize the two Prxs compared in their study, Wood et al. (2003) showed that the sensitivity to oxidative inactivation is caused by structural features that slow disulfide formation (step 3 of Fig. 2.1), lengthening the lifetime of the C_p-SOH intermediate, thereby allowing it more time to react with a second peroxide (Wood et al., 2003). It was further proposed that the sensitivity of certain eukaryotic Prxs has been selected for during evolution because it allows these enzymes to remove hydrogen peroxide under normal conditions, but (by being inactivated) allows hydrogen peroxide to locally build up and oxidize specific target proteins when the peroxide is purposefully produced as a part of a signaling pathway. Such a "floodgate" type of behavior has recently been shown to occur for the mitochondrial enzyme Prx3 in the regulation of cortisone production (Kil et al., 2012). Although much still remains to be learned about the physiological role(s) of Prx hyperoxidation, the existence of sulfiredoxin, an ATP-dependent reactivator of hyperoxidized Prx (Biteau, Labarre, & Toledano, 2003; Lowther & Haynes, 2011) that is inducible (Jeong, Bae, Toledano, & Rhee, 2012) and important for the survival of cells under low (10–20 μM) peroxide exposure (Baek et al., 2012), underscores the physiological relevance of Prx hyperoxidation.

By using antibodies specific for sulfinic and sulfonic acid, numerous examples have been found of Prx hyperoxidation in biologically relevant systems. Recently, in one such study, Prx hyperoxidation was seen to occur as part of what was proposed to be an ancient circadian rhythm that occurs in very diverse organisms (Edgar et al., 2012). Given the need to better understand the physiological roles of Prx hyperoxidation, it is important to be able to characterize this property for any given Prx. Furthermore, since sensitivity to hyperoxidation is a continuous rather than a binary property, it is an oversimplification to characterize enzymes as simply "sensitive" or "robust"; instead, what is needed is a set of approaches to quantify and compare the level of sensitivity of Prxs. Studies using Western blots specific for hyperoxidized Cys residues are of limited use both because they provide qualitative rather than quantitative data and because the antibodies available typically recognize sulfonic acid more strongly than sulfinic acid. Mass spectrometry (MS) and isoelectric focusing are in principle useful but have not yet been applied to provide information about the relative sensitivity to hyperoxidation of various proteins or variants. Here, we describe three methods for making such quantitative or semiquantitative comparisons. We apply these methods variously to demonstrate the high degree of insensitivity of bacterial AhpC toward H_2O_2 inactivation, to quantify the relative sensitivities of *Escherichia coli* thiol peroxidase (Tpx) toward inactivation by cumene hydroperoxide (CHP) versus hydrogen peroxide, and also to explore the previously uncharacterized sensitivity of Cp20, an unusual glutaredoxin (Cp9)-dependent Prx from *Clostridium pasteurianum*.

1.2. Theory of the peroxide dependence of sensitivity

In our earlier work (Wood et al., 2003), we briefly presented a model for how the rate of inactivation observed would be expected to vary with the concentration of peroxide used in the assay, and we explain this more fully here. When any Prx reacts with a peroxide, the sulfenic acid formed at the active site can then either form a disulfide bond with a resolving SH group (Fig. 2.1, steps 2a and 3) or react with a second peroxide to undergo irreversible inactivation (Fig. 2.1, step 2b). The partitioning of the sulfenic acid into the two pathways during turnover depends on several rates associated with chemistry and conformational changes, because disulfide bond formation requires a conformational change from fully folded (FF) to locally unfolded (LU), and the hyperoxidation reaction occurs in the fully folded enzyme. The estimated rates of conformational change are quite large for one Prx (Perkins, Gretes, Nelson, Poole, & Karplus, 2012), suggesting that this can be treated as a rapid equilibration step with the equilibrium constant for locally unfolded relative to fully folded enzyme defined as K_{LU}. Defining rate constants for disulfide bond formation and hyperoxidation as k_{SS} and k_{SO_2H}, respectively (Fig. 2.1), the rate of disulfide bond formation is $k_{SS}[E_{LU}]$ (where $[E_{LU}]$ is the concentration of locally unfolded enzyme) and the rate of inactivation is $k_{SO_2H}[ROOH]E_{FF}$. Neither reaction is chemically reversible (back to the sulfenic acid) under physiological conditions. Combining these, the fraction of protein that is inactivated during each catalytic cycle can be described by

$$f_{inact} = \frac{k_{SO_2H}[ROOH][E_{FF}]}{k_{SO_2H}[ROOH][E_{FF}] + k_{SS}[E_{LU}]}$$

Dividing the numerator and denominator by $[E_{FF}]$, and assuming that the FF and LU forms are in rapid equilibrium so that the ratio $[E_{LU}]/[E_{FF}]$ can be replaced by K_{LU}, we get

$$f_{inact} = \frac{k_{SO_2H}[ROOH]}{k_{SO_2H}[ROOH] + k_{SS}K_{LU}}$$

Then, as was described in our earlier paper (Wood et al., 2003), when f_{inact} is <0.1, the first term in the denominator is small compared with the second one, and so the above relationship can be approximated as

$$f_{inact} \approx \frac{k_{SO_2H}[ROOH]}{k_{SS}K_{LU}}$$

Thus, f_{inact} under these limiting conditions is expected to be linearly dependent on peroxide concentration. This also makes clear that for given values of k_{SO_2H} and k_{SS}, the stabilization of the FF conformation of Prxs (i.e., a lowering of K_{LU}) will have the effect of increasing f_{inact}, making the enzyme more sensitive. That f_{inact} only depends on the parameters noted means that it should not depend on the identity of the reductant used for the recycling of the disulfide form back to the dithiol form. With this in mind, one can potentially carry out the kinds of studies described here using reductants ranging from something as general as dithiothreitol (DTT) to something as specific as the true physiological partner protein. Also, we note that by reporting the sensitivity in terms of the fraction of molecules inactivated per catalytic cycle, given a constant amount of peroxide substrate, the total level of inactivation will depend directly on how many catalytic cycles have occurred.

As can be seen, the value of f_{inact} itself for a given peroxide is not an intrinsic property of the enzyme but depends on the peroxide concentration used. Previously (Wood et al., 2003), to establish a value that depends just on the enzyme and the type of peroxide, we reported the sensitivity as the slope of the $f_{inact}/[H_2O_2]$ line, which approximates $k_{SO_2H}/(k_{SS}K_{LU})$; for human PrxI reactivity toward H_2O_2, the value was 162 M^{-1}. This value implies that given a 6.2 mM solution of H_2O_2 (the reciprocal of the slope), all Prx molecules will inactivate in a single catalytic cycle (i.e., $f_{inact} = 1$). Unfortunately, this manner of reporting the sensitivity is not especially meaningful, in part because the equation used to derive it is an approximation only valid up to an f_{inact} value of 0.1. For this reason, we here suggest that a more intuitively and physiologically meaningful way to report the sensitivity would be to report the concentration of peroxide at which f_{inact} would have a value of 0.01 (i.e., 1%). This corresponds to the peroxide concentration at which 1 out of every 100 Prx molecules would be inactivated per turnover and can be calculated as the inverse of the above-mentioned slope divided by 100. For human PrxI, this works out to 6.2×10^{-5} M or 62 μM H_2O_2. Here, we will call this quantity $C_{hyp1\%}$, as in $C_{hyp1\%} = 62$ μM for human PrxI hyperoxidation by H_2O_2. At this concentration, the half-life for activity is about 70 turnovers, which, given the k_{cat} for these enzymes near 60 s^{-1}, could be as fast as ~1.2 s under conditions where the catalytic reaction with H_2O_2 is the rate-limiting step.

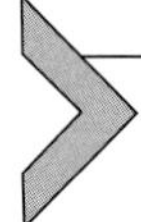

2. MATERIALS

2.1. Solutions

25 mM potassium phosphate, pH 7.0, 1 mM EDTA, with 100 mM ammonium sulfate (assay buffer)

25 mM potassium phosphate, pH 7.0, 1 mM EDTA (standard buffer)

15 mM NADPH dissolved in 10 mM Tris–SO_4, pH 8.5

100 mM CHP, dissolved first in dimethyl sulfoxide (DMSO) to make a 100 mM stock, standardized (Nelson & Parsonage, 2011)

100 mM hydrogen peroxide, standardized (Nelson & Parsonage, 2011; Poole & Ellis, 2002)

100 mM DTT

5× nonreducing sample buffer with 100 mM *N*-ethylmaleimide, 250 mM Tris/HCl pH 6.8, 50% (v/v) glycerol, 10% sodium dodecyl sulfate, 0.2% bromophenol blue

2.2. Proteins

All proteins used were recombinantly expressed and purified as described previously:

E. coli thioredoxin (Trx1 and Trx2) (Reeves, Parsonage, Nelson, & Poole, 2011)

E. coli thioredoxin reductase (TrxR) (Poole, Godzik, Nayeem, & Schmitt, 2000)

E. coli Tpx (Baker & Poole, 2003)

Salmonella typhimurium AhpC (Poole & Ellis, 1996)

C. pasteurianum Cp20 (Reynolds, Meyer, & Poole, 2002)

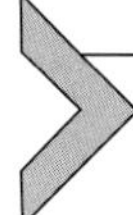

3. MEASURING INACTIVATION SENSITIVITY BY STEADY-STATE NADPH-LINKED ASSAYS

3.1. Evaluation of peroxide sensitivity of *E. coli* thiol peroxidase with thioredoxin and thioredoxin reductase

3.1.1 Demonstration of linear relationship between reaction rate and Prx concentration and establishment of the range of peroxide concentrations where inactivation is observed

As has been shown previously, the inactivation of sensitive Prx proteins can be identified by observation of the kinetic traces during turnover in the presence of micromolar to millimolar concentrations of peroxide substrates and a suitable reductant (Baker & Poole, 2003; Wood et al., 2003; Yang et al., 2002). As the Prx protein becomes hyperoxidized over the course of the reaction (with a fraction inactivated during each turnover), the kinetic trace increasingly deviates from the linear kinetics expected for the portion of the reaction having peroxide concentrations significantly above the apparent K_M. The example Prx analyzed in detail here is the Tpx protein from *E. coli*, which was reported to be sensitive to inactivation during turnover with CHP and 15-hydroperoxyeicosatetraenoic acid, but not hydrogen peroxide (Baker & Poole, 2003). The previous enzymological analysis of Tpx established the kinetic parameters for catalytic turnover using Trx1, TrxR, and NADPH. By varying peroxide and Trx1 concentrations in a full bisubstrate analysis, K_M values for CHP and H_2O_2 were found to be 9.1 ± 1.8 µ*M* and 1730 ± 360 µ*M*, respectively, while the k_{cat} values with these two substrates were about the same, at ~73 s^{-1}. The K_M value for Trx1 was ~24 µ*M* and did not vary with different peroxides. While the inactivation during turnover of Tpx with CHP was noted to occur at concentrations of 100 µ*M* and higher under the assay conditions used, the kinetics of inactivation was not further assessed.

In the present work, assays were designed to monitor the absorbance of the NADPH in the system containing TrxR, Trx1, Tpx, and CHP over multiple turnovers. Briefly, a "master mix" of TrxR, Trx, and Tpx was made in standard assay buffer and then divided into multiple cuvettes (270 µL each). NADPH was added (20 µL) and the solution was mixed, and then a 30 × peroxide solution (10 µL) was added and mixed as quickly as possible to start the assay. The 30 × stock CHP solutions all contained an equal amount of DMSO (final concentration in the assays was 0.14%). For all peroxide concentrations, the rate of NADPH oxidation in the absence of Prx was measured and TrxR concentrations were kept very low (0.05 μ*M*) in these assays to minimize the Tpx-independent loss of NADPH over these extended incubation periods. From the data shown in Fig. 2.2A using 50 μ*M* CHP, an average of 500 turnovers for Tpx in each of the

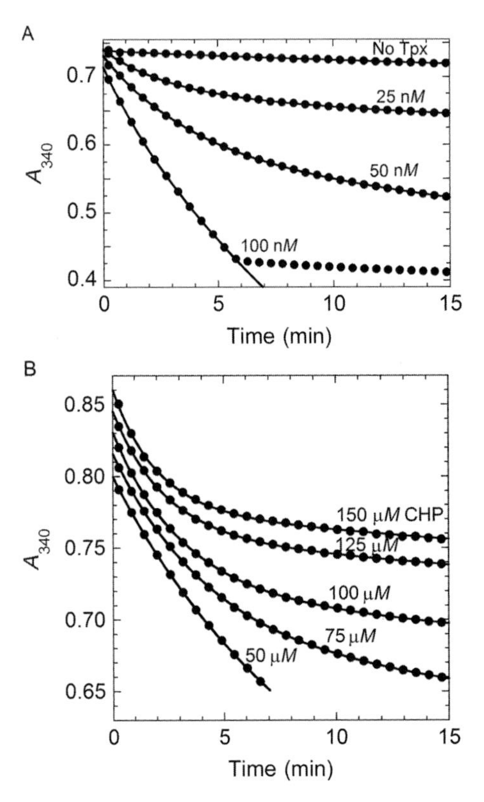

Figure 2.2 Turnover and inactivation of *E. coli* Tpx in the presence of varying Tpx and cumene hydroperoxide (CHP) concentrations. The absorbance of NADPH was monitored at 340 nm in the presence of 150 μ*M* NADPH, 0.05 μ*M E. coli* thioredoxin reductase, 1 μ*M E. coli* Trx1, and specified amounts of Tpx and CHP in standard assay buffer. Only one out of every four data points is shown. During the course of multiple turnovers, Tpx becomes increasingly inactivated and the slopes deviate from linear. Data were fit directly to a combined linear and exponential decay model as described in the text. In panel (A), assays included 50 μ*M* CHP and varying amounts of Tpx. In panel (B), assays included 0.1 μ*M* Tpx and varying CHP concentrations. Starting absorbances were intentionally staggered (by adding or subtracting a constant A_{340} to all data points in each set) in order to readily visualize the constant initial rate across CHP concentrations. Inactivation becomes more pronounced as the CHP concentration increases.

enzymes samples was complete by ~5 min; assays were typically monitored over 15–20 min.

For experiments like this one monitoring the Prx reaction rate indirectly through absorbance changes related to NADPH oxidation, it is important to establish at the outset of the studies that the rate is directly dependent on the concentration of the active Prx rather than being limited by some other component of the system. This was established in Fig. 2.2A by evaluating the initial rates at the very beginning of each reaction (the first ~4 points, collected in <0.5 min), which were indeed shown to be directly dependent on Tpx concentrations from 0.025 to 0.2 μM. The abrupt transition from turnover to background TrxR oxidase rates just before 6 min in Fig. 2.2A is due to depletion of substrate and confirms that under these conditions the $K_{M,app}$ for CHP is quite low, as we do not observe a significant change in rate until the CHP is almost completely depleted.

Given our results from Fig. 2.2A, we chose to study turnover and inactivation of Tpx with CHP using 0.1 μM Tpx and 50–150 μM CHP. As emphasized in Fig. 2.2B, the initial rate of turnover with Tpx is the same at all concentrations (reflecting the amount of active Tpx), but the divergence from linearity due to inactivation occurs to a greater extent at the higher CHP concentrations where the rate of inactivation is expected to increase.

3.1.2 Data fitting and analysis

In order to analyze the data obtained with Tpx to determine f_{inact}, we began with an analysis of the "instantaneous" rates of reaction over a short slice of time (from slopes across five time points) using a linear fit and a moving window (sliding by one point each time) of the data in an Excel spreadsheet. Rates of NADPH oxidation were first corrected for the background rate in the absence of Tpx and then converted to micromolar NADPH oxidized per minute by dividing the ΔA_{340}/min value by the molar extinction coefficient for NADPH, 6200 M^{-1} cm^{-1}. Using a logarithmic scale for the *rate* axis demonstrates that the data obey pseudo-first-order decay kinetics as expected, given that CHP is in significant excess and changing very little during the assay (Fig. 2.3). The slopes of these lines yield k_{inact} (equal to k_{SO_2H}[CHP]), which can then be divided by the initial (turnover) rate in order to obtain f_{inact} at each CHP concentration.

For this work, we were also able to develop and validate a complementary approach whereby the exponential decay rate is determined using a

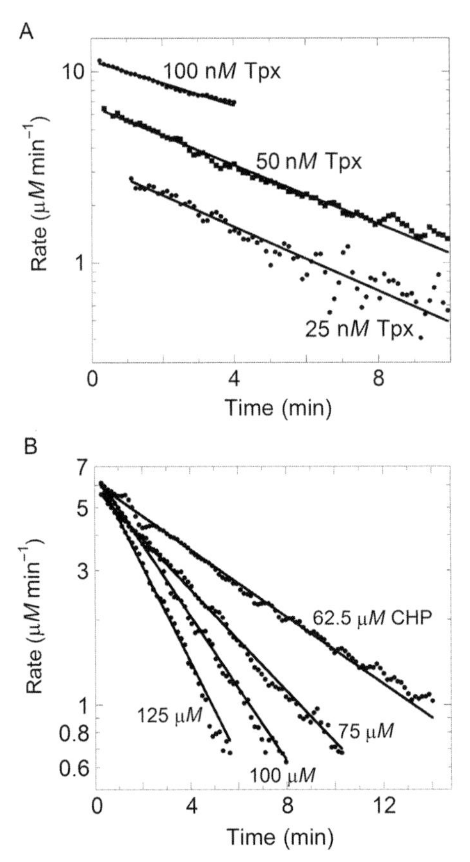

Figure 2.3 Changing rates of NADPH oxidation over time at various Tpx (A) or CHP (B) concentrations. First, slopes to give "instantaneous" rates were obtained by linear fitting, over a narrow window of five points at a time, of the data shown in Fig. 2.2; the window was advanced by one point and the process repeated across the time series. Rates were calculated from the 340 nm data using the molar extinction coefficient for NADPH as described in the text. Rates were then replotted on a logarithmic scale versus time and fit to an exponential decay model. Fits did not include data beyond 90% inactivation due to the noise in the data at these levels.

direct fit to an equation including both linear and exponential decay terms (fits by this approach are shown in Fig. 2.2).

$$y = ae^{-kt} + bt + c$$

where y is absorbance at time t, k is the exponential rate of decay in activity, and a, b, and c are coefficients that are floated during the fit. We found, in practice, that a reliable value for k can only be obtained where there is sufficient curvature in the line so that a simple linear fit of the data gives $R < 0.995$. We also determined through many repetitions that the fit to yield k was of high quality only if the error on this fit value was <2–3%. Either raw absorbance data or data converted to units of rates can be used for these fits, and background rates of absorbance changes in the absence of Tpx do not have to be subtracted prior to fitting the data as long as the background loss in absorbance over time is linear (as observed in our data, Fig. 2.2A, no Tpx).

Using multiple determinations and the NADPH–TrxR–Trx1–Tpx system, the slope of the f_{inact}/[CHP] line, which approximates $k_{SO_2H}/(k_{SS}K_{LU})$, is 64 M^{-1} (Fig. 2.4). Given this value, $C_{hyp1\%} = 156$ μM for *E. coli* Tpx with CHP. Comparing this to the $C_{hyp1\%}$ of 62 μM with H_2O_2 for human Prx I, one can see that Tpx is not quite as sensitive toward CHP as human Prx I is toward H_2O_2.

3.2. Evaluation of substrate specificity for inactivation

3.2.1 Tpx inactivation by CHP versus H_2O_2

In the original kinetic analysis of Tpx, K_M values for CHP and H_2O_2 were shown to be very different (9.1 and 1730 μM, respectively), and only the

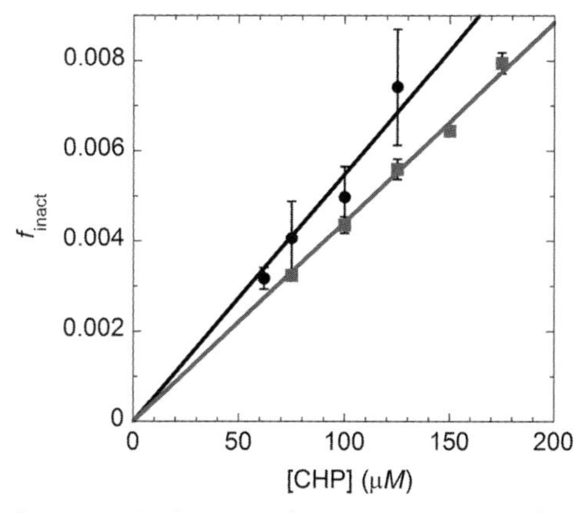

Figure 2.4 Effect of cumene hydroperoxide concentration on fraction of inactivation per turnover of Tpx. Assays as shown in Fig. 2.2 (including 150 μM NADPH, 0.05 μM TrxR, 0.1 μM Tpx, and varying CHP concentrations) were conducted using either 1 μM Trx1 (black circles) or 2 μM Trx2 (gray squares) as the direct Tpx reductant. The fraction of inactivation per turnover (f_{inact}) was calculated by dividing the pseudo-first order rate of inactivation (determined as described in the text) by the turnover rate (the rate of NADPH oxidation divided by the Tpx concentration). Data (from three or more determinations at each CHP concentration) were averaged (shown with standard error bars) and fit to a line through the origin based on the equations given in the text. The slopes of the lines using Trx1 or Trx2 as reductant were 64 and 46 M^{-1}, respectively.

CHP was reported to cause inactivation during turnover (Baker & Poole, 2003). In order to further explore the differences between these two substrates under conditions more suitable to measuring inactivation, we conducted the assays of Tpx with H_2O_2 over the same concentrations as CHP. Even though initial rates were somewhat different with the two peroxides at 100 and 250 μ*M* (reflecting the higher K_M for H_2O_2), the H_2O_2 traces are quite linear making it clear that, unlike CHP, H_2O_2 does not significantly inactivate Tpx during turnover under these conditions (Fig. 2.5). The inactivating effect of H_2O_2 was difficult to detect below m*M* levels, and above 1 m*M* H_2O_2, the background (Tpx independent) NADPH oxidation becomes high and obfuscates the analysis. With a single determination at 1 m*M* H_2O_2, an approximate value of 0.5 M^{-1} was obtained for $f_{inact}/[H_2O_2]$, corresponding to $C_{hyp1\%} \sim 20{,}000$ μ*M* indicating (by comparison with the $C_{hyp1\%} = 156$ μ*M* for CHP) that there is a >100-fold difference in sensitivity of Tpx for inactivation by these two hydroperoxides. Further evaluation of these two substrates as inactivators of Tpx was also conducted using complementary methods, as described in Section 4.

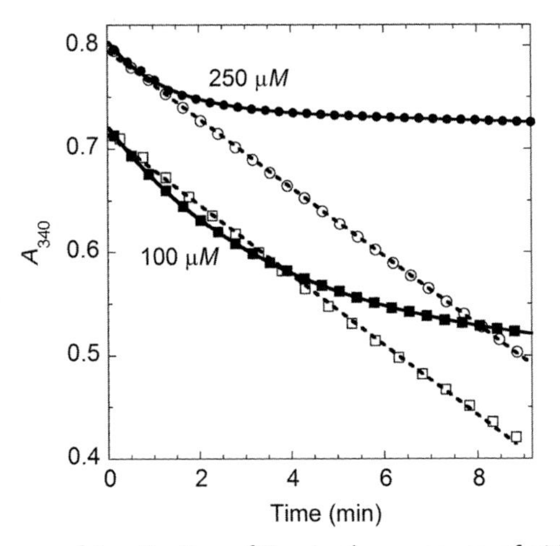

Figure 2.5 Turnover and inactivation of Tpx in the presence of either CHP or H_2O_2. Assays were conducted with NADPH, TrxR, Trx, and Tpx concentrations as in Fig. 2.2 using hydroperoxide concentrations of 100 μ*M* (squares) and 250 μ*M* (circles) and either CHP (closed symbols) or H_2O_2 (open symbols). Starting absorbances were intentionally staggered (by adding or subtracting a constant A_{340} to all data points in each set) for clarity. Data with CHP were fit directly to a combined linear and exponential decay model as described in the text, whereas data with H_2O_2 were fit to a linear equation. Only one out of every four data points is shown.

3.2.2 *Tpx inactivation by CHP in the presence of different reductants (E. coli Trx1 vs. Trx2)*

As the analysis of f_{inact} involves the determination of the pseudo-first-order rate of decay in activity adjusted for turnover rates, we hypothesized that a sensitive Prx would exhibit its characteristic peroxide sensitivity with even a relatively poor reductant as long as a sufficient number of turnovers are measured. In order to test this hypothesis, we conducted the assays described above with Tpx and CHP swapping in *E. coli* Trx2 in place of Trx1. In earlier studies of Tpx, Trx1 was shown to be a much better substrate than Trx2 (Baker & Poole, 2003). We evaluated the activity of Tpx with Trx2 using assay conditions similar to those above (0.3–0.6 μ*M* Tpx, 100 μ*M* NADPH, 0.05 μ*M* TrxR, and 50 μ*M* CHP) and found that the rate of turnover with Tpx and 10 μ*M* Trx2 was 5.4-fold slower than with Tpx and 10 μ*M* Trx1.

Upon evaluation of the inactivation and turnover rates of Tpx with Trx2 (using 2 μ*M* Trx2 rather than 1 μ*M* to partially compensate for the difference in rates between the two), the $C_{hyp1\%}$ was 217 μ*M* for CHP (Fig. 2.4). Thus, the inactivation profiles observed for Tpx with CHP with either the more efficient reductant, Trx1, or the poorer reductant, Trx2, differ by less than a factor of 1.4.

To also ask whether the generic chemical reductant DTT could be used in similar assays, we conducted a set of studies to evaluate inactivation with this relatively slow reductant of Tpx. Unfortunately, there is no strong spectral feature that changes between reduced and oxidized DTT that could be used to monitor reaction rates, so an endpoint assay utilizing ferrous ammonium sulfate and Xylenol Orange (a FOX assay) was conducted to monitor peroxide levels over time (Nelson & Parsonage, 2011). Due to the low rate of reaction between Tpx and DTT, a high concentration of Tpx (5 μ*M*) was required to observe significant changes in peroxide concentration. Because of the noise in the experimental data, which is relatively large, and the small fraction of Tpx inactivated with so few turnovers, we were unable to use this assay to measure any significant inactivation of Tpx with CHP.

3.2.3 *Inactivation of* S. typhimurium *AhpC by different hydroperoxide substrates*

It was clear in our previous work with *S. typhimurium* AhpC, a rather robust Prx (Wood et al., 2003), that the quality of the inactivation data used to assess AhpC sensitivity by this method was marginal due to the 10–30 m*M* levels of H_2O_2 required to collect these data; at these peroxide concentrations, Prx-independent background rates of NADPH oxidation are significant.

In later work by our group, the acquisition time was increased to better observe the curvature associated with inactivation and to allow comparisons of the relatively robust AhpC with even less sensitive proteins (Parsonage et al., 2005, 2010). Reexamining our previously published data investigating the sensitivity of *S. typhimurium* AhpC toward different peroxide substrates (Parsonage, Karplus, & Poole, 2008), we were able to fit data at 10 and 30 m*M* ethyl hydroperoxide using our new direct fit to the combined linear and exponential decay model, although only the 30 m*M* data yielded a rate of inactivation that matched our criteria above for extracting reliable rates as described in Section 3.1.2. Using this value and the initial rate of the reaction, and assuming a linear relationship through the origin, the f_{inact}/[ethyl hydroperoxide] value was calculated to be 0.98 M^{-1}. We note that this is quite similar to our previously reported value for H_2O_2 of $1.4 \pm 0.3\ M^{-1}$ (Wood et al., 2003). Thus, $C_{hyp1\%}$ is ~10,000 μ*M* for both ethyl hydroperoxide and H_2O_2 with *S. typhimurium* AhpC. Exploring inactivation with bulky hydroperoxides like CHP and *t*-butyl hydroperoxide, which exhibited higher K_M values with AhpC than did ethyl hydroperoxide and H_2O_2, we found that these substrates were not prone to hyperoxidize AhpC during turnover to any detectable degree based on a qualitative evaluation of the data (Parsonage et al., 2008). Thus, like with Tpx, peroxide substrates of AhpC exhibiting high K_M values do not significantly promote inactivation during turnover.

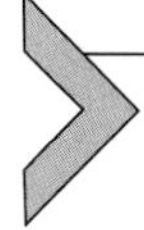

4. MEASURING INACTIVATION SENSITIVITY BY MULTITURNOVER CYCLING WITH ROOH AND DTT FOLLOWED BY MASS SPECTROMETRY ANALYSIS

As described above, monitoring inactivation by the decrease in activity during continuously monitored assays has many advantages, including quantitative evaluation of $C_{hyp1\%}$ as a measure of inactivation sensitivity. When this value is very high, however, its accurate determination becomes more difficult as shown above. However, even very robust Prxs subjected to many turnovers in the presence of high DTT and peroxide levels can eventually be inactivated through hyperoxidation, as observed by MS. Thus, we evaluated inactivation based on levels of $—SO_2H$ and $—SO_3H$ as assessed by MS after addition of varying ratios of peroxide to Prx protein in order to further evaluate peroxide sensitivity of sensitive and robust enzymes. This assay is particularly convenient in having very few components (just the enzyme, the peroxide of interest and an excess of DTT) and an incubation time that is long enough to allow for full reaction of the peroxide (overnight incubations seemed to be sufficient for this in the cases studied). Thus, it can be used even when the identity of the physiological reductant is unknown. Further, we were able to simply exchange the protein (after incubation with DTT and peroxide) into a MS-friendly buffer, 10 m*M* ammonium bicarbonate, before addition of acetonitrile and formic acid for MS analysis without first blocking free thiol groups by alkylation (avoiding additional sample manipulations and potential for incomplete reactions), although this requires that buffers be completely free of peroxides or metals which can promote oxidation in air. It should be noted that the degree of Prx hyperoxidation is dependent on both the number of turnovers and the peroxide concentration; in the assays conducted herein, we held the starting peroxide concentration constant (at 1 m*M*) and modified the number of potential turnovers by varying the Prx concentration (Fig. 2.6). Also, given that the peroxide is the limiting reagent and the incubation is extended to allow for all of it to react, the peroxide concentration is changing throughout the incubation, and we did not pursue a rigorous mathematical treatment of the data as described above.

Consistent with the results presented above, this approach demonstrates that Tpx is much more sensitive toward CHP than toward H_2O_2, with half the molecules of Tpx becoming hyperoxidized in assays where the Tpx concentration is only 50- to 100-fold less than the CHP concentration (Fig. 2.6). AhpC hyperoxidation, like that of Tpx with H_2O_2, is not significant even at a ratio of 2000:1 for peroxide to protein using either CHP or H_2O_2 (Fig. 2.6B).

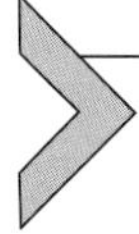

5. MEASURING INACTIVATION SENSITIVITY OF Prx1/ AhpC Prxs UNDER SINGLE TURNOVER CONDITIONS FOLLOWED BY GEL ELECTROPHORESIS

A distinct way to assess sensitivity toward inactivation by peroxides was also developed where the reduced Prx is subjected to very high peroxide concentrations in the absence of reductant, allowing only one opportunity during this partial turnover experiment for each active site to partition between either the disulfide-bonded form or the hyperoxidized (sulfinic or sulfonic acid) forms (following steps 2a or 2b, respectively, in Fig. 2.1). This analysis is particularly straightforward with *S. typhimurium* AhpC, where a simple gel separation (after alkylation of free thiol groups with NEM during denaturation to prepare samples) distinguishes readily between proteins with zero, one, and two disulfide bonds per dimer (Fig. 2.7). Because this readout reflects the competition between

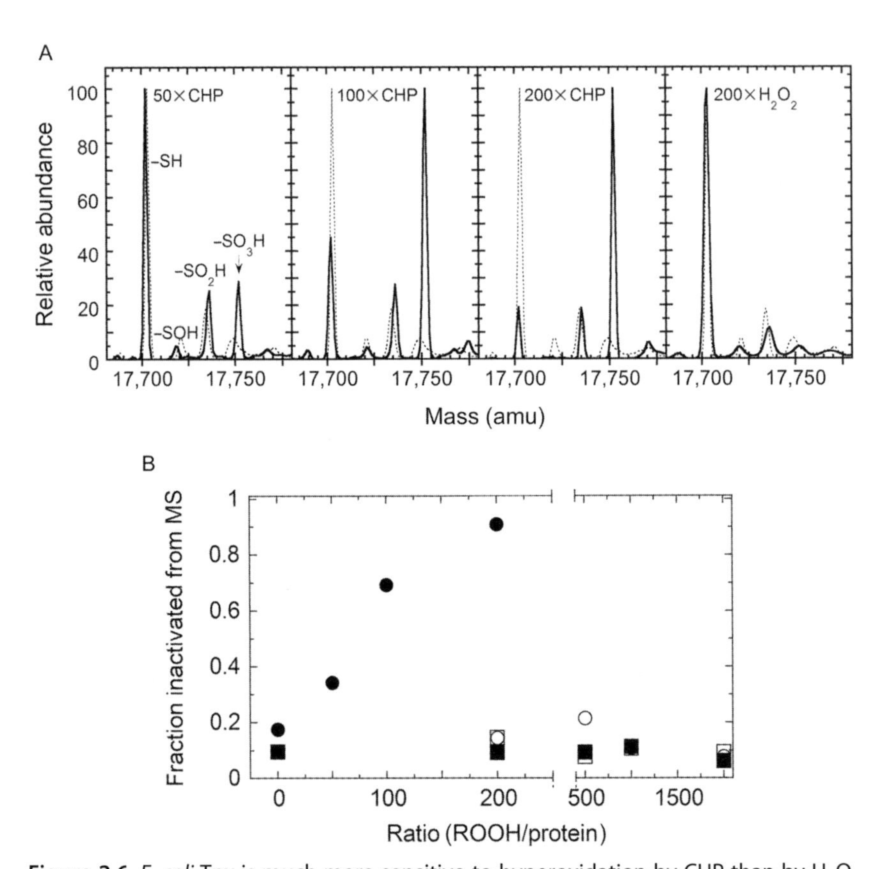

Figure 2.6 *E. coli* Tpx is much more sensitive to hyperoxidation by CHP than by H_2O_2, and *S. typhimurium* AhpC is insensitive to both, based on MS after incubation with peroxide and excess DTT. Tpx or AhpC was treated overnight (in standard buffer) with 10 m*M* DTT and 1 m*M* peroxide (CHP or H_2O_2) using protein dilutions as needed in order to obtain ratios of peroxide to protein of 50 to 2000. Samples were then exchanged into 10 m*M* ammonium bicarbonate using ultrafiltration (with Centricon YM-30 filtration units) for an effective dilution of the original buffer of ~100,000-fold prior to MS analysis. Shortly prior to MS analysis, samples were mixed 1:1 with acetonitrile and supplemented with 1% formic acid or further diluted into 50% acetonitrile with 1% formic acid as needed to obtain high quality MS data. Samples were analyzed through direct infusion and electrospray ionization using a Micromass Quattro II triple quadrupole mass spectrometer equipped with a Z-spray source. Data were processed and analyzed using MassLynx, version 3.5. Deconvoluted data shown in panel (A) include peaks at 17,701.6, 17,718.8, 17,737.1, and 17,051.2 representing the reduced thiol (—SH), sulfenic acid (—SOH), sulfinic acid (—SO_2H), and sulfonic acid (—SO_3H) forms, respectively (as indicated in the left panel). In each spectrum of panel (A), the profile of untreated Tpx is shown as a dotted line. A summary of the data with Tpx (circles) or AhpC (squares) incubated with CHP (closed symbols) or H_2O_2 (open symbols) is shown in panel (B). The approximate fractions of protein in sulfinic (—SO_2H) or sulfonic (—SO_3H) acid forms relative to all forms was calculated from the relative intensities.

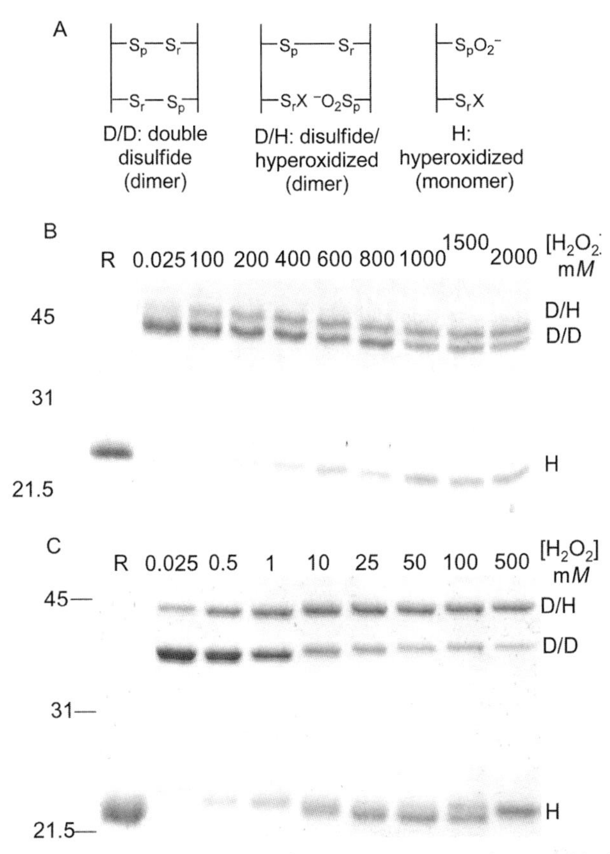

Figure 2.7 *C. pasteurianum* Cp20 is much more sensitive to hyperoxidation by H_2O_2 than *S. typhimurium* AhpC based on partial turnover experiments and subsequent mobility in gel electrophoresis. Protein in standard assay buffer was reduced with 15–20 m*M* DTT for 30–60 min at room temperature. Excess DTT was removed using a PD10 desalting column (GE Healthcare). Protein was diluted to 10 μ*M* and 25 μL aliquots were mixed with an equal volume of various concentrations of H_2O_2. After 10 min, the protein mixtures were denatured and alkylated by adding 12.5 μL of 5 × non-reducing sample loading buffer containing *N*-ethylmaleimide (final concentrations of 20 m*M* NEM and 2% SDS). Samples were incubated at room temperature for a further 10 min before heating to 95 °C for 5 min to complete the denaturation. Samples of each reaction mixture (15 μL) were loaded on a 12% polyacrylamide gel, electrophoresed, and stained for protein with GelCode Blue. Because of the sensitivity of Cp20 to oxidation, all buffers used in the preparation of the Cp20 samples were bubbled for at least 20 min with high-purity argon, and the Cp20 solution was made anaerobic by cycles of evacuation and equilibration with argon, and exposure to air was kept to a minimum. The positions of protein markers, with sizes in kDa, are shown to the left of the gels and peroxide concentrations (in mM) are given above each lane. Oxidized and hyperoxidized species as depicted in panel A are labeled to the right in panels B and C. (B) Data for *S. typhimurium* AhpC. (C) Data for *C. pasteurianum* Cp20.

hyperoxidation and disulfide bond formation, and AhpC is quite significantly resistant to hyperoxidation, partitioning is not readily observed until nearly molar levels of H_2O_2 (Fig. 2.7B).

Using this approach to investigate other Prxs in the Prx1/AhpC group (also known as typical 2-Cys Prxs), we found that the glutaredoxin (Cp9)-dependent Prx1/AhpC-like Cp20 protein from *C. pasteurianum* (Reynolds et al., 2002) was, in fact, highly sensitive to inactivation by H_2O_2 (Fig. 2.7C). This high degree of sensitivity was also observed by MS analysis after treatment with peroxide and DTT as described in the previous section. Unfortunately, the recombinantly expressed, purified Cp20 protein is a mixture of two species due to a partial codon misread in the third position (substituting Lys for Arg in 36% of the molecules), making full detailed analysis of the various redox states by MS difficult (Reynolds et al., 2002). However, it can be readily concluded that more than half of the protein is irreversibly oxidized to —SO_2H and —SO_3H upon addition of a 50-fold excess of either CHP or H_2O_2 in the presence of excess DTT, indicating that this protein is more sensitive than Tpx toward inactivation by peroxides. Assays using the *E. coli* TrxR and Trx1 system as described in Section 3 further confirmed this result and indicated that, unlike Tpx, Cp20 exhibited a similar sensitivity toward both H_2O_2 and CHP (not shown).

It should be noted that this gel approach takes significant advantage of the dimeric nature of disulfide-bonded AhpC and Cp20, given that the peroxidatic and resolving Cys residues are on different subunits. However, this method would need to be adapted to study Prxs from other groups, since outside the Prx1/AhpC group nearly all catalytic disulfide bonds are intrasubunit disulfides and covalent dimers would not be observed. Therefore, an additional method to distinguish disulfide bonded from nondisulfide bonded species would likely be required (e.g., AMS modification to cause an observable increase in apparent molecular weight with each thiol group modified) (Åslund et al., 1999).

6. CONCLUSIONS/SUMMARY

Here, we describe multiple methods that allow for the quantitation and analysis of Prx inactivation for both sensitive and robust Prx proteins. In addition to providing a simplified method of calculating f_{inact} values from steady-state turnover assays in the presence of Trx, TrxR, and NADPH, we also present complementary methods to extract quantitative or semiquantitative data using MS analysis and polyacrylamide gel shift experiments.

Because the fraction of Prx inactivated per turnover is dependent on the peroxide concentration, we propose that data for Prx sensitivity would be better presented as the peroxide concentration at which 1 of every 100 active sites is hyperoxidized per turnover, which we designate here as $C_{hyp1\%}$. Our analysis here provides data consistent with the expectation that the sensitivity of a given Prx to inactivation can be strongly dependent on the peroxide substrate but does not vary significantly with different reductants.

ACKNOWLEDGMENT

This work was supported in part by U. S. Public Health Service Grant GM050389 from the National Institutes of Health, and by a 2011 Sparkdrug discovery grant from Wake Forest School of Medicine. We thank Mike Samuel for help with the mass spectrometry analyses reported in figure 2.6.

REFERENCES

Åslund, F., Zheng, M., Beckwith, J., & Storz, G. (1999). Regulation of the OxyR transcription factor by hydrogen peroxide and the cellular thiol-disulfide status. *Proceedings of the National Academy of Sciences of the United States of America, 96*, 6161–6165.

Baek, J. Y., Han, S. H., Sung, S. H., Lee, H. E., Kim, Y. M., Noh, Y. H., et al. (2012). Sulfiredoxin protein is critical for redox balance and survival of cells exposed to low steady-state levels of H_2O_2. *The Journal of Biological Chemistry, 287*, 81–89.

Baker, L. M., & Poole, L. B. (2003). Catalytic mechanism of thiol peroxidase from *Escherichia coli*. Sulfenic acid formation and overoxidation of essential CYS61. *The Journal of Biological Chemistry, 278*, 9203–9211.

Biteau, B., Labarre, J., & Toledano, M. B. (2003). ATP-dependent reduction of cysteine-sulphinic acid by *S. cerevisiae* sulphiredoxin. *Nature, 425*, 980–984.

Chae, H. Z., Chung, S. J., & Rhee, S. G. (1994). Thioredoxin-dependent peroxide reductase from yeast. *The Journal of Biological Chemistry, 269*, 27670–27678.

Edgar, R. S., Green, E. W., Zhao, Y., van Ooijen, G., Olmedo, M., Qin, X., et al. (2012). Peroxiredoxins are conserved markers of circadian rhythms. *Nature, 485*, 459–464.

Hall, A., Karplus, P. A., & Poole, L. B. (2009). Typical 2-Cys peroxiredoxins—Structures, mechanisms and functions. *FEBS Journal, 276*, 2469–2477.

Jeong, W., Bae, S. H., Toledano, M. B., & Rhee, S. G. (2012). Role of sulfiredoxin as a regulator of peroxiredoxin function and regulation of its expression. *Free Radical Biology and Medicine, 53*, 447–456.

Kil, I. S., Lee, S. K., Ryu, K. W., Woo, H. A., Hu, M. C., Bae, S. H., et al. (2012). Feedback control of adrenal steroidogenesis via H(2)O(2)-dependent, reversible inactivation of peroxiredoxin III in mitochondria. *Molecular Cell, 46*, 584–594.

Lowther, W. T., & Haynes, A. C. (2011). Reduction of cysteine sulfinic acid in eukaryotic, typical 2-Cys peroxiredoxins by sulfiredoxin. *Antioxidants and Redox Signaling, 15*, 99–109.

Nelson, K. J., & Parsonage, D. (2011). Measurement of peroxiredoxin activity. *Current Protocols in Toxicology, 49*, 7.10.1–7.10.28, Chapter 7, Unit7.10.

Parsonage, D., Desrosiers, D. C., Hazlett, K. R., Sun, Y., Nelson, K. J., Cox, D. L., et al. (2010). Broad specificity AhpC-like peroxiredoxin and its thioredoxin reductant in the sparse antioxidant defense system of *Treponema pallidum*. *Proceedings of the National Academy of Sciences of the United States of America, 107*, 6240–6245.

Parsonage, D., Karplus, P. A., & Poole, L. B. (2008). Substrate specificity and redox potential of AhpC, a bacterial peroxiredoxin. *Proceedings of the National Academy of Sciences of the United States of America*, *105*, 8209–8214.

Parsonage, D., Youngblood, D. S., Sarma, G. N., Wood, Z. A., Karplus, P. A., & Poole, L. B. (2005). Analysis of the link between enzymatic activity and oligomeric state in AhpC, a bacterial peroxiredoxin. *Biochemistry*, *44*, 10583–10592.

Perkins, A., Gretes, M. C., Nelson, K. J., Poole, L. B., & Karplus, P. A. (2012). Mapping the active site helix-to-strand conversion of CxxxxC Peroxiredoxin Q enzymes. *Biochemistry*, *51*, 7638–7650.

Poole, L. B., & Ellis, H. R. (1996). Flavin-dependent alkyl hydroperoxide reductase from *Salmonella typhimurium*. 1. Purification and enzymatic activities of overexpressed AhpF and AhpC proteins. *Biochemistry*, *35*, 56–64.

Poole, L. B., & Ellis, H. R. (2002). Identification of cysteine sulfenic acid in AhpC of alkyl hydroperoxide reductase. *Methods in Enzymology*, *348*, 122–136.

Poole, L. B., Godzik, A., Nayeem, A., & Schmitt, J. D. (2000). AhpF can be dissected into two functional units: Tandem repeats of two thioredoxin-like folds in the N-terminus mediate electron transfer from the thioredoxin reductase-like C-terminus to AhpC. *Biochemistry*, *39*, 6602–6615.

Reeves, S. A., Parsonage, D., Nelson, K. J., & Poole, L. B. (2011). Kinetic and thermodynamic features reveal that *Escherichia coli* BCP is an unusually versatile peroxiredoxin. *Biochemistry*, *50*, 8970–8981.

Reynolds, C. M., Meyer, J., & Poole, L. B. (2002). An NADH-dependent bacterial thioredoxin reductase-like protein in conjunction with a glutaredoxin homologue form a unique peroxiredoxin (AhpC) reducing system in *Clostridium pasteurianum*. *Biochemistry*, *41*, 1990–2001.

Winterbourn, C. C. (2008). Reconciling the chemistry and biology of reactive oxygen species. *Nature Chemical Biology*, *4*, 278–286.

Wood, Z. A., Poole, L. B., & Karplus, P. A. (2003). Peroxiredoxin evolution and the regulation of hydrogen peroxide signaling. *Science*, *300*, 650–653.

Yang, K. S., Kang, S. W., Woo, H. A., Hwang, S. C., Chae, H. Z., Kim, K., et al. (2002). Inactivation of human peroxiredoxin I during catalysis as the result of the oxidation of the catalytic site cysteine to cysteine-sulfinic acid. *The Journal of Biological Chemistry*, *277*, 38029–38036.

CHAPTER THREE

Peroxiredoxins as Preferential Targets in H_2O_2-Induced Signaling

Lía M. Randall[*,†], Gerardo Ferrer-Sueta[*,†,1], Ana Denicola[*,†,1]

[*]Laboratorio de Fisicoquímica Biológica, Instituto de Química Biológica, Facultad de Ciencias, Universidad de la República, Montevideo, Uruguay
[†]Center for Free Radical and Biomedical Research, Facultad de Medicina, Universidad de la República, Montevideo, Uruguay
[1]Corresponding authors: e-mail address: gfe@fmed.edu.uy; denicola@fcien.edu.uy

Contents

1. Introduction 42
2. Reaction of H_2O_2 with Cellular Thiols 43
3. H_2O_2 Diffusion Versus Reaction with Cellular Thiols 47
4. Sulfenic Acids as Signal Transduction Intermediates 49
5. Prxs as Preferential Targets for H_2O_2 51
6. Prxs as Primary H_2O_2 Sensors and Transducers 52
7. Prx–Protein Interactions are Needed to Transmit the Signal 52
8. Posttranslational Regulation of Prxs 55
9. Summary 57
Acknowledgments 58
References 58

Abstract

Evidence has accumulated showing that hydrogen peroxide (H_2O_2) acts as a signaling molecule via oxidation of critical cysteine residues on target proteins. The reaction of H_2O_2 with thiols is thermodynamically favored, but its selectivity is imposed by differences in reaction kinetics. Previously proposed signal relaying mechanisms, such as the floodgate hypothesis and widespread protein sulfenylation, appear inconsistent with kinetic and diffusion considerations. Among all cellular thiols, the peroxidatic cysteines of peroxiredoxins (Prxs) represent preferential targets considering their high rate constants and their cellular abundance that place them as the first step in the H_2O_2-induced signaling pathways. The oxidized Prxs could transfer the signal either via thiol-disulfide redox reactions or through nonredox protein–protein interactions. Recent studies evidence Prxs interactions with protein tyrosine kinases and phosphatases, indicating a potential connection between redox and phosphorylation signaling pathways that does not need the direct reaction of H_2O_2 with phosphatase or kinase critical cysteines. Posttranslational modifications of Prxs have been observed *in vivo* (mainly overoxidation of cysteines and phosphorylation of threonines) that affect their peroxidase activity,

Methods in Enzymology, Volume 527
ISSN 0076-6879
http://dx.doi.org/10.1016/B978-0-12-405882-8.00003-9

redox state, and/or oligomeric structure, and likely impact on H_2O_2 signaling. More focus on kinetic data and redox-sensitive protein–protein interactions are needed to unravel the molecular mechanisms of H_2O_2 signaling.

1. INTRODUCTION

Biological oxidants such as hydrogen peroxide (H_2O_2) have long been recognized as responsible for oxidative modifications of macromolecules causing cellular dysfunction as the final step of oxidative stress. Now experimental evidence has accumulated indicating that production of low levels of H_2O_2 has a physiological effect, participating in redox signaling and regulating important cellular events such as proliferation, growth, migration, apoptosis, and survival (Dickinson & Chang, 2011; Gough & Cotter, 2011; Veal, Day, & Morgan, 2007). H_2O_2 is continuously produced by metabolic reactions, oxidases or oxidoreductases, and leakage from mitochondrial respiratory chain and peroxisomes. In addition, several stimuli (such as growth factors, cytokines, insulin) can trigger the assembly of different subunits to form the active membrane NADPH oxidase (Nox), or induce the expression of Nox genes, resulting in the production of superoxide, major precursor of endogenous H_2O_2. It has long been established that phagocytes activate NADPH oxidase complexes (Nox 2) to generate superoxide (thus H_2O_2) as a cytotoxic agent against pathogens. More recently, different Nox isoforms have been identified as well in nonphagocytic cell types, and not only in plasma membrane but also in membrane organelles (Cheng, Cao, Xu, van Meir, & Lambeth, 2001; Suh et al., 1999). There is a growing list of stimuli capable of increasing the expression of Nox in nonphagocytic cells: PDGF, EGF, IFN-γ, HIF-1α, TNF-α, LPS, insulin, angiotensin II, plasminogen, Th1 and Th2 cytokines that appear to be Nox isoform and cell type specific (Gough & Cotter, 2011). Nox 1, 2, 3, and 5 produce superoxide that dismutates to H_2O_2 (spontaneously or enzymatically) whereas Duox 1 and 2, and more recently, Nox 4 (Takac et al., 2011) directly form H_2O_2. Cells with increased Nox 1 expression have an increased level of endogenous H_2O_2 (Arnold et al., 2001).

Although it is now accepted that, at low physiological levels, H_2O_2 can act as an intracellular signaling molecule, the molecular mechanism by which H_2O_2 participates in redox signaling is still under intense investigation. How H_2O_2 modulates signaling by oxidation needs mechanistic insights. Unlike other second messengers (i.e., cAMP), H_2O_2 has such a simple chemical structure that its molecular recognition represents a challenge; as a matter of fact, no protein is known to bind H_2O_2 reversibly. However, it acts as an oxidant that can react with cysteine residues of target proteins in order to transmit the signal. It is widely accepted that the mechanism by which H_2O_2 participates in redox signaling is via oxidation of reactive cysteine residues; now, what are the preferential cellular thiols that react with H_2O_2?

2. REACTION OF H_2O_2 WITH CELLULAR THIOLS

In approaching the question of thiol involvement in H_2O_2-mediated redox signaling, researchers face an apparent contradiction. On one hand, all thiols can undergo oxidation by H_2O_2, and on the other hand, an all thiol-reacting scenario is unworkable as a signaling mechanism.

On thermodynamic grounds, all biological thiols have extremely favorable reactions with H_2O_2; reduction potentials span a range from −325 for PrxQ (Rouhier et al., 2004) to −89 mV/NHE for DsbA (Wunderlich & Glockshuber, 1993), whereas $E^{\circ\prime}$ for H_2O_2 is +1350 mV/NHE (Koppenol, Stanbury, & Bounds, 2010). The reactions are therefore favored with $\Delta G^{\circ\prime}$ values from −77 to −66 kcal/mol or equilibrium constants ranging from 10^{56} to 10^{49}, always favoring the oxidation of the thiol.

$$\text{P-(SH)}_2 + H_2O_2 \rightleftharpoons \text{P-S}_2 + 2H_2O \tag{3.1}$$

These estimates extracted from reduction potentials indicate a seemingly inexorable fate: Given enough H_2O_2, all thiols will eventually be oxidized.

The thermodynamic outlook of the preceding paragraph is obviously different from any biologically relevant situation. The fact that any thiol will react with H_2O_2 implies a number of experimental setups that could lead to erroneous conclusions, particularly, if mass balance is not taken into account. For instance, if a thiol protein is treated *in vitro* with excess H_2O_2 and its oxidation is observed, it does not imply that such oxidation necessarily happens *in vivo*. As we shall see below, there is a kinetic order of precedence in the reaction of thiols with H_2O_2. If cultured cells are exposed to exogenous H_2O_2, the observation of oxidized protein thiols is, once again, inexorable. However, the degree and identity of the oxidized thiols depend on a number of factors such as extracellular H_2O_2 concentration, time of exposure, cell membrane permeability, mass balance between intracellular reductants and extracellular oxidant equivalents, and the availability of repair systems.

In the most drastic scenario, excess H_2O_2 over a long time leads to indiscriminate oxidation of all intracellular thiols, what we could call the "thermodynamic scenario" and lacks any biological relevance. A more limited exposure (in amount and/or time) will yield partial oxidation, in a pattern dictated by reaction kinetics and target availability, but even in that case, the evidence of which thiols end up oxidized has to be considered critically. The H_2O_2 surpassing the first line of defense (i.e., peroxidases and catalases) could cause the oxidation of secondary targets, but there is also the possibility of second-hand oxidation due to the reaction of disulfides or sulfenic acids formed in the first place. The overall result of these "kinetic scenarios" depends on a combination of concentrations of H_2O_2 and targets, rate constants, and biophysical properties of the system, such as compartmentalization, membrane permeability, and diffusion.

Finally, even the endogenous cellular formation of H_2O_2 by oxygen-dependent systems such as Nox might be artificially enhanced in cell culture experiments where the enzyme has practically unlimited supply of O_2. This situation presents two potential pitfalls: first, H_2O_2 production could be augmented because of substrate availability (Chen, Gill, & Welch, 2005) and second, NADPH depletion could compromise the reduction and repair systems such as thioredoxin reductase and glutathione reductase, thus disrupting redox homeostasis beyond the effect of H_2O_2 per se.

If reaction (3.1) is thermodynamically so favorable, how can it be harnessed to affect signaling? The answer lies in chemical kinetics. As with other mechanisms of signal transduction such as phosphorylation, thermodynamically unstable (but kinetically inert) intermediates can be guided through the desired pathway via enzymatic catalysis and other kinetically favored reactions. We have previously complained about the scarcity of hard kinetic data in the area of thiols in redox signaling (Ferrer-Sueta et al., 2011); only a few rate constants of reaction (3.1) with biological thiols have been determined. In trying to overcome the paucity of actual numbers, several assumptions are usually made. One of the most common (and erroneous) assumptions relates thiol reactivity with its pK_a and has been repeated very often, particularly, in the PTPs literature (Held & Gibson, 2012; Karisch et al., 2011; Melvin et al., 2011; Nagahara, Matsumura, Okamoto, & Kajihara, 2009; Tanner, Parsons, Cummings, Zhou, & Gates, 2011). The line of reasoning followed is

- thiolates are more reactive than thiols,
- acidic (low pK_a) cysteines exist as thiolates at neutral pH,
- therefore, acidic cysteines are more reactive.

This reasoning is not correct because the relevant comparison is not between thiols and thiolates, but among thiolates, which differ not only in pK_a but also in many other reactivity parameters. The list of rate constants extracted from the literature in Table 3.1 should suffice to demonstrate that there is no observable correlation between thiol pK_a and reactivity toward H_2O_2. The sad truth is that we do not have any good predictor of thiol reactivity toward H_2O_2 and, so far, the only possibility is to determine the rate constant of every thiol candidate and compare it with other known targets. Such an approach has yielded extremely clear results for mitochondria (Cox, Winterbourn, & Hampton, 2010), where four peroxidases (Prx 3, GPx 1, Prx 5, and GPx 4) account for 99.9% of the H_2O_2 consumed. However, as only a few rate constants have been experimentally determined, the literature on H_2O_2 signaling tends to be largely descriptive and speculative about which are the preferred oxidation targets.

Rate constants alone are a limited descriptor of the competition between potential targets. Mass action dictates that the rate of each specific reaction involves the concentration of reactants. Since reaction (3.1) is first order in both thiolate and H_2O_2, the rate is equal to k^{app}[Target][H_2O_2], where k^{app} is the apparent rate constant at a given pH and [Target] is the total thiol concentration. If we consider a number of thiols competing for a limiting concentration of H_2O_2, the term [H_2O_2] would be equal in all cases and we can

Table 3.1 Rate constants of H_2O_2 reaction with protein cysteines

RSH	Cys pK_a	Rate constant with H_2O_2 (M^{-1} s^{-1}, pH 7.4)
Papain (Lewis, Johnson, & Shafer, 1976; Stone, 2004)	3.4	62
Prx 6 (Toledo et al., 2011)	5.2	34,000,000
Prx 5 (Trujillo et al., 2007)	5.2	300,000
Prx 2 (Manta et al., 2009)	5.3	100,000,000
DJ-1 (Andres-Mateos et al., 2007)	5.4	0.56
PTP1B (Barrett et al., 1999; LaButti, Chowdhury, Reilly, & Gates, 2007)	5.5	9 42.8
AhpC (Jaeger et al., 2004; Nelson, Parsonage, Hall, Karplus, & Poole, 2008)	5.8	39,000,000
Methionine sulfoxide reductase A (Lim et al., 2012)	7.2	14.7
GSH[a] (Winterbourn & Metodiewa, 1999)	8.7	0.87

[a]GSH is included in this table for comparison purposes.

compare the product k^{app}[Target] as the relative rate of each process (Table 3.2).

To provide a more detailed picture in time and space, we can consider that the product k^{app}[Target] behaves like a pseudo first-order rate constant; therefore, the apparent half-life of the reaction can be calculated as:

$$t_{1/2} = \frac{\ln 2}{k^{app}[\text{Target}]} \tag{3.2}$$

Even using a very conservative estimate of 20 μM Prx2 (concentrations as high as 410 μM have been measured in erythrocytes; Cho, Kato, et al., 2010), the relative rate of peroxide consumption by Prx2 is eight orders of magnitude higher than that of a typical PTP such as PTP1B (Table 3.2). To overcome this overwhelming kinetic superiority, a temporary inactivation of Prx2 and Prx1 via overoxidation (or phosphorylation) has been proposed as a mechanism of redox signaling (Woo et al., 2010; Wood, Schröder, Robin Harris, & Poole, 2003b). Thence, the so-called "floodgate hypothesis" proposes that, under Prx inactivation, the intracellular H_2O_2 levels would increase, allowing direct oxidation of other critical cysteine residues in a redox-sensitive signaling protein such as PTP1B, less abundant

Table 3.2 Comparison of reaction kinetics and diffusion distances of H_2O_2 for several thiol targets

Target	[Target] (M)	k^{app} (pH 7.4, M^{-1} s^{-1})	Relative rate (k^{app}[Target], s^{-1})	$t_{1/2}$ (s)	Distance to $C/C_0 = 0.5$ in $t_{1/2}$ (μm)[a]
Prx2	20×10^{-6}	1×10^{8} (Manta et al., 2009)	2000	0.000347	0.424
GPx1	2×10^{-6}	6×10^{7} (Flohe, Loschen, Gunzler, & Eichele, 1972)	120	0.00578	1.7
Prx5	20×10^{-6}	3×10^{5} (Trujillo et al., 2007)	6	0.116	7.7
GSH	5×10^{-3}	0.87 (Winterbourn & Metodiewa, 1999)	0.00435	159	288
PTP1B	5×10^{-6}	42.8 (Barrett et al., 1999)	2.14×10^{-4}	3239	1280

[a]Calculated using Eq. (3.3) and using a value of $D = 5.7 \times 10^{-10}$ m^2/s for H_2O_2, that is, assuming that molecular crowding inside the cell causes a 60% decrease in the diffusion coefficient of H_2O_2.

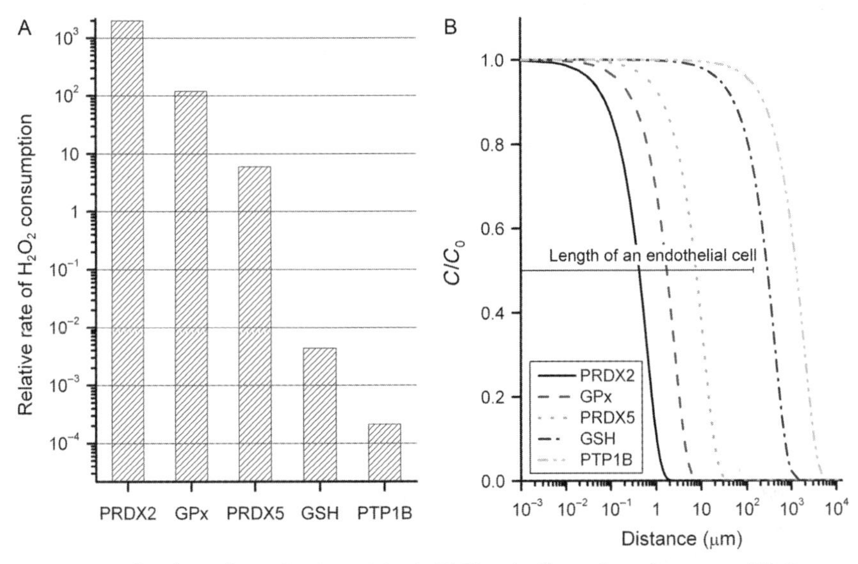

Figure 3.1 Floodgate hypothesis revisited. (A) Kinetically preferred targets of H_2O_2 represent barriers impeding it to react with putative signaling proteins, such as PTP1B. Not only enzymes such as Prx2 (potentially inhibited or inactivated) represent kinetic floodgates but also nonenzymatic targets such as GSH. (B) The presence of each target molecule has the possibility of generating a H_2O_2 gradient assuming localized formation of the oxidant. The length and steepness of the gradient depends on the competition between rate of reaction and diffusion, so 20 μ*M* Prx2 generates a gradient of roughly 1 μm where half the H_2O_2 is consumed. As a comparison, the gradient generated by 5 μ*M* PTP1B is several millimeters wide. In other words, no change in H_2O_2 inside a cell will be observed depending on slow and scarce intracellular targets such as PTP1B. (For color version of this figure, the reader is referred to the online version of this chapter.)

and less reactive toward H_2O_2. However, several other thiol targets have notoriously faster relative rates of reaction with H_2O_2 than PTP1B.

So, in reanalyzing the floodgate hypothesis, we can conclude that there are in fact *several* possible floodgates, all of which will advantageously compete with phosphatases for reacting with H_2O_2, including some nonenzymatic (therefore noninhibitable) reactions of abundant thiols such as GSH (Table 3.2; Fig. 3.1A).

3. H_2O_2 DIFFUSION VERSUS REACTION WITH CELLULAR THIOLS

H_2O_2 is a very small and polar molecule and, as such, it can freely diffuse in aqueous compartments. Its diffusion coefficient in water is

1.43×10^{-9} m^2 s^{-1} (van Stroe-Blezen, Everaerts, Janssen, & Tacken, 1993), although its permeation through biomembranes could be hindered by its polar nature. Experiments in *Saccharomyces cerevisiae* have shown that transport across yeast plasma membrane has an activation energy of 58 kJ mol^{-1}, that permeation depends on the composition of the membrane, and that it is facilitated by the presence of aquaporins (Bienert et al., 2007; Branco, Marinho, Cyrne, & Antunes, 2004). Intracellular diffusion is also limited by molecular crowding, so the effective diffusion coefficient could be reduced by up to 60% for small molecules in highly crowded environments (Straube & Ridgway, 2009). One very important aspect of diffusion that is often overlooked in signaling model schemes is that diffusion is a random process and if H_2O_2 is produced extracellularly, particularly, in cell culture where the volume outside cells is orders of magnitude larger than inside, it will most probably remain outside the cells just for statistical reasons.

Anyway, once H_2O_2 is inside a cellular compartment, two sets of events occur: chemical reactions and physical diffusion (Fig. 3.2).

It is often mentioned that if H_2O_2 is locally generated at close quarters of a target, its only choice will be reacting, independent of kinetic considerations (Rhee, Woo, Kil, & Bae, 2012; Woo et al., 2010). This "localized peroxide" assumption ignores diffusion, which is a very fast process, particularly, when dealing with short distances such as subcellular compartments. Assuming a focal formation of H_2O_2, diffusion will cause its concentration to change in space and time according to the solution of Fick's second law (Eq. 3.3).

$$C(x,t) = C_0 \operatorname{erfc}\left(\frac{x}{\sqrt{4Dt}}\right) \tag{3.3}$$

where C is the concentration of the molecule, C_0 is the concentration at the point of formation, x is the distance, D is the diffusion coefficient, and erfc is the complementary error function. Using the times calculated as $t_{1/2}$ of each reaction in Table 3.2 and substituting them in Eq. (3.3), we can estimate how far the boundary condition of $C/C_0 = 0.5$ has moved from the point of H_2O_2 formation. Figure 3.1B provides a visual aid of how far would H_2O_2 diffuse by the time half of it has been consumed by reaction with the selected targets. It is clear that fast targets such as peroxidases (Prx2, GPx, or Prx5) can confine the diffusion of H_2O_2 whereas slow targets such as GSH or PTP1B will see the oxidant diffusing out of the cell before having the chance to react appreciably with it.

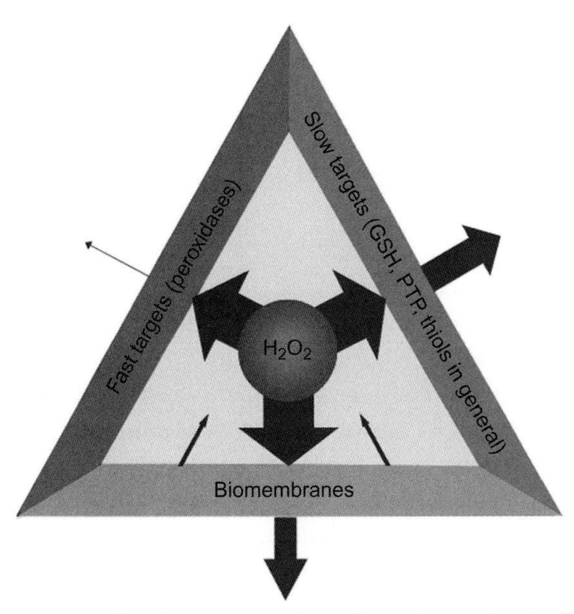

Figure 3.2 The options of hydrogen peroxide. Diffusion (arrows) is the fastest process for H_2O_2 within a biological compartment, even considering molecular crowding. After collision with some fast targets, mainly Prxs and GPxs, a very small fraction of H_2O_2 molecules survive the chemical reaction. By contrast, encounters with slow targets such as nonperoxidatic thiols and other reductants leave H_2O_2 mostly unscathed. Small rate constants indicate that the chemical reaction is highly improbable unless a large number of encounters happen, and that is not likely to occur if H_2O_2 can diffuse. Finally, if diffusion takes H_2O_2 to a biomembrane, permeation can happen by passive diffusion or through aquaporins but also a fraction of H_2O_2 molecules will bounce back to its original compartment. (For color version of this figure, the reader is referred to the online version of this chapter.)

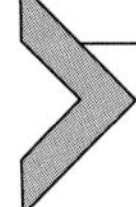

4. SULFENIC ACIDS AS SIGNAL TRANSDUCTION INTERMEDIATES

Direct oxidation of thiols by H_2O_2 yields the corresponding sulfenic acid (RSOH) as the first intermediate. As a matter of fact, reaction (3.1) is stoichiometrically correct, but to describe the course of the reaction we should write it as Eq. (3.4).

$$\text{P-(SH)}_2 + H_2O_2 \rightarrow \text{P-(SH)(SOH)} + H_2O \rightarrow \text{P-S}_2 + 2H_2O \quad (3.4)$$

As sulfenic acids are mandatory intermediates of the oxidation of thiols by peroxides, we should consider their role in H_2O_2-mediated signal transduction.

We mentioned before that the first part of reaction (3.4) has been characterized only for a small number of thiols. For the second part, the number is even smaller. The reaction between cysteine sulfenic and cysteine (free amino acids) is so fast that the intermediate is not observed during the reaction (Table 3.3). Kinetically characterized reactions between protein cysteine sulfenic acids and thiols, either intra or intermolecular, are scarce (Table 3.3). What we know from the few studies available is that:

- Solvent-exposed sulfenic acids react rapidly with thiols to form disulfides.
- In very specific cases, protein sulfenic acids have been shown to react with amides to form relatively inert sulfenylamides (Salmeen et al., 2003).
- Dimedone-based probes have successfully been used to detect sulfenic acids in cell culture (Klomsiri et al., 2010; Paulsen et al., 2012) despite slow kinetics.

Notwithstanding the inherent lability of sulfenic acids (that argues against its role as signaling intermediates), detected protein sulfenic acids are surprisingly abundant in cells subjected to exogenous H_2O_2 or to growth factors

Table 3.3 Rate constants of sulfenic acids reacting with thiols

Protein sulfenic	**Thiol**	
Intermolecular reactions		k (M^{-1} s^{-1})
HSA C34 (Turell et al., 2008)	Cys	21.6
HSA C34 (Turell et al., 2008)	GSH	2.9
HSA C34 (Turell et al., 2008)	CysGly	55
AhpE C45 (Hugo et al., 2009)	Thionitrobenzoate	1500
AhpE C45 (Reyes et al., 2011)	DTT	90
Intramolecular reactions		k (s^{-1})
Prx5 C47 (Trujillo et al., 2007)	Prx5 C151	12
Cdc25B C437 (Sohn & Rudolph, 2003)	Cdc25B C426	0.16
Cdc25C C377 (Sohn & Rudolph, 2003)	Cdc25C C330	0.012
Detection reaction		k (M^{-1} s^{-1})
HSA C34 (Turell et al., 2008)	Dimedone[a]	0.027
Papain (Klomsiri et al., 2010)	Biotinylated dimedone	28
fRMsr (Klomsiri et al., 2010)	Biotinylated dimedone	2.2
AhpC (Klomsiri et al., 2010)	Biotinylated dimedone	0.05

[a]Dimedone is not a thiol, it is included in this table for comparison purposes.

that promote the formation of endogenous H_2O_2, such as TNF-α or EGF (Barelier et al., 2010; Klomsiri et al., 2010; Paulsen et al., 2012; Saurin, Neubert, Brennan, & Eaton, 2004). Positive detection of protein sulfenic acids is usually obtained with probes based in 5,5-dimethylcyclohexane-1,3-dione (dimedone) derivatives, whereas other approaches that attempt to quantify the modified residues have failed to observe such abundant modification (Held et al., 2010). Dimedone as a probe for detecting sulfenic acids is in itself somewhat counterintuitive. The few rate constants available indicate a very slow reaction (Table 3.3) that should not compete with endogenous thiols.

Based on the kinetic and diffusional considerations mentioned in the preceding sections, in order to form a sulfenic acid in a nonperoxidatic thiol, one should get rid of other thiols that act as kinetic competitors and also abolish diffusion. One way this could be accomplished is direct transfer of the metabolite (H_2O_2) formed in the active site of one enzyme to the oxidizable target without being released to the bulk medium. Evidently, before such direct transfer can be invoked as a signaling mechanism, the specific and intimate interaction involved has to be identified. A second possibility would be a reaction in which the sulfenic acid group is transferred from one cysteine protein to another (Eq. 3.5).

$$R^1SH + R^2SOH \rightleftharpoons R^1SSR^2 + H_2O \rightleftharpoons R^1SOH + R^2SH \quad (3.5)$$

Although going from R^2SOH to R^1SOH is not impossible (only highly improbable as the disulfide is the most stable species), once again, such transfer needs to be experimentally demonstrated before it can be used as an explanation for redox signaling.

5. Prxs AS PREFERENTIAL TARGETS FOR H_2O_2

As discussed above, the mechanism of H_2O_2 redox signaling involves the initial reaction of H_2O_2 with unique cysteine residues. The specificity of the signal lies on kinetics: the most reactive and abundant cysteine residues will be the preferential targets. In that sense, with the kinetic data available today, thiol peroxidases (peroxiredoxins, Prxs) and selenocysteine-glutathione peroxidases (GPxs) emerge as the preferential targets. While most protein thiols react with H_2O_2 with rate constants $<100\ M^{-1}\ s^{-1}$, the active-site residue of Prxs and GPxs reacts with $k = 10^5$–$10^8\ M^{-1}\ s^{-1}$ (Cox et al., 2009; Flohe, Loschen, Gunzler, & Eichele, 1972; Manta et al., 2009; Peskin et al., 2007; Toledo et al., 2011; Trujillo et al., 2007). The high reactivity of Prxs Cp is due to the particular cationic protein environment of the active site that stabilizes the transition state and also the anionic leaving group in a concerted manner (Ferrer-Sueta et al., 2011; Hall, Parsonage, Poole, & Karplus, 2010). Peroxiredoxins are antioxidant enzymes but their specificity for hydroperoxides makes them ideal sensors of endogenous H_2O_2.

6. Prxs AS PRIMARY H_2O_2 SENSORS AND TRANSDUCERS

Considering the extraordinary reactivity of Prxs with H_2O_2 and their cellular abundance (15–60 μ*M*; Chae, Kim, Kang, & Rhee, 1999) and even higher than 200 μ*M* in erythrocytes (Cho, Lee et al., 2010b; Moore, Mankad, Shriver, Mankad, & Plishker, 1991), peroxide-induced oxidative modification of mammalian signaling proteins can occur via an indirect process, beginning with Prxs as the H_2O_2 receptors that transduce the peroxide bond into disulfide-exchange networks that affect signal responses (Fig. 3.3).

This H_2O_2 sensor/transducer role of Prxs is well documented in yeast. For example, Orp1 (also known as Gpx3), a *S. cerevisiae* thiol peroxidase, senses H_2O_2 forming a sulfenic intermediate that reacts with a cysteine from a redox transcription factor (Yap1) forming a mixed disulfide that rearranges into an intramolecular disulfide bond, and finally, the disulfide Yap1 interacts with DNA (Delaunay, Pflieger, Barrault, Vinh, & Toledano, 2002).

Moreover, a recent article has elegantly shown that the regulation of gene expression in response to H_2O_2 in *S. cerevisiae* is abolished in cells lacking all peroxiredoxins and glutathione peroxidases, and that direct oxidation of redox proteins by H_2O_2 is a minor event (Fomenko et al., 2011).

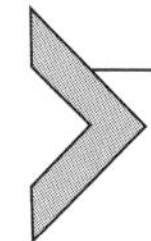

7. Prx–PROTEIN INTERACTIONS ARE NEEDED TO TRANSMIT THE SIGNAL

If Prxs serve as the initial H_2O_2 receptor, further interactions of the oxidized Prxs with other proteins are essential and determinant for the transduction of the signal (Fig. 3.3).

In particular, interactions of Prxs with kinases or phosphatases represent an interesting link between redox and phosphorylation signaling pathways. In that sense, recent studies have reported the interaction of Prxs with protein tyrosine kinases and protein tyrosine phosphatases.

It has been observed that activation of MST1 kinase (Mammalian Ste20-like kinase-1) by H_2O_2 requires its interaction with Prx1 (Morinaka, Funato, Uesugi, & Miki, 2011). After exposure of cells or the recombinant

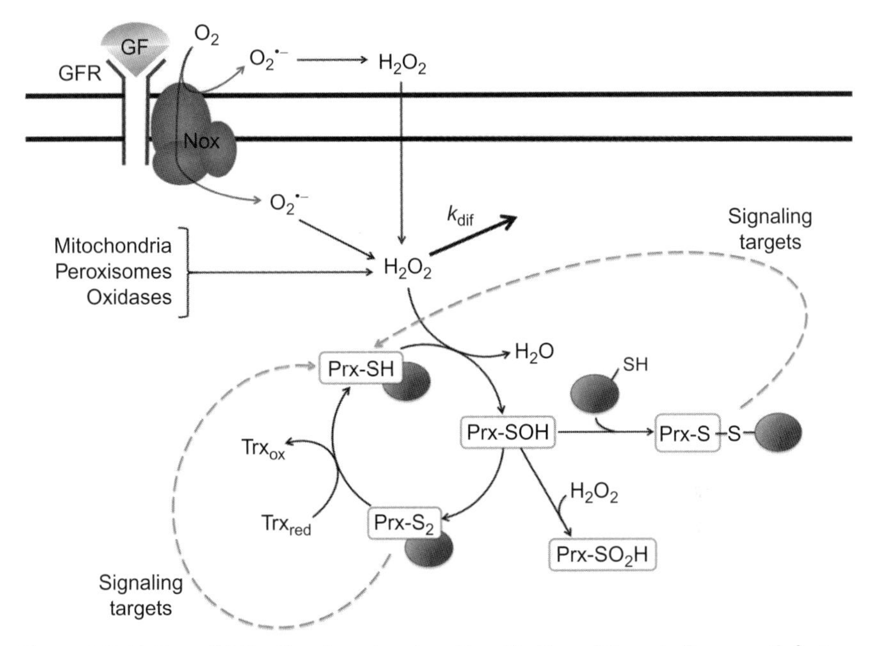

Figure 3.3 H_2O_2 and 2-Cys Prxs in redox signaling. Binding of ligands like growth factors (GF) to their receptors (GFR) stimulates NADPH oxidase (Nox) assembly, source of extra- and intracellular superoxide that dismutates to H_2O_2. Intracellular H_2O_2 is also generated in mitochondria and peroxisomes and by cytosolic oxidases. The fate of this H_2O_2 is to diffuse and/or react with its preferential targets inside the cell, Prxs, initiating their catalytic cycle. The resulting peroxidatic cysteine sulfenic acid can react with the Prx-resolving cysteine, closing the catalytic cycle, or with a second H_2O_2 molecule, getting overoxidized, or alternatively with other reduced protein thiols to generate mixed disulfide bonds. Protein–protein interactions can occur between different Prx redox forms and other signaling proteins (purple circles), as a way to amplify the response to downstream effectors. (For interpretation of the references to color in this figure legend, the reader is referred to the online version of this chapter.)

enzyme to H_2O_2, MST1 kinase is activated by autophosphorylation only in the presence of oligomeric Prx1. Coimmunoprecipitation shows the association between the two proteins, and RNA interference knockdown experiments demonstrate that Prx1 is an essential intermediate in H_2O_2-induced MST1 activation.

Another example is the kinase ASK1 (apoptosis signaling kinase 1), which is known to form a complex with Trx *in vivo*. Reduced Trx was shown to constitutively interact with ASK1, directly inhibiting its kinase activity to prevent stress- and cytokine-induced apoptosis. When Trx is oxidized (e.g., by the reduction of oxidized Prx), it dissociates from ASK1, which recovers its kinase activity and stimulates apoptosis (Lu & Holmgren, 2012; Ray, Huang, & Tsuji, 2012; Saitoh et al., 1998).

A recent study shows the interaction of Prx2 with the complex ASK1-Trx (reduced) (Hu et al., 2011). Moreover, overexpression of Prx2 in dopaminergic neurons prevented apoptosis via ASK1 inhibition after exposure to 6-hydroxydopamine. The authors observed that overoxidation of Prx2 preserved Trx in its reduced state and associated with ASK1, thus keeping the kinase inactive. However, other studies showed that oxidation of ASK1 is required to activate the cascade of phosphorylation that ends up in apoptosis (Nadeau, Charette, Toledano, & Landry, 2007) and that Trx activity is required for reducing back ASK1. It has even been proposed that Prx1 catalyzes the oxidation of ASK1 and transmits a phosphorylation signal to p38 in response to peroxide (Jarvis, Hughes, & Ledgerwood, 2012). The authors suggest the formation of a transient mixed disulfide between Prx and ASK1 that is resolved into a disulfide-linked homomultimeric (active) ASK1.

The interaction of Prx with phosphatases has been reported as well. PTEN (tumor suppressor phosphatase with sequence homology to tensin) is transiently inactivated by cysteine oxidation when cells are stimulated with growth factors or exogenous H_2O_2 (Kwon et al., 2004). A C124–C71 disulfide is formed after H_2O_2 treatment, which is reduced back in the cell via Trx (coimmunoprecipitation of Trx and PTEN was observed; Lee et al., 2002). Interestingly, association between PTEN and Prx1 was reported (Cao et al., 2009). The reduced form of Prx1 associates with PTEN and the complex dissociates under oxidative stress when Prx1 is overoxidized. The authors propose a protective role of Prx1 against PTEN oxidation. This interaction might also stabilize PTEN binding to the plasma membrane, required for PTEN phosphatase activity (Cao et al., 2009).

More recently, it was found that Prx1 also binds to MAPK phosphatases, MKP-1 and MKP-5, but only dissociates from MKP-1 after overoxidation (Turner-Ivey et al., 2013).

Other redox-regulated phosphatases, such as Cdc25 and low-molecular weight PTP, are also known to form a disulfide bond between the active-site cysteine and another nearby cysteine residue during H_2O_2 oxidation (Cho et al., 2004). By analogy with what has been observed for the yeast redox transcription factor Yap1 (Delaunay, Pflieger, Barrault, Vinh, & Toledano, 2002), a similar mechanism of disulfide formation in these mammalian PTPs has been proposed, involving thiol peroxidases as intermediate H_2O_2 sensors (Cho et al., 2004). Thus, the active-site cysteine of PTPs is not directly oxidized by H_2O_2, but the sulfenic intermediate of Prxs, Cp-SOH, forms a mixed disulfide that is eventually resolved as an intramolecular PTP disulfide.

As suggested by the results of Gutscher et al. (2009), the interaction of Prxs with the oxidizable target protein is essential to give the right specificity to H_2O_2 signaling. Using Prx-roGFP fusion proteins in living cells exposed to H_2O_2, they demonstrated that some thiol peroxidases function as redox relays when in close proximity to redox-regulated proteins.

Altogether, these recent results point to the role of Prxs as obligatory intermediates in H_2O_2 signaling and address the importance of protein–protein interactions to pass on the oxidation signal. Prx–protein interactions may also represent a key point to modulate redox signaling pathways.

8. POSTTRANSLATIONAL REGULATION OF Prxs

Considering Prxs as critical-point regulators of intracellular redox signaling, posttranslational modifications (PTMs) affecting their peroxidase activity, redox state, and/or oligomeric structure will undoubtedly impact on H_2O_2 signaling. Several PTMs have been reported on Prxs, not only in the catalytic residues but in other residues as well, affecting their structure and function.

A particular feature of Prxs is their turnover-induced inactivation by overoxidation of the Cp residue to sulfinic acid (Fig. 3.3), specifically reversed by sulfiredoxin (Srx) through an ATP-dependent mechanism (Biteau, Labarre, & Toledano, 2003; Woo et al., 2003, 2005). Given the abundance of these enzymes, it seems likely that the biological significance of Cp overoxidation could be the modulation of Prx–protein interactions (Cao et al., 2009), rather than the inhibition of peroxidase activity, as proposed in support of the floodgate hypothesis (Wood, Poole, & Karplus, 2003). Moreover, this overoxidation could be regarded as a release of reduced Trx from the Prx catalytic cycle, allowing Trx-mediated reduction of other secondary intracellular targets (Day et al., 2012).

Typical 2-Cys Prxs' catalytic cycle depends on their Cp redox state, which, *in vitro*, was shown to be related to the oligomerization state of these proteins, showing a prevalence of the decamer for the reduced and overoxidized enzymes, and mainly a dimer for the disulfide form (Wood, Poole, Hantgan, & Karplus, 2002). Although a dynamic equilibrium between these species is always established, oxidation of the Cp could be a way of modulating Prx–protein interactions through these oligomeric changes, as seen for MST1 kinase and Prx1 (Morinaka, Funato,

Uesugi, & Miki, 2011). Interestingly, when typical 2-Cys Prxs overoxidize, they tend to form high-molecular weight oligomers that have been shown to gain molecular chaperone-like activity (Jang et al., 2004; Lim et al., 2008; Moon et al., 2005).

Another cysteine modification commonly observed in biological systems and particularly elevated under oxidative stress is glutathionylation (Fratelli et al., 2002).

$$\text{P-SOH} + \text{GSH} \rightleftharpoons \text{P-SSG} + H_2O \tag{3.6}$$

Protein glutathionylation has been described as a protective mechanism of reactive cysteines against overoxidation. For Prx1, glutathionylation on the noncatalytic C83, which is responsible for stabilizing the interface between dimers, disrupts the decameric form of the enzyme (Park, Piszczek, Rhee, & Chock, 2011), thus potentially affecting its interaction with other signaling proteins.

Although Srx was first described as an enzyme that reduces Cp sulfinic acid back to its sulfenic state, it was also shown to specifically deglutathionylate Prx1's C83 and the resolving C172 (Jonsson, Murray, Johnson, & Lowther, 2008; Park, Mieyal, Rhee, & Chock, 2009), while glutaredoxin 1 (Grx1) catalyzes the deglutathionylation of the peroxidatic cysteine (Park, Mieyal, Rhee, & Chock, 2009).

S-nitros(yl)ation, a nitric oxide-dependent PTM on cysteine residues, has been detected in many biological systems under physiological and pathological conditions. However, its cellular mechanism of formation is still under investigation (Foster, Hess, & Stamler, 2009). Prx2 was found S-nitrosylated on its peroxidatic and resolving cysteines in the brains of patients with Parkinson's disease (Fang, Nakamura, Cho, Gu, & Lipton, 2007), suggesting that S-nitrosylation could be a potential modulator of Prx peroxidase activity *in vivo*.

Noncatalytic Prx residues have also shown to be susceptible to modifications that affect the structure and activity of the enzyme. Woo et al. (2010) observed in cells stimulated with growth factors that Prx1 (but not Prx2) was transiently inactivated via a Src kinase-mediated tyrosine phosphorylation at Y194, suggesting a connection between redox and phosphorylation pathways. Prx1 was also shown to get inactivated by phosphorylation at T90 by cyclin B-dependent kinase 1 (Cdk1) (Chang et al., 2002). Phosphorylation of T90 or T89 of Prx1 and Prx2, respectively, adds a negative charge close to the Cp that leads to structural changes and oligomerization of these enzymes, switching their peroxidase activity to a molecular chaperone-like activity (Chang et al., 2004; Jang et al., 2006). While phosphorylation of Prx1 on Y194 or T90 inactivates the enzyme, phosphorylation on S32 was shown to increase its peroxidase activity (Woo et al., 2010; Zykova et al., 2010). Phosphorylation of noncatalytic residues could be seen as an alternative mechanism for the regulation of Prxs. Although a nonredox modification, it appears as an interplay between H_2O_2- and kinase-driven signaling pathways.

Finally, lysine N-acetylation has been reported as another post-translational modification of Prxs. N-acetylation of Prx1 and Prx2 was found to increase their peroxidase activity as well as resistance to overoxidation by H_2O_2 (Parmigiani et al., 2008). These authors found K197 and K196 N-acetylated in Prx1 and Prx2, respectively, and HDAC6 responsible for their deacetylation in prostate cancer cells. However, other authors working on HeLa cells exclusively found Prx1 acetylated (Seo et al., 2009).

The cross-talk between all Prxs PTMs and their effect on Prx–protein interactions and signaling has yet to be elucidated, but it seems pretty clear that many regulation mechanisms have coevolved to regulate Prxs activity, suggesting a key role of these peroxidases in H_2O_2-dependent redox signaling.

9. SUMMARY

H_2O_2 acts as a signaling molecule by oxidizing critical cysteine residues in redox-regulated proteins. On thermodynamic grounds, all thiols can be oxidized by H_2O_2, so the specificity is given by differences in kinetics. As suggested before by others (Brigelius-Flohe & Flohe, 2011; Winterbourn & Hampton, 2008), Prxs and GPxs emerge as the preferential targets for endogenous H_2O_2 signal, based on their cellular abundance and high reactivity toward H_2O_2, allowing them to compete with the fast process of diffusion of H_2O_2 molecules out of the cell. Direct oxidation by H_2O_2 of cysteine residues other than the peroxidatic cysteine Cp forming the corresponding sulfenic acid intermediates is hard to visualize with the available data. However, less reactive and less abundant protein thiols as PTP1B have been proved to get oxidized in a H_2O_2-dependent manner in stimulated cells. A more plausible mechanism is direct reaction of H_2O_2 with Prxs Cp and transmission of the signal from the oxidized Prx to a redox-sensitive protein. The molecular mechanism could involve the formation of a mixed disulfide from Cp-SOH with a cysteine of the oxidizable target, or with the resolving cysteine, followed by disulfide-exchange reactions. A nonredox transduction of the signal is also possible, depending on specific interactions between oxidized Prx and nonredox-sensitive proteins. Therefore, a close look into Prx–protein interactions is needed to get a better insight into the mechanisms of H_2O_2 signaling.

ACKNOWLEDGMENTS

The authors want to acknowledge the financial support from CSIC (Comisión Sectorial de Investigación Científica), Universidad de la República, PEDECIBA (Programa de Desarrollo de las Ciencias Básicas), and ANII (Agencia Nacional de Investigación e Innovación), Uruguay.

REFERENCES

Andres-Mateos, E., Perier, C., Zhang, L., Blanchard-Fillion, B., Greco, T. M., Thomas, B., et al. (2007). DJ-1 gene deletion reveals that DJ-1 is an atypical peroxiredoxin-like peroxidase. *Proceedings of the National Academy of Sciences of the United States of America, 104*(37), 14807–14812.

Arnold, R. S., Shi, J., Murad, E., Whalen, A. M., Sun, C. Q., Polavarapu, R., et al. (2001). Hydrogen peroxide mediates the cell growth and transformation caused by the mitogenic oxidase Nox1. *Proceedings of the National Academy of Sciences of the United States of America, 98*(10), 5550–5555.

Barelier, S., Linard, D., Pons, J., Clippe, A., Knoops, B., Lancelin, J. M., et al. (2010). Discovery of fragment molecules that bind the human peroxiredoxin 5 active site. *PLoS One, 5*(3), e9744.

Barrett, W. C., DeGnore, J. P., Keng, Y. F., Zhang, Z. Y., Yim, M. B., & Chock, P. B. (1999). Roles of superoxide radical anion in signal transduction mediated by reversible regulation of protein-tyrosine phosphatase 1B. *Journal of Biological Chemistry, 274*(49), 34543–34546.

Bienert, G. P., Moller, A. L., Kristiansen, K. A., Schulz, A., Moller, I. M., Schjoerring, J. K., et al. (2007). Specific aquaporins facilitate the diffusion of hydrogen peroxide across membranes. *Journal of Biological Chemistry, 282*(2), 1183–1192.

Biteau, B., Labarre, J., & Toledano, M. B. (2003). ATP-dependent reduction of cysteine-sulphinic acid by *S. cerevisiae* sulphiredoxin. *Nature, 425*(6961), 980–984.

Branco, M. R., Marinho, H. S., Cyrne, L., & Antunes, F. (2004). Decrease of H_2O_2 plasma membrane permeability during adaptation to H_2O_2 in Saccharomyces cerevisiae. *Journal of Biological Chemistry, 279*(8), 6501–6506.

Brigelius-Flohe, R., & Flohe, L. (2011). Basic principles and emerging concepts in the redox control of transcription factors. *Antioxidants & Redox Signaling, 15*(8), 2335–2381.

Cao, J., Schulte, J., Knight, A., Leslie, N. R., Zagozdzon, A., Bronson, R., et al. (2009). Prdx1 inhibits tumorigenesis via regulating PTEN/AKT activity. *EMBO Journal, 28*(10), 1505–1517.

Chae, H. Z., Kim, H. J., Kang, S. W., & Rhee, S. G. (1999). Characterization of three isoforms of mammalian peroxiredoxin that reduce peroxides in the presence of thioredoxin. *Diabetes Research and Clinical Practice, 45*(2–3), 101–112.

Chang, T. S., Jeong, W., Choi, S. Y., Yu, S., Kang, S. W., & Rhee, S. G. (2002). Regulation of peroxiredoxin I activity by Cdc2-mediated phosphorylation. *Journal of Biological Chemistry, 277*(28), 25370–25376.

Chang, T. S., Jeong, W., Woo, H. A., Lee, S. M., Park, S., & Rhee, S. G. (2004). Characterization of mammalian sulfiredoxin and its reactivation of hyperoxidized peroxiredoxin through reduction of cysteine sulfinic acid in the active site to cysteine. *Journal of Biological Chemistry, 279*(49), 50994–51001.

Chen, Y., Gill, P. S., & Welch, W. J. (2005). Oxygen availability limits renal NADPH-dependent superoxide production. *American Journal of Physiology. Renal Physiology, 289*(4), F749–F753.

Cheng, G., Cao, Z., Xu, X., van Meir, E. G., & Lambeth, J. D. (2001). Homologs of gp91phox: Cloning and tissue expression of Nox3, Nox4, and Nox5. *Gene, 269*(1–2), 131–140.

Cho, C. S., Kato, G. J., Yang, S. H., Bae, S. W., Lee, J. S., Gladwin, M. T., et al. (2010). Hydroxyurea-induced expression of glutathione peroxidase 1 in red blood cells of individuals with sickle cell anemia. *Antioxidants & Redox Signaling, 13*(1), 1–11.

Cho, S. H., Lee, C. H., Ahn, Y., Kim, H., Kim, H., Ahn, C. Y., et al. (2004). Redox regulation of PTEN and protein tyrosine phosphatases in H(2)O(2) mediated cell signaling. *FEBS Letters, 560*(1–3), 7–13.

Cho, C. S., Lee, S., Lee, G. T., Woo, H. A., Choi, E. J., & Rhee, S. G. (2010). Irreversible inactivation of glutathione peroxidase 1 and reversible inactivation of peroxiredoxin II by H_2O_2 in red blood cells. *Antioxidants & Redox Signaling, 12*(11), 1235–1246.

Cox, A. G., Pearson, A. G., Pullar, J. M., Jonsson, T. J., Lowther, W. T., Winterbourn, C. C., et al. (2009). Mitochondrial peroxiredoxin 3 is more resilient to hyperoxidation than cytoplasmic peroxiredoxins. *The Biochemical Journal, 421*(1), 51–58.

Cox, A. G., Winterbourn, C. C., & Hampton, M. B. (2010). Mitochondrial peroxiredoxin involvement in antioxidant defence and redox signalling. *The Biochemical Journal, 425*(2), 313–325.

Day, A. M., Brown, J. D., Taylor, S. R., Rand, J. D., Morgan, B. A., & Veal, E. A. (2012). Inactivation of a peroxiredoxin by hydrogen peroxide is critical for thioredoxin-mediated repair of oxidized proteins and cell survival. *Molecular Cell, 45*(3), 398–408.

Delaunay, A., Pflieger, D., Barrault, M. B., Vinh, J., & Toledano, M. B. (2002). A thiol peroxidase is an H_2O_2 receptor and redox-transducer in gene activation. *Cell, 111*(4), 471–481.

Dickinson, B. C., & Chang, C. J. (2011). Chemistry and biology of reactive oxygen species in signaling or stress responses. *Nature Chemical Biology*, 7(8), 504–511.

Fang, J., Nakamura, T., Cho, D. H., Gu, Z., & Lipton, S. A. (2007). S-nitrosylation of peroxiredoxin 2 promotes oxidative stress-induced neuronal cell death in Parkinson's disease. *Proceedings of the National Academy of Sciences of the United States of America, 104*(47), 18742–18747.

Ferrer-Sueta, G., Manta, B., Botti, H., Radi, R., Trujillo, M., & Denicola, A. (2011). Factors affecting protein thiol reactivity and specificity in peroxide reduction. *Chemical Research in Toxicology, 24*(4), 434–450.

Flohe, L., Loschen, G., Gunzler, W. A., & Eichele, E. (1972). Glutathione peroxidase, V. The kinetic mechanism. *Hoppe-Seyler's Zeitschrift für Physiologische Chemie, 353*(6), 987–999.

Fomenko, D. E., Koc, A., Agisheva, N., Jacobsen, M., Kaya, A., Malinouski, M., et al. (2011). Thiol peroxidases mediate specific genome-wide regulation of gene expression in response to hydrogen peroxide. *Proceedings of the National Academy of Sciences of the United States of America, 108*(7), 2729–2734.

Foster, M. W., Hess, D. T., & Stamler, J. S. (2009). Protein S-nitrosylation in health and disease: A current perspective. *Trends in Molecular Medicine, 15*(9), 391–404.

Fratelli, M., Demol, H., Puype, M., Casagrande, S., Eberini, I., Salmona, M., et al. (2002). Identification by redox proteomics of glutathionylated proteins in oxidatively stressed human T lymphocytes. *Proceedings of the National Academy of Sciences of the United States of America, 99*(6), 3505–3510.

Gough, D. R., & Cotter, T. G. (2011). Hydrogen peroxide: A Jekyll and Hyde signalling molecule. *Cell Death & Disease, 2*, e213.
Gutscher, M., Sobotta, M. C., Wabnitz, G. H., Ballikaya, S., Meyer, A. J., Samstag, Y., et al. (2009). Proximity-based protein thiol oxidation by H_2O_2-scavenging peroxidases. *Journal of Biological Chemistry, 284*(46), 31532–31540.
Hall, A., Parsonage, D., Poole, L. B., & Karplus, P. A. (2010). Structural evidence that peroxiredoxin catalytic power is based on transition-state stabilization. *Journal of Molecular Biology, 402*(1), 194–209.
Held, J. M., Danielson, S. R., Behring, J. B., Atsriku, C., Britton, D. J., Puckett, R. L., et al. (2010). Targeted quantitation of site-specific cysteine oxidation in endogenous proteins using a differential alkylation and multiple reaction monitoring mass spectrometry approach. *Molecular & Cellular Proteomics, 9*(7), 1400–1410.
Held, J. M., & Gibson, B. W. (2012). Regulatory control or oxidative damage? Proteomic approaches to interrogate the role of cysteine oxidation status in biological processes. *Molecular & Cellular Proteomics, 11*(4) R111.013037.
Hu, X., Weng, Z., Chu, C. T., Zhang, L., Cao, G., Gao, Y., et al. (2011). Peroxiredoxin-2 protects against 6-hydroxydopamine-induced dopaminergic neurodegeneration via attenuation of the apoptosis signal-regulating kinase (ASK1) signaling cascade. *The Journal of Neuroscience, 31*(1), 247–261.
Hugo, M., Turell, L., Manta, B., Botti, H., Monteiro, G., Netto, L. E., et al. (2009). Thiol and sulfenic acid oxidation of AhpE, the one-cysteine peroxiredoxin from Mycobacterium tuberculosis: Kinetics, acidity constants, and conformational dynamics. *Biochemistry, 48*(40), 9416–9426.
Jaeger, T., Budde, H., Flohe, L., Menge, U., Singh, M., Trujillo, M., et al. (2004). Multiple thioredoxin-mediated routes to detoxify hydroperoxides in Mycobacterium tuberculosis. *Archives of Biochemistry and Biophysics, 423*(1), 182–191.
Jang, H. H., Kim, S. Y., Park, S. K., Jeon, H. S., Lee, Y. M., Jung, J. H., et al. (2006). Phosphorylation and concomitant structural changes in human 2-Cys peroxiredoxin isotype I differentially regulate its peroxidase and molecular chaperone functions. *FEBS Letters, 580*(1), 351–355.
Jang, H. H., Lee, K. O., Chi, Y. H., Jung, B. G., Park, S. K., Park, J. H., et al. (2004). Two enzymes in one; two yeast peroxiredoxins display oxidative stress-dependent switching from a peroxidase to a molecular chaperone function. *Cell, 117*(5), 625–635.
Jarvis, R. M., Hughes, S. M., & Ledgerwood, E. C. (2012). Peroxiredoxin 1 functions as a signal peroxidase to receive, transduce, and transmit peroxide signals in mammalian cells. *Free Radical Biology & Medicine, 53*(7), 1522–1530.
Jonsson, T. J., Murray, M. S., Johnson, L. C., & Lowther, W. T. (2008). Reduction of cysteine sulfinic acid in peroxiredoxin by sulfiredoxin proceeds directly through a sulfinic phosphoryl ester intermediate. *Journal of Biological Chemistry, 283*(35), 23846–23851.
Karisch, R., Fernandez, M., Taylor, P., Virtanen, C., St-Germain, J. R., Jin, L. L., et al. (2011). Global proteomic assessment of the classical protein-tyrosine phosphatome and "Redoxome" *Cell, 146*(5), 826–840.
Klomsiri, C., Nelson, K. J., Bechtold, E., Soito, L., Johnson, L. C., Lowther, W. T., et al. (2010). Use of dimedone-based chemical probes for sulfenic acid detection evaluation of conditions affecting probe incorporation into redox-sensitive proteins. *Methods in Enzymology, 473*, 77–94.
Koppenol, W. H., Stanbury, D. M., & Bounds, P. L. (2010). Electrode potentials of partially reduced oxygen species, from dioxygen to water. *Free Radical Biology & Medicine, 49*(3), 317–322.
Kwon, J., Lee, S. R., Yang, K. S., Ahn, Y., Kim, Y. J., Stadtman, E. R., et al. (2004). Reversible oxidation and inactivation of the tumor suppressor PTEN in cells stimulated with

peptide growth factors. *Proceedings of the National Academy of Sciences of the United States of America, 101*(47), 16419–16424.
LaButti, J. N., Chowdhury, G., Reilly, T. J., & Gates, K. S. (2007). Redox regulation of protein tyrosine phosphatase 1B by peroxymonophosphate (=O3POOH). *Journal of the American Chemical Society, 129*(17), 5320–5321.
Lee, S. R., Yang, K. S., Kwon, J., Lee, C., Jeong, W., & Rhee, S. G. (2002). Reversible inactivation of the tumor suppressor PTEN by H_2O_2. *Journal of Biological Chemistry, 277*(23), 20336–20342.
Lewis, S. D., Johnson, F. A., & Shafer, J. A. (1976). Potentiometric determination of ionizations at the active site of papain. *Biochemistry, 15*(23), 5009–5017.
Lim, J. C., Choi, H. I., Park, Y. S., Nam, H. W., Woo, H. A., Kwon, K. S., et al. (2008). Irreversible oxidation of the active-site cysteine of peroxiredoxin to cysteine sulfonic acid for enhanced molecular chaperone activity. *Journal of Biological Chemistry, 283*(43), 28873–28880.
Lim, J. C., Gruschus, J. M., Kim, G., Berlett, B. S., Tjandra, N., & Levine, R. L. (2012). A low pKa cysteine at the active site of mouse methionine sulfoxide reductase A. *Journal of Biological Chemistry, 287*(30), 25596–25601.
Lu, J., & Holmgren, A. (2012). Thioredoxin system in cell death progression. *Antioxidants & Redox Signaling, 17*(12), 1738–1747.
Manta, B., Hugo, M., Ortiz, C., Ferrer-Sueta, G., Trujillo, M., & Denicola, A. (2009). The peroxidase and peroxynitrite reductase activity of human erythrocyte peroxiredoxin 2. *Archives of Biochemistry and Biophysics, 484*(2), 146–154.
Melvin, J. A., Murphy, C. F., Dubois, L. G., Thompson, J. W., Moseley, M. A., & McCafferty, D. G. (2011). Staphylococcus aureus sortase A contributes to the Trojan horse mechanism of immune defense evasion with its intrinsic resistance to Cys184 oxidation. *Biochemistry, 50*(35), 7591–7599.
Moon, J. C., Hah, Y. S., Kim, W. Y., Jung, B. G., Jang, H. H., Lee, J. R., et al. (2005). Oxidative stress-dependent structural and functional switching of a human 2-Cys peroxiredoxin isotype II that enhances HeLa cell resistance to H_2O_2-induced cell death. *Journal of Biological Chemistry, 280*(31), 28775–28784.
Moore, R. B., Mankad, M. V., Shriver, S. K., Mankad, V. N., & Plishker, G. A. (1991). Reconstitution of Ca(2+)-dependent K+transport in erythrocyte membrane vesicles requires a cytoplasmic protein. *Journal of Biological Chemistry, 266*(28), 18964–18968.
Morinaka, A., Funato, Y., Uesugi, K., & Miki, H. (2011). Oligomeric peroxiredoxin-I is an essential intermediate for p53 to activate MST1 kinase and apoptosis. *Oncogene, 30*(40), 4208–4218.
Nadeau, P. J., Charette, S. J., Toledano, M. B., & Landry, J. (2007). Disulfide bond-mediated multimerization of Ask1 and its reduction by thioredoxin-1 regulate H(2)O(2)-induced c-Jun NH(2)-terminal kinase activation and apoptosis. *Molecular Biology of the Cell, 18*(10), 3903–3913.
Nagahara, N., Matsumura, T., Okamoto, R., & Kajihara, Y. (2009). Protein cysteine modifications: (2) reactivity specificity and topics of medicinal chemistry and protein engineering. *Current Medicinal Chemistry, 16*(34), 4490–4501.
Nelson, K. J., Parsonage, D., Hall, A., Karplus, P. A., & Poole, L. B. (2008). Cysteine pK(a) values for the bacterial peroxiredoxin AhpC. *Biochemistry, 47*(48), 12860–12868.
Park, J. W., Mieyal, J. J., Rhee, S. G., & Chock, P. B. (2009). Deglutathionylation of 2-Cys peroxiredoxin is specifically catalyzed by sulfiredoxin. *Journal of Biological Chemistry, 284*(35), 23364–23374.
Park, J. W., Piszczek, G., Rhee, S. G., & Chock, P. B. (2011). Glutathionylation of peroxiredoxin I induces decamer to dimers dissociation with concomitant loss of chaperone activity. *Biochemistry, 50*(15), 3204–3210.

Parmigiani, R. B., Xu, W. S., Venta-Perez, G., Erdjument-Bromage, H., Yaneva, M., Tempst, P., et al. (2008). HDAC6 is a specific deacetylase of peroxiredoxins and is involved in redox regulation. *Proceedings of the National Academy of Sciences of the United States of America, 105*(28), 9633–9638.

Paulsen, C. E., Truong, T. H., Garcia, F. J., Homann, A., Gupta, V., Leonard, S. E., et al. (2012). Peroxide-dependent sulfenylation of the EGFR catalytic site enhances kinase activity. *Nature Chemical Biology, 8*(1), 57–64.

Peskin, A. V., Low, F. M., Paton, L. N., Maghzal, G. J., Hampton, M. B., & Winterbourn, C. C. (2007). The high reactivity of peroxiredoxin 2 with H(2)O(2) is not reflected in its reaction with other oxidants and thiol reagents. *Journal of Biological Chemistry, 282*(16), 11885–11892.

Ray, P. D., Huang, B. W., & Tsuji, Y. (2012). Reactive oxygen species (ROS) homeostasis and redox regulation in cellular signaling. *Cellular Signalling, 24*(5), 981–990.

Reyes, A. M., Hugo, M., Trostchansky, A., Capece, L., Radi, R., & Trujillo, M. (2011). Oxidizing substrate specificity of *Mycobacterium tuberculosis* alkyl hydroperoxide reductase E: kinetics and mechanisms of oxidation and overoxidation. *Free Radical Biology and Medicine, 51*(2), 464–473.

Rhee, S. G., Woo, H. A., Kil, I. S., & Bae, S. H. (2012). Peroxiredoxin functions as a peroxidase and a regulator and sensor of local peroxides. *Journal of Biological Chemistry, 287*(7), 4403–4410.

Rouhier, N., Gelhaye, E., Gualberto, J. M., Jordy, M. N., De Fay, E., Hirasawa, M., et al. (2004). Poplar peroxiredoxin Q. A thioredoxin-linked chloroplast antioxidant functional in pathogen defense. *Plant Physiology, 134*(3), 1027–1038.

Saitoh, M., Nishitoh, H., Fujii, M., Takeda, K., Tobiume, K., Sawada, Y., et al. (1998). Mammalian thioredoxin is a direct inhibitor of apoptosis signal-regulating kinase (ASK) 1. *The EMBO Journal, 17*(9), 2596–2606.

Salmeen, A., Andersen, J. N., Myers, M. P., Meng, T. C., Hinks, J. A., Tonks, N. K., et al. (2003). Redox regulation of protein tyrosine phosphatase 1B involves a sulphenyl-amide intermediate. *Nature, 423*(6941), 769–773.

Saurin, A. T., Neubert, H., Brennan, J. P., & Eaton, P. (2004). Widespread sulfenic acid formation in tissues in response to hydrogen peroxide. *Proceedings of the National Academy of Sciences of the United States of America, 101*(52), 17982–17987.

Seo, J. H., Lim, J. C., Lee, D. Y., Kim, K. S., Piszczek, G., Nam, H. W., et al. (2009). Novel protective mechanism against irreversible hyperoxidation of peroxiredoxin: Nalpha-terminal acetylation of human peroxiredoxin II. *Journal of Biological Chemistry, 284*(20), 13455–13465.

Sohn, J., & Rudolph, J. (2003). Catalytic and chemical competence of regulation of cdc25 phosphatase by oxidation/reduction. *Biochemistry, 42*(34), 10060–10070.

Stone, J. R. (2004). An assessment of proposed mechanisms for sensing hydrogen peroxide in mammalian systems. *Archives of Biochemistry and Biophysics, 422*(2), 119–124.

Straube, R., & Ridgway, D. (2009). Investigating the effects of molecular crowding on Ca^{2+} diffusion using a particle-based simulation model. *Chaos, 19*(3), 037110.

Suh, Y. A., Arnold, R. S., Lassegue, B., Shi, J., Xu, X., Sorescu, D., et al. (1999). Cell transformation by the superoxide-generating oxidase Mox1. *Nature, 401*(6748), 79–82.

Takac, I., Schroder, K., Zhang, L., Lardy, B., Anilkumar, N., Lambeth, J. D., et al. (2011). The E-loop is involved in hydrogen peroxide formation by the NADPH oxidase Nox4. *Journal of Biological Chemistry, 286*(15), 13304–13313.

Tanner, J. J., Parsons, Z. D., Cummings, A. H., Zhou, H., & Gates, K. S. (2011). Redox regulation of protein tyrosine phosphatases: Structural and chemical aspects. *Antioxidants & Redox Signaling, 15*(1), 77–97.

Toledo, J. C., Jr., Audi, R., Ogusucu, R., Monteiro, G., Netto, L. E., & Augusto, O. (2011). Horseradish peroxidase compound I as a tool to investigate reactive protein-cysteine residues: From quantification to kinetics. *Free Radical Biology & Medicine, 50*(9), 1032–1038.

Trujillo, M., Clippe, A., Manta, B., Ferrer-Sueta, G., Smeets, A., Declercq, J. P., et al. (2007). Pre-steady state kinetic characterization of human peroxiredoxin 5: Taking advantage of Trp84 fluorescence increase upon oxidation. *Archives of Biochemistry and Biophysics, 467*(1), 95–106.

Turell, L., Botti, H., Carballal, S., Ferrer-Sueta, G., Souza, J. M., Duran, R., et al. (2008). Reactivity of sulfenic acid in human serum albumin. *Biochemistry, 47*(1), 358–367.

Turner-Ivey, B., Manevich, Y., Schulte, J., Kistner-Griffin, E., Jezierska-Drutel, A., Liu, Y., et al. (2013). Role for Prdx1 as a specific sensor in redox-regulated senescence in breast cancer. *Oncogene*, http://dx.doi.org/10.1038/onc.2012.624.

van Stroe-Blezen, S. A. M., Everaerts, F. M., Janssen, L. J. J., & Tacken, R. A. (1993). Diffusion coefficients of oxygen, hydrogen peroxide and glucose in a hydrogel. *Analytica Chimica Acta, 273*, 553–560.

Veal, E. A., Day, A. M., & Morgan, B. A. (2007). Hydrogen peroxide sensing and signaling. *Molecular Cell, 26*(1), 1–14.

Winterbourn, C. C., & Hampton, M. B. (2008). Thiol chemistry and specificity in redox signaling. *Free Radical Biology & Medicine, 45*(5), 549–561.

Winterbourn, C. C., & Metodiewa, D. (1999). Reactivity of biologically important thiol compounds with superoxide and hydrogen peroxide. *Free Radical Biology & Medicine, 27*(3–4), 322–328.

Woo, H. A., Chae, H. Z., Hwang, S. C., Yang, K. S., Kang, S. W., Kim, K., et al. (2003). Reversing the inactivation of peroxiredoxins caused by cysteine sulfinic acid formation. *Science, 300*(5619), 653–656.

Woo, H. A., Jeong, W., Chang, T. S., Park, K. J., Park, S. J., Yang, J. S., et al. (2005). Reduction of cysteine sulfinic acid by sulfiredoxin is specific to 2-cys peroxiredoxins. *Journal of Biological Chemistry, 280*(5), 3125–3128.

Woo, H. A., Yim, S. H., Shin, D. H., Kang, D., Yu, D. Y., & Rhee, S. G. (2010). Inactivation of peroxiredoxin I by phosphorylation allows localized H(2)O(2) accumulation for cell signaling. *Cell, 140*(4), 517–528.

Wood, Z. A., Poole, L. B., Hantgan, R. R., & Karplus, P. A. (2002). Dimers to doughnuts: Redox-sensitive oligomerization of 2-cysteine peroxiredoxins. *Biochemistry, 41*(17), 5493–5504.

Wood, Z. A., Poole, L. B., & Karplus, P. A. (2003). Peroxiredoxin evolution and the regulation of hydrogen peroxide signaling. *Science, 300*(5619), 650–653.

Wood, Z. A., Schröder, E., Robin Harris, J., & Poole, L. B. (2003). Structure, mechanism and regulation of peroxiredoxins. *Trends in Biochemical Sciences, 28*(1), 32–40.

Wunderlich, M., & Glockshuber, R. (1993). Redox properties of protein disulfide isomerase (DsbA) from Escherichia coli. *Protein Science, 2*(5), 717–726.

Zykova, T. A., Zhu, F., Vakorina, T. I., Zhang, J., Higgins, L. A., Urusova, D. V., et al. (2010). T-LAK cell-originated protein kinase (TOPK) phosphorylation of Prx1 at Ser-32 prevents UVB-induced apoptosis in RPMI7951 melanoma cells through the regulation of Prx1 peroxidase activity. *Journal of Biological Chemistry, 285*(38), 29138–29146.

CHAPTER FOUR

Selenium in the Redox Regulation of the Nrf2 and the Wnt Pathway

Regina Brigelius-Flohé[1], Anna Patricia Kipp
Department Biochemistry of Micronutrients, German Institute of Human Nutrition Potsdam-Rehbruecke, Nuthetal, Germany
[1]Corresponding author: e-mail address: flohe@dife.de

Contents

1. Some Historical Background for Introduction 66
2. Selenium Status and Selenoprotein Synthesis 68
3. Selenium Status and the Keap1/Nrf2 System 69
4. Selenium and the Wnt Pathway 72
5. Common Players and Events in Nrf2 and Wnt Signaling 75
6. How Does Selenium Come into Play? 78
References 80

Abstract

Selenium deficiency is known to increase cancer risk by so far unclear mechanisms. Selenium exerts its biological effects via selenocysteine as an integral part of selenoproteins. Certain selenoproteins have redox properties, thereby providing a tool to regulate hydroperoxide-mediated signaling. Selenium deficiency does not only reduce synthesis of selenoproteins but also affects the expression of other proteins and even pathways. A moderate Se deficiency activates the Nrf2 and the Wnt pathways. The link between both pathways appears to be GSK3β which in the active state prepares Nrf2 as well as β-catenin, the key player in Wnt signaling, for ubiquitination and proteasomal degradation, thus silencing their transcriptional activity. Upon stimulation by Wnt signals, GSK3β becomes inactivated and transcription factors are stabilized. Many intermediate steps in both pathways can be modulated by hydroperoxides, making them predestined to be regulated by selenoproteins. Oxidation sensors are (i) Keap1 which keeps Nrf2 in the cytosol unless it is modified by hydroperoxides/electrophiles and (ii) nucleoredoxin (Nrx) which is associated with disheveled (Dvl). NOX1-derived H_2O_2 oxidizes Nrx leading to the liberation of Dvl and the activation of Wnt signaling. Selenium deficiency can support oxidation of both sensors and activate both pathways. The consequences are dual: while the Keap1/Nrf2 system is generally believed to protect against oxidative stress, diverse xenobiotics, inflammation, and carcinogenesis, the Wnt response is considered rather a risky one in these respects. However, not only healthy cells but also malignant ones benefit from intact Keap1/Nrf2 signaling, making a dysregulated hydroperoxide signaling a plausible explanation for the increased cancer risk in selenium deficiency.

Methods in Enzymology, Volume 527
ISSN 0076-6879
http://dx.doi.org/10.1016/B978-0-12-405882-8.00004-0

1. SOME HISTORICAL BACKGROUND FOR INTRODUCTION

In 1973, glutathione peroxidase (GPx) from red blood cells, now known as GPx1, became the first selenoprotein to be identified by Rotruck et al. (1973) and Flohé, Günzler, and Schock (1973). Just 3 years later, Lawrence and Burk (1976) described a second GPx that preferentially reduced organic hydroperoxides and, surprisingly, persisted even in severe selenium deficiency. The novel "selenium-independent GPx" turned out to be a glutathione-*S*-transferase (GST) (Lawrence, Parkhill, & Burk, 1978). This GST, now known as GST1, was found in many organs and increased under selenium deficiency in the liver, kidney, and in the duodenal section of the intestine of rats (Masukawa, Nishimura, & Iwata, 1984). Refeeding of selenium led to a return of the GST activity to normal lower values. In 1983, a group working on "selenium and drug metabolism" observed an upregulation of several enzymes by selenium deprivation, not only of GSTs but also of UDP glucuronyl transferases (UGTs) and sulfotransferases (SULTs) (Reiter & Wendel, 1983). Two years later, heme oxygenase was added to this list of selenium-responsive proteins (Reiter & Wendel, 1985). By now, it has become known that all these enzymes are targets of the transcription factor Nrf2 (Surh, Kundu, & Na, 2008). Thus, Nrf2 appeared to be activated by selenium deficiency, and this hypothesis was finally proved in 2008 by Burk et al. (2008): selenium deprivation caused an increase of liver GST only in wild-type (WT) mice but not in Nrf2 knockout (KO) mice.

Nrf2 is a transcription factor that orchestrates the adaptive response to an oxidative challenge or potentially harmful exposure to xenobiotics by transcriptional upregulation of antioxidant and detoxifying phase 2 enzymes. It is activated by electrophiles and oxidants. Therefore it is not surprising that it is also activated in selenium deficiency, because under these conditions the decline of selenium-dependent peroxidase systems shifts the cellular redox state toward a more oxidized one. It is not surprising, either, that the expression of some peroxidases was observed to depend on Nrf2 activation. One of these peroxidases is the selenoprotein GPx2 (Banning, Deubel, Kluth, Zhou, & Brigelius-Flohé, 2005) (see below), and another one is peroxiredoxin1 (Prx1) (Ishii et al., 2004), which is not a selenoprotein but depends on its cosubstrate thioredoxin (Trx) which is maintained in the reduced state by the selenoprotein thioredoxin reductase (TrxR)

(Tamura & Stadtman, 1996). The functionality of these and other peroxidase systems is guaranteed by two other Nrf2 targets: γ-glutamylcysteine synthetase (Wild, Moinova, & Mulcahy, 1999), the key enzyme for glutathione (GSH) biosynthesis, and sulfiredoxin (Srx), which reverses the oxidative inactivation of Prx-type peroxidases (Biteau, Labarre, & Toledano, 2003).

The Nrf2 pathway is not the only pathway that can be activated by oxidation. Another intriguing example appears to be the Wnt pathway. Its key transcription factor β-catenin controls the expression of the very same selenoprotein that is also induced by Nrf2: GPx2. β-Catenin further regulates the expression of two more selenoproteins, TrxR2 and TrxR3 (Kipp, Müller, Göken, Deubel, & Brigelius-Flohé, 2012). Activation of Wnt signaling recently turned out to be regulated via the thiol-disulfide relais nucleoredoxin (Nrx) (Funato, Michiue, Asashima, & Miki, 2006) (see below). The pathway stimulates mitogenesis and differentiation and is pivotal in embryonic morphogenesis. Constitutively activated Wnt signaling is associated with many tumors (Logan & Nusse, 2004), while Nrf2 activation is generally believed to prevent carcinogenesis.

Selenium deficiency has for long been suspected to increase the risk of cancer development (Shamberger & Frost, 1969), and its role in the reduction of mutagenic hydroperoxides as an integral constituent of GPx offered a plausible explanation for the presumed anticarcinogenic effect of the trace element (Flohé, 1989). In fact, selenium supplementation significantly reduced the incidence of virally induced mammary tumors in mice (Medina & Shepherd, 1980) but did not influence the growth of transplanted tumors (Medina & Shepherd, 1981). More recently, selenium acted anticarcinogenic in a chemically induced esophageal cancer model in rats in the early stage but had no effect in the late stage (Yang, Jia, Chen, Yang, & Li, 2012). A Janus-faced role was also shown for GPx2 in carcinogenicity studies with KO mice (Krehl et al., 2012), and a similar "split personality" has been ascribed to other selenoproteins (Yoo et al., 2012). Collectively these observations indicate that the effect of selenium differs with the stage of carcinogenesis. Carcinogenicity studies with selenium are further complicated by the circumstance that many selenium compounds, if administered beyond the dosage required for proper selenoprotein biosynthesis, tend to autoxidize with the initiation of free radical chain reactions (Spallholz, 1994), which may trigger mutagenesis as well as apoptosis and overt cytotoxicity.

We are, thus, facing several puzzles: (i) A seemingly protective signaling system (Keap1/Nrf2) and a procarcinogenic one (Wnt) induce expression of the same (GPx2) or closely related selenoproteins (TrxR1 or TrxR2 and 3, respectively). (ii) These gene products could be presumed to be protective by constituting detoxifying peroxidase systems but might equally promote tumor growth. (iii) The essential trace element selenium, depending on dosage and chemical form of administration, certainly sustains an appropriate redox metabolism but may also cause adverse effects resulting from nonenzymatic free radical reactions. It was, therefore, deemed rewarding to compile the present knowledge on the effect of the selenium status on the two signaling pathways and their mutual cross-talk.

2. SELENIUM STATUS AND SELENOPROTEIN SYNTHESIS

Selenium as selenocysteine (Sec) is a constituent of selenoproteins which is integrated in the peptide chain. The human selenoproteome comprises five GPxs from which at least four (GPx1–4) are able to efficiently reduce H_2O_2 and other hydroperoxides at the expense of GSH. The members of the GPx family differ dramatically in their position in the hierarchy. GPx2 ranks highest in the hierarchy of all selenoproteins which means that it is synthesized even when selenium becomes limiting, and its mRNA was even increased by selenium restriction (Wingler, Böcher, Flohé, Kollmus, & Brigelius-Flohé, 1999). In contrast, GPx1 and 3 occupy the lower ranks, and disappearance of GPx1 mRNA in selenium deficiency has been amply documented (reviewed in Sunde, 2012).

The exceptional position of GPx2 in the hierarchy prompted a series of (in part ongoing) investigations aimed at a better understanding of its potential role in the prevention of inflammation and carcinogenesis. GPx2 is preferentially expressed in the gastrointestinal system and in cancer cells of epithelial origin (reviewed in Brigelius-Flohé & Kipp, 2009). *Gpx2*$^{-/-}$ mice had been shown to display a mild phenotype, but when GPx1 was also knocked out, the mice developed ileocolitis and finally gastrointestinal tumors (Chu et al., 2004). With this background in mind, a study was designed to mimic the impact of a moderate selenium deficiency, as it might be expected from common variations in nutritional habits (Kipp et al., 2009). Four week-old mice were fed a diet containing 0.086 mg Se/kg for 6 weeks, which is about half the recommended daily intake for mice (0.150 mg/kg diet) (Ritskes-Hoitinga, 2004). For control, selenium was added in the form of selenomethionine up to the recommended dosage.

After 6 weeks, global gene expression was analyzed by microarray analyses in the colon and changes confirmed by qPCR in the colon (Kipp et al., 2009) and the duodenum (Müller, Banning, Brigelius-Flohé, & Kipp, 2010). The success of the feeding regimen was indicated by highly decreased plasma selenium and a reduction of GPx activity in liver and colon by about 50%. Selenoprotein W (SelW) was the most affected selenoprotein in the colon and also in the leukocytes of these mice (Kipp, Banning, van Schothorst, Meplan, Coort, et al., 2012). Other significantly affected selenoproteins were GPx1, selenoprotein H (SelH), and selenoprotein M (SelM) which, together with SelW, belong to the stress-related selenoproteins (Carlson, Xu, Gladyshev, & Hatfield, 2005). Surprisingly, the mRNA of TrxR1 and GPx2 was decreased in selenium-poor mice when measured in the homogenate of the entire colon, whereas it was increased in the duodenum, as expected for GPx2 and shown for the first time for TrxR1 (Müller et al., 2010). However, GPx2 mRNA was clearly increased by the moderate selenium deficiency when measured in isolated colonic crypt base cells, which is the site of its preferential expression in the colon (Kipp, Müller, et al., 2012). The interference of these changes with regulatory systems will be discussed in the following chapters.

3. SELENIUM STATUS AND THE Keap1/Nrf2 SYSTEM

The Keap1/Nrf2 system is briefly compiled in Fig. 4.1. In short, in the "off" state, Nrf2 is kept in the cytosol by Keap1. Keap1 acts as an adaptor for the E3 ubiquitin ligase Cullin 3/Ring/ligase (CRL) in the CRL/Keap1 complex (Cullinan, Gordan, Jin, Harper, & Diehl, 2004) by which Nrf2 is ubiquitinated and presented for proteasomal degradation (McMahon, Itoh, Yamamoto, & Hayes, 2003). Activation of the Nrf2 pathway is mediated by a conformational change of Keap1, which is associated with an abrogation of its adapter function for Cul3 and an accumulation of Nrf2 at the degradation complex (Rachakonda et al., 2008; Zhang, Lo, Cross, Templeton, & Hannink, 2004). Newly synthesized Nrf2 can no longer bind to Keap1 and takes over transcriptional activation (for reviews, see Kobayashi et al., 2006; Sekhar, Rachakonda, & Freeman, 2010; Taguchi, Motohashi, & Yamamoto, 2011). The most important redox-sensitive event of the signaling cascade is the conformational change of Keap1. Keap1 contains distinct cysteine residues from which Cys273, 288, and 151 are preferentially and selectively modified by electrophiles or oxidants (Takaya et al., 2012). For an in-depth discussion of the redox regulation of the system, we

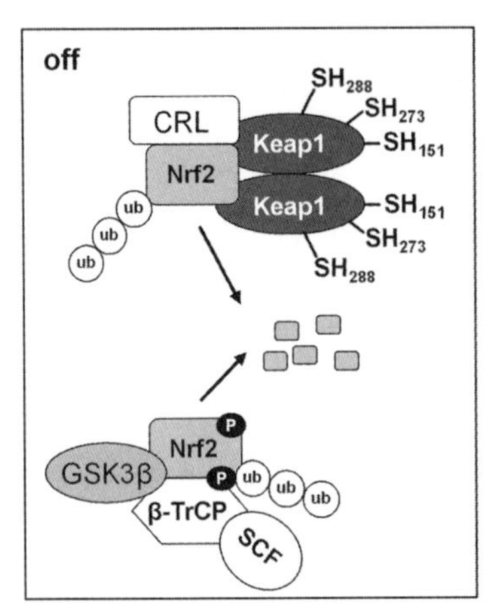

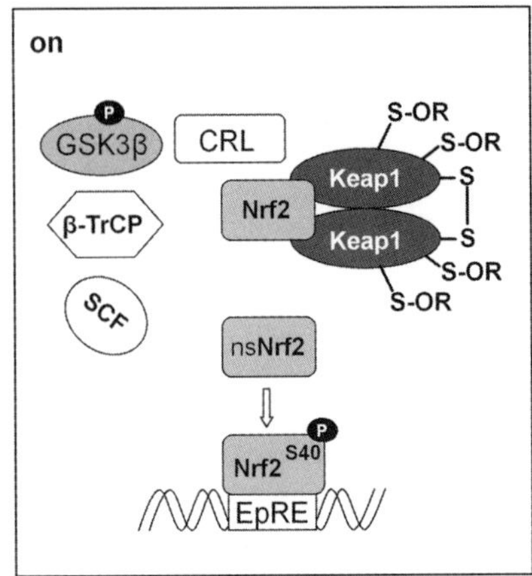

Figure 4.1 Overview of Nrf2 signaling. Left part: in the "off" state, Nrf2 binds to Keap1, which acts as an adapter to present Nrf2 to the Cul3-based E3 ligase complex (Cullin3/Ring/ligase, CRL) for ubiquitination and proteasomal degradation. Alternatively, a Keap1 independent degradation of Nrf2 is initiated by phosphorylation of Nrf2 by GSK3β. Phosphorylated Nrf2 is recognized by the adapter β-TrCP (β-transducin repeat containing protein) in the ubiquitin E3 ligase complex SCF (Skp1-Cullin-F-box), presented for ubiquitination and degraded. Right part: in the "on" state, Keap1 is modified by oxidants/electrophiles either at cysteines 288 or 273 or by disulfide formation between cysteines 151 of the two Keap1 molecules bound to Nrf2. The interaction between Nrf2 and Keap1 is disturbed and Nrf2 stays at Keap1 undegraded. Newly synthesized (ns) Nrf2 can be phosphorylated by PKC isoforms at Ser40 (Bloom & Jaiswal, 2003), enters the nucleus, and activates target genes by binding to its responsive element (EpRE). R in S-OR in the "on" state can be H coming from H_2O_2 or any residue coming from the adding electrophile. In the Keap1-independent pathway, Nrf2 is stabilized by inhibition of GSK3β. For further details see text and references therein. (For color version of this figure, the reader is referred to the online version of this chapter.)

refer readers to Brigelius-Flohé and Flohé (2011). In addition, Nrf2 can be directly phosphorylated by glycogen synthase kinase 3 beta (GSK3β), leading to a Keap1-independent degradation (see below, chapter on common players).

When screening the literature for regulation of GPx2 expression, GPx2 was detected on a list of genes upregulated in a model of pulmonary fibrosis. Surprisingly, however, upregulation of GPx2 was not observed in Nrf2 KO mice (Cho et al., 2002). Nrf2 stimulates the expression of target genes via binding to an electrophile-responsive element (EpRE) (Friling, Bensimon, Tichauer, & Daniel, 1990), first named antioxidant-responsive element (ARE) (Rushmore & Pickett, 1990). An EpRE was, indeed, found in the GPx2 promoter and its functionality was confirmed by the inducibility of GPx2 by electrophiles such as sulforaphane (SFN) at the level of promoter activity, mRNA, and protein (Banning et al., 2005). Independently, TrxR1 was identified as the target of Nrf2 (Hintze, Wald, Zeng, Jeffery, & Finley, 2003; Sakurai et al., 2005). The thioredoxin system (Trx/TrxR), apart from its pivotal function in providing deoxynucleotides, plays a major role in the redox regulation of cellular processes by maintaining cysteines of involved proteins in the thiol form (Arnér, 2009). Thus, the expression of at least two selenoproteins involved in redox regulation is controlled by Nrf2. Enhanced RNA levels would guarantee a fast regeneration of the respective selenoprotein when selenium is repleted. Moreover, the induction of certain selenoproteins by Nrf2 activation appears to be essential for the overall protective role of the Keap1/Nrf2 system. In a mouse model of inflammation-induced carcinogenesis, the classical Nrf2 activator SFN *enhanced* inflammation in selenium deficiency, while it displayed its expected anti-inflammatory action under adequate selenium supply (Krehl et al., 2012).

The feeding experiment mentioned above (Kipp et al., 2009) further revealed that the moderate selenium deficiency activated the Keap1/Nrf2 system. Microarray analyses revealed a large variety of nonselenium protein genes that are known to be Nrf2 targets, indeed, upregulated in moderate Se deficiency (Müller et al., 2010). Among those were genes of phase 2 enzymes such as GSTs or UGTs, of redox-active enzymes such as Prx or Srx, and of the sequestosome (p62). Further, upregulation of genes for NAD(P)H quinone oxidoreductase-1, Sult1b1, and Ugt1a6 a/b was confirmed by qPCR in the colon as well as in the duodenum, as was upregulation of heme oxygenase-1, Prx1, Srx1, and γ-glutamylcysteine synthetase-c. Generally, the response was larger in the duodenum than in the colon (Müller et al., 2010).

In line with the feeding experiments, upregulation of Nrf2 targets, especially phase 2 enzymes, was also observed in mice with a liver-specific KO of all selenoproteins by deletion of a functional Sec $tRNA^{ser(sec)}$ (gene name *Trsp*) (Sengupta et al., 2008). Replacement of housekeeping selenoproteins but not of stress-related selenoproteins normalized phase 2 enzyme expression. A link between housekeeping selenoproteins and phase 2 enzyme induction was thereby suggested. As TrxR1 belongs to the housekeeping selenoproteins (Carlson et al., 2007), this enzyme may play the major role in the proposed interplay (see Fig. 4.3). On the other hand, the stress-responsive selenoproteins, in particular GPx1, decline fastest under selenium deficiency, and the resulting impairment of hydroperoxide metabolism may well contribute to an oxidative activation of the Keap1/Nrf2 system.

In summary so far, the Keap1/Nrf2 system controls the expression of the selenoproteins GPx2 and TrxR1, which may complement the adaptive response thus triggered. Inversely, impaired redox homeostasis resulting from the decline of selenium-dependent peroxidases leads to oxidative activation of Keap1/Nrf2, and a moderate selenium deficiency is sufficient to trigger the activation.

4. SELENIUM AND THE WNT PATHWAY

So far only a few studies have described an association between selenium and the Wnt pathway. Wnt signaling was inhibited by high concentrations of certain selenium compounds (see below) and responded to selenium restriction with activation (Kipp et al., 2009). Further, an intriguing colocalization of GPx2 (Florian et al., 2001) and the Wnt system was observed in the intestinal tract. Wnt activity is particularly high at intestinal crypt bases, where it regulates proliferation and differentiation of intestinal stem cells. The finding that GPx2 is also regulated by Wnt signals (Kipp, Banning, & Brigelius-Flohé, 2007) suggested that GPx2 might be involved in the regulation of proliferation.

Wnt signaling is depicted in Fig. 4.2. When unstimulated ("off" state), its transcription factor β-catenin is kept in the cytosol, where it is associated with a "degradation complex" that consists of adenomatous polyposis coli (APC), Axin, and the kinases GSK3β and casein kinase-1 (CK1). Phosphorylation of β-catenin promotes ubiquitination by the ubiquitin E3 ligase complex SCF (Skp1-Cullin-F-box)/β-transducing repeat containing protein (β-TrCP), SCF/β-TrCP, and subsequent proteasomal degradation of β-catenin. As a result, levels of free β-catenin remain low, and the formation of the DNA-binding T cell factor (TCF)/β-catenin heterodimer in the nucleus is prevented. Instead, transcription factors of the TCF/lymphocyte enhancer factor (LEF) family are associated to the transcriptional corepressors transducin-like enhancer of split (TLE)/groucho and, thus, act as transcriptional repressors (van Es, Barker, & Clevers, 2003).

Upon binding of Wnt proteins ("on" state), the Wnt receptor Frizzled (Frz) interacts with its coreceptor LDL receptor-like proteins (LRP5 or LRP6) to initiate Wnt signaling. Several models have been postulated for subsequent down-stream events to follow (reviewed by Metcalfe & Bienz, 2011).

The *first* model suggests the formation of a signalosome at the plasma membrane. Activated Frz recruits the adapter protein disheveled (Dvl) to

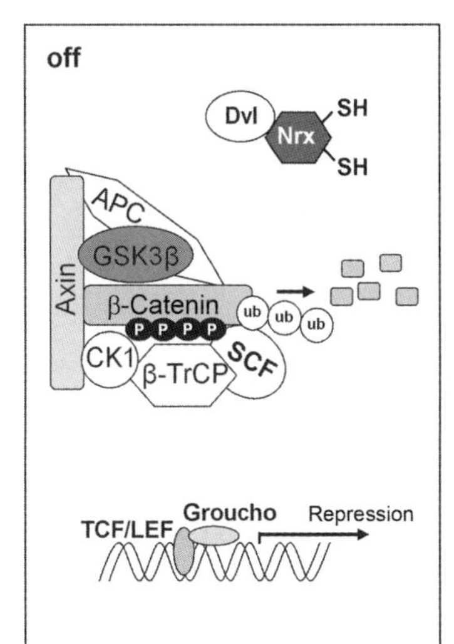

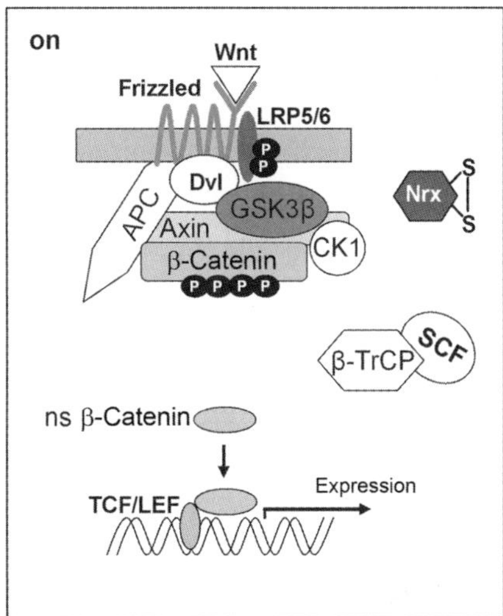

Figure 4.2 Simplified scheme of Wnt signaling. Left part: in the silent state ("off"), β-catenin is included in the so-called degradation complex consisting of the anchor proteins adenomatous polyposis coli (APC) and axis inhibiting protein (Axin), the Ser/Thr kinases glycogen synthase 3β (GSK3β) and casein kinase-1 (CK1), and the ubiquitin E3 ligase complex SCF (Skp1-Cullin-F-box)/β-TrCP (β-transducin repeat containing protein). Continuous phosphorylation by CK1 and GSK3β leads to ubiquitination and proteasomal degradation of β-catenin. In the nucleus, transcription factors of the TCF/LEF (T cell factor/transducin-like enhancer) family repress Wnt target genes by heterodimerization with the transcriptional corepressor groucho. Disheveled (Dvl) is associated to nucleoredoxin (Nrx). Right part: in the "on" state, Wnt proteins bind to their receptor complex consisting of Frizzled (Frz) and LDL receptor-related protein 5 (LRP5) or LRP6 and the signaling cascade starts. LRP becomes phosphorylated by CK1 and GSK3β, and Axin is recruited and associated via Dvl to the plasma membrane. Deliberation of Dvl is achieved by oxidative modification of Nrx. Depending on the model used (see text), GSK3β is inactivated by its interaction with phosphorylated LRP, β-catenin is deliberated, or β-TrCP is released during recruitment of the complex to the plasma membrane. All alternatives allow newly synthesized β-catenin (ns β-catenin) to enter the nucleus. There it binds to TCF/LEF and activates target gene expression. For further details see text and references therein. (For color version of this figure, the reader is referred to the online version of this chapter.)

the membrane. In turn, Dvl interacts with Axin in the destruction complex and recruits it together with associated proteins to the membrane forming the signalosome. GSK3β activity is blocked in the signalosome by its interaction with LRP, which is phosphorylated by CK1 (Bilic et al., 2007) thus allowing release and nuclear translocation of β-catenin.

The *second* model is based on the internalization of signalosome compounds, for example, Axin and GSK3β, into multivesicular bodies

(MVBs), thereby separating GSK3β from the other cellular compartments (Taelman et al., 2010). This "insulated" GSK3β is not able to phosphorylate newly synthesized β-catenin, which now can enter the nucleus unhampered. In both models, inhibition of GSK3β is a key event in Wnt signaling. In model 1, this is achieved by a direct catalytic inhibition of Axin-associated GSK3β by the interaction with phosphorylated LRP. In model 2, phosphorylation of β-catenin is indirectly inhibited by sequestration of LRP/Axin and GSK3β into MVBs.

Both models require the inhibition of GSK3β activity but whether this is achieved in the Wnt pathway by phosphorylation as in insulin signaling is still a matter of debate (Ding, Chen, & McCormick, 2000; McManus et al., 2005; Ng et al., 2009). Before any clarification could be offered, a *third* model has been developed. According to this model, Axin, upon binding of Wnt signals to its receptor, associates with LRP6 and recruits the destruction complex to the plasma membrane (Fig. 4.2 "on"). This results in an inhibition of β-catenin degradation due to the release of SCF/β-TrCP (Li et al., 2012). GSK3β remains active, and phosphorylated β-catenin accumulates and saturates the complex. Newly synthesized β-catenin can no longer associate to the complex and is able to enter the nucleus to activate target genes.

Present views on the redox regulation of the pathway center around the role of Dvl, which appears to be not yet finally clarified. In all three models, Dvl acts as a link between the membrane and the complex. It binds Axin1 directly (Malbon & Wang, 2006) and might, thus, provide a docking site for the interaction between Axin and Frz (Zeng et al., 2008). Dvl was first described as inhibitor of GSK3β (Cadigan & Liu, 2006). As mentioned, Dvl is bound to Nrx and deliberated upon oxidation of Nrx (Funato et al., 2006), whereby signaling can proceed (Korswagen, 2006). The oxidation equivalents for Nrx oxidation are assumed to be provided by H_2O_2 generated from NADPH oxidase-1 (NOX1)-derived superoxide anions. Interestingly, NOX1 is activated by Wnt signals (Kajla et al., 2012), which thus can guide signaling through the canonical Wnt pathway.

The presence of multiple H_2O_2-dependent steps suggests that the Wnt pathway is also affected by selenium. Inhibition of the pathway due to increased degradation of β-catenin was observed in CaCo2 cells by pXSC [1,4-phenylene bis(methylene)selenocyanate] (Narayanan, Narayanan, Desai, Pittman, & Reddy, 2004). Increased degradation of β-catenin by methylseleninic acid (MSeA) in human esophageal squamous cell carcinoma (Zhang et al., 2010) and HT29 cell lines (Saifo, Rempinski, Rustum, & Azrak, 2010) was accompanied by activation of GSK3β. Sodium selenite inhibited tumor formation in Muc2/p21 double KO mice and decreased cellular β-catenin levels as well as COX-2 expression. The effects were associated with an activation of JNK1 (Fang, Han, Bi, Xiong, & Yang, 2010). In the $APC^{min/+}$ mouse model of familial adenomatous polyposis, pXSC decreased the incidence of small intestinal tumors and protein levels of β-catenin and COX-2 activity in these tumors (Rao et al., 2000). In colorectal cancer cells (HCT116 and SW480) and xenograft tumors, high concentrations of selenite (10 μ*M*) inhibited β-catenin signaling, as was indicated by inhibition of Akt, suppression of β-catenin transcriptional activity, and a decrease in the expression of its targets cyclin D1 and survivin, finally resulting in apoptosis (Luo et al., 2012). In this case, Wnt inhibition was attributed to "ROS" generation resulting from the excessive concentration of selenite. However, the role of selenium in most of the quoted experiments remains to be elucidated, because its potentially controversial functions, that is, either supporting the synthesis of antioxidant selenoprotein or causing oxidative damage by autoxidation and redox cycling, had not been systematically addressed.

Activation of the Wnt pathway, as observed in moderate selenium deficiency (Kipp et al., 2009), appears to be more revealing. Under low-selenium feeding not only selenoproteins (see above) but also genes for β-catenin, Dvl, GSK3β, LEF1, and c-Myc responded significantly. A significant upregulation of β-catenin, Dvl2, Lef1, and c-Myc genes in the colon of Se-poor mice was observed in microarray analyses and confirmed by qPCR, as was a downregulation of GSK3β (Kipp et al., 2009). Especially the upregulation of the Wnt target c-Myc showed that the Wnt pathway responded to limited selenium supply with activation under *in vivo* conditions. The most likely reason underlying this phenomenon is an oxidative activation of the pathway due to impaired H_2O_2 metabolism in selenium deficiency.

5. COMMON PLAYERS AND EVENTS IN Nrf2 AND Wnt SIGNALING

Nrf2 cross-talks with a number of other transcription factors (reviewed in Wakabayashi, Slocum, Skoko, Shin, & Kensler, 2010) such as the aryl hydrocarbon receptor, NF-κB, p53, and Notch. A cross-talk with the Wnt pathway has so far hardly been considered. Both pathways, however, share certain components, and it would therefore be surprising if they operated independently. Such shared components are protein kinases and phosphatases, in particular the serine/threonine kinase GSK3β.

In the Keap1/Nrf2 system, GSK3β (Jain & Jaiswal, 2007; Salazar, Rojo, Velasco, de Sagarra, & Cuadrado, 2006) is in charge of termination of signaling by enabling the export of Nrf2 from the nucleus. More specifically, GSK3β phosphorylates and, thus, activates Fyn, a tyrosine phosphorylase belonging to the Src-A family. Upon its own phosphorylation, Fyn phosphorylates Nrf2 at Y568 which is a prerequisite for Nrf2 export from the nucleus. Consequently, inhibition of GSK3β leads to an accumulation of Nrf2 in the nucleus and perpetuation of Nrf2-dependent gene expression (Rojo et al., 2008) (Fig. 4.3). In addition, GSK3β also phosphorylates Nrf2 at a cluster of serines. These so-called degrons serve as recognition sequences for adaptors mediating the interaction of ubiquitin ligases with their substrates. The ubiquitin E3 ligase complex SCF/β-TrCP associates to these phosphorylated degrons and presents Nrf2 for proteasomal degradation in a Keap1-independent manner (Fig. 4.1) (Chowdhry et al., 2012; Rada et al., 2011, 2012). In both ways, either supporting nuclear export or degradation in the cytosol, GSK3β turns off Nrf2 signaling.

Also in the Wnt pathway, active GSK3β turns off signaling, and inhibition of GSK3β is a prerequisite for Wnt signaling. As outlined above, several models to achieve GSK3β inhibition are being discussed: (i) inhibition of complexed GSK3 by phosphorylated LRP, (ii) spatial separation (insulation) of the GSK3β complex in MVBs, (iii) liberation of the GSK3β inhibitor Dvl by oxidation of Nrx, and (iv) inhibition of GSK3β by phosphorylation at Ser-9 by, for example, Akt (Cohen & Frame, 2001).

Thus, GSK3β appears to be a common player in both pathways. However, whether the turning off signal in the Keap1-independent Nrf2 pathway comes from the GSK3β associated to the β-catenin degradation complex or from the free pool is not known. Whereas a number of studies report on an activation of the Wnt pathway by the phosphorylation and thereby inhibition of complexed GSK3β by the PI3K/Akt pathway (Naito et al., 2005; Rochat et al., 2004), others did not confirm this (Ding et al., 2000; McManus et al., 2005). Separation of GSK3β in the Axin complex was made responsible for the failure to activate the Wnt pathway by Akt (Ng et al., 2009). This means that two different pools of GSK3β exist in the cell. In fact only 3–5% of total GSK3β appear to be associated to Axin. Thus, a stabilization of Nrf2 by activation of the Wnt pathway needs further investigation. However, the possibility remains, as the GSK3β-mediated Keap1-independent degradation of Nrf2 could be inhibited by the very

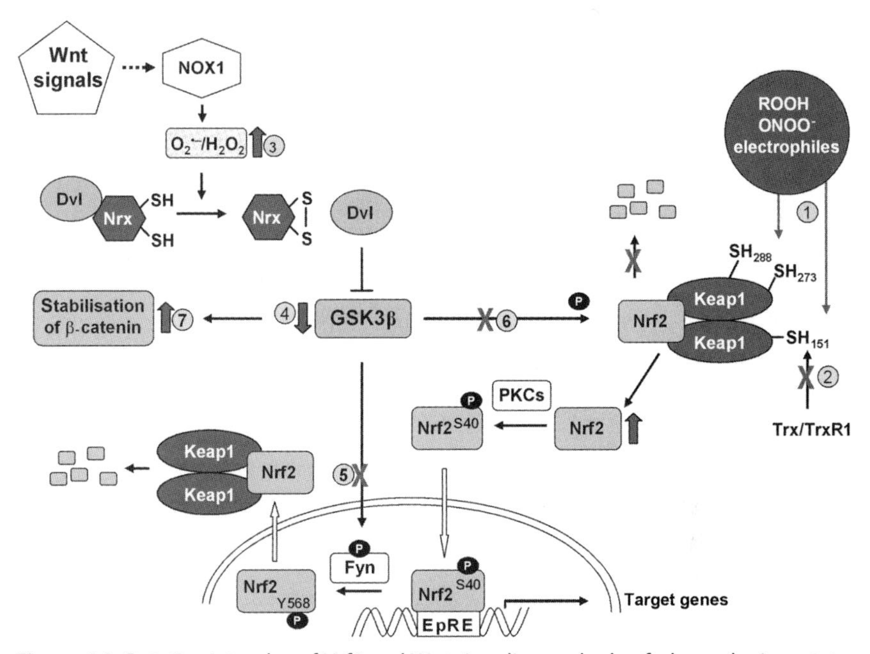

Figure 4.3 Putative interplay of Nrf2 and Wnt signaling and role of a low selenium state. Processes influenced by selenium deficiency are indicated by 1–7 and explained in detail in the text. Blocks in processes or up/downregulation by a low-selenium state are indicated by brown crosses or arrows, respectively. Abbreviations (not explained in Figs. 4.1 and 4.2): TrxR1, thioredoxin reductase-1; NOX1, NADPH oxidase-1; Fyn, member of the nonreceptor protein-tyrosine kinase subfamily Src-A. (1) Lack of hydroperoxide removal by, for example, GPxs leads to an enhanced stabilization of Nrf2 in the cytosol and nuclear translocation. (2) Inhibition of rereduction of Keap1 disulfides by a low activity of the Trx/TrxR1 system (Fourquet, Guerois, Biard, & Toledano, 2010) or by peroxiredoxins and their subsequent reduction by Trx/TrxRs, prevents Keap1 restoration. Newly synthesized Nrf2 can enter the nucleus. (3) Enhanced production of H_2O_2 via NOX1 upon stimulation with Wnt signals (Kajla et al., 2012) by lack of GPxs or TrxR1-mediated reduction of Prxs leads to persistent oxidation of Nrx and release of Dvl (Funato et al., 2006). (4) Inhibition of GSK3β activity by deliberation of Dvl (Funato et al., 2006) or decreased expression (Kipp et al., 2009) prevents steps 5–7. (5) Suppressed GSK3β activity prevents phosphorylation of Fyn and thus the export of Nrf2 out of the nucleus (Jain & Jaiswal, 2007). (6) Inhibition of phosphorylation of Nrf2 by GSK3β prevents Keap1-independent Nrf2 degradation (Chowdhry et al., 2012; Rada et al., 2011, 2012). (7) Putative stabilization of β-catenin due to decreased action of GSK3β (needs to be validated). (See Color Insert.)

same inhibitor of GSK3β, SB216763 (Rada et al., 2011), which otherwise activated Wnt signaling, as demonstrated by GSK3β phosphorylation and translocation of β-catenin into the nucleus (Die et al., 2012).

The most straightforward explanation for a link between both pathways is an oxidative inactivation of GSK3β resulting from massively impaired H_2O_2 reduction via, for example, GPx1. An impaired hydroperoxide metabolism facilitates Nrx oxidation and subsequent recruitment of the degradation complex to the membrane. On the other hand, it would stabilize inactive phosphorylated GSK3β, as its reactivation by dephosphorylation appears to depend on redox-sensitive phosphatases such as PP1 and PP2A (Hernandez, Langa, Cuadros, Avila, & Villanueva, 2010), which are candidates to be inhibited by oxidative modification of susceptible cysteines (Brigelius-Flohé & Flohé, 2011).

6. HOW DOES SELENIUM COME INTO PLAY?

A number of selenoproteins are redox-active enzymes and, thus, are predestined to regulate redox-sensitive signaling pathways. Obviously, Nrf2 is activated already under marginal selenium deficiency, a situation where selenoprotein synthesis is decreased. A decrease in GPx activity would facilitate the oxidative modification of Keap1 due to increased hydroperoxide concentrations (Fig. 4.3, 1). For the Keap1/Nrf2 system, the TrxR1/Trx couple is implicated as turn-off device (Fig. 4.3, 2). An intermolecular disulfide formation between the two Cys151 residues of the two Nrf2-bound Keap1 molecules has been detected in H_2O_2-treated HeLa cells (Fourquet et al., 2010). Silencing of TrxR1 led to an increased formation of these disulfide-linked forms of Keap1. Blocking TrxR activity by aurothioglucose led to a stimulation of Nrf2 responses, which points in the same direction (Locy et al., 2012). Deletion of TrxR1 in parenchymal hepatocytes of mice resulted in a compensatory upregulation of Nrf2 targets including GSTs, GPx2, and Srx (Suvorova et al., 2009). All these observations make TrxR1 an attractive candidate for delivering the turn-off signal for the Nrf2 pathway at the level of Keap1 and, accordingly, a low TrxR1 activity, as is observed in marginal selenium deficiency (Krehl et al., 2012), would equally block the turn-off (Fig. 4.3, 2). Whether the TrxR/Trx couple plays a similar role in protein disulfide formation in the Wnt pathway remains to be investigated.

The role of GPx2 is less clear. NOX1-derived $O_2^{\cdot -}/H_2O_2$ (Fig. 4.3, 3) mediates oxidation of Nrx leading to a release of Dvl and inhibition of GSK3β (Fig. 4.3, 4). H_2O_2 is stabilized in selenium deficiency due to lack of sufficient GPx1 and probably GPx2 activity. The observed increase in the levels of GPx2 RNA in colon crypt bases of selenium-poor mice (Kipp, Müller, et al., 2012) does not necessarily mean that protein and its activity are also enhanced, and GPx2 also needs selenium for being synthesized. Increased RNA in selenium deficiency only guarantees fast resynthesis upon refeeding. Removal of specific hydroperoxides in specific pathways by specific GPxs is attractive and could explain their specific function as recently demonstrated for GPx4 in the inhibition of apoptosis stimulated by 12,15-LOX-derived products (Seiler et al., 2008). Whether GPx1 or 2 counteracts NOX1 activity in Wnt signaling remains to be elucidated. However, there are other options than impaired H_2O_2 removal to interpret the selenium dependency of the pathways. As has been amply reviewed elsewhere (Brigelius-Flohé & Maiorino, 2013; Ferrer-Sueta et al., 2011; Forman, Maiorino, & Ursini, 2010) peroxidases of the Prx and the GPx family can interfere with signaling cascades in controversial ways: by removing H_2O_2 and thereby preventing oxidative modifications of signaling proteins or by acting as sensors for H_2O_2 and promoting oxidative modification of specific targets. The latter action has so far been reported only for non-mammalian systems. But "redoxins" such as Trx and Nrx may well be considered as potential substrates of thiol peroxidases. In any case, whether the reversal of protein disulfide formation is for termination or for triggering signal transduction deserves attention in this context.

Inhibition of GSK3β in the destruction complex or in a free pool, as well as its observed downregulation in marginal selenium deficiency, will prevent phosphorylation of Fyn (Fig. 4.3, 5) and subsequent stabilization of Nrf2 in the nucleus. Inhibition of GSK3β will also prevent phosphorylation of Nrf2 and its Keap1-independent proteasomal degradation (Fig. 4.3, 6), and in this way, stabilize β-catenin and its nuclear translocation (Fig. 4.3, 7).

In conclusion, the Keap1/Nrf2 system and the Wnt pathway similarly respond to oxidizing conditions. The primary oxidant in both cases is H_2O_2 produced by dismutation of superoxide anions which are derived from NOX-type enzymes: NOX1 in case of Wnt signaling. The oxidative challenge exerted by NOX activation obviously can be mimicked by a moderate selenium restriction, which leads to impaired hydroperoxide reduction by selenium-dependent peroxidase systems and prevention of signaling shut off due to shortage of reduced Trx. The direct and indirect targets of H_2O_2 or Trx are not entirely clear for any of the pathways (for review, see Brigelius-Flohé & Flohé, 2011; Miki & Funato, 2012), but the two pathways share at least one common redox-sensitive event, the inactivation of

GSK3β by phosphorylation or other mechanisms. Although these pathways are similarly regulated, their overall profile of regulated proteins is rather contrasting: While the Keap1/Nrf2 system is generally believed to promote a hormetic response that protects against oxidative stress, diverse xenobiotics, inflammation, and carcinogenesis, the Wnt response is rather considered a risky one in these respects (Logan & Nusse, 2004). It is therefore tempting to speculate that the increased risk of carcinogenesis associated with selenium deficiency is attributable to enhanced Wnt signaling. However, a continuous activation of the Keap1/Nrf2 system cannot be rated as benign either. As compiled elsewhere (Brigelius-Flohé, Müller, Lippmann, & Kipp, 2012), mutations in the system leading to its constitutive activation as well as loss-of-function mutations have been found associated with clinical cancers, and xenograft experiments revealed that not only healthy cells but also malignant ones benefit from intact Keap1/Nrf2 signaling. Findings that under certain conditions high as well as low selenium can contribute to carcinogenesis (Felix et al., 2004; Irmak, Ince, Ozturk, & Cetin-Atalay, 2003; Novoselov et al., 2005) are controversial enough to preclude any generalized recommendation on the use of selenium for cancer prevention.

REFERENCES

Arnér, E. S. (2009). Focus on mammalian thioredoxin reductases—Important selenoproteins with versatile functions. *Biochimica et Biophysica Acta*, *1790*, 495–526.

Banning, A., Deubel, S., Kluth, D., Zhou, Z., & Brigelius-Flohé, R. (2005). The GI-GPx gene is a target for Nrf2. *Molecular and Cellular Biology*, *25*, 4914–4923.

Bilic, J., Huang, Y. L., Davidson, G., Zimmermann, T., Cruciat, C. M., Bienz, M., et al. (2007). Wnt induces LRP6 signalosomes and promotes dishevelled-dependent LRP6 phosphorylation. *Science*, *316*, 1619–1622.

Biteau, B., Labarre, J., & Toledano, M. B. (2003). ATP-dependent reduction of cysteine-sulphinic acid by *S. cerevisiae* sulphiredoxin. *Nature*, *425*, 980–984.

Bloom, D. A., & Jaiswal, A. K. (2003). Phosphorylation of Nrf2 at Ser40 by protein kinase C in response to antioxidants leads to the release of Nrf2 from INrf2, but is not required for Nrf2 stabilization/accumulation in the nucleus and transcriptional activation of antioxidant response element-mediated NAD(P)H:quinone oxidoreductase-1 gene expression. *Journal of Biological Chemistry*, *278*, 44675–44682.

Brigelius-Flohé, R., & Flohé, L. (2011). Basic principles and emerging concepts in the redox control of transcription factors. *Antioxidants & Redox Signaling*, *15*, 2335–2381.

Brigelius-Flohé, R., & Kipp, A. (2009). Glutathione peroxidases in different stages of carcinogenesis. *Biochimica et Biophysica Acta*, *1790*, 1555–1568.

Brigelius-Flohé, R., & Maiorino, M. (2013). Glutathione peroxidases. *Biochimica et Biophysica Acta*, *1830*, 3289–3303.

Brigelius-Flohé, R., Müller, M., Lippmann, D., & Kipp, A. P. (2012). The yin and yang of nrf2-regulated selenoproteins in carcinogenesis. *International Journal of Cell Biology*, *2012*, 486147. http://dx.doi.org/10.1155/2012/486147.

Burk, R. F., Hill, K. E., Nakayama, A., Mostert, V., Levander, X. A., Motley, A. K., et al. (2008). Selenium deficiency activates mouse liver Nrf2-ARE but vitamin E deficiency does not. *Free Radical Biology & Medicine, 44*, 1617–1623.

Cadigan, K. M., & Liu, Y. I. (2006). Wnt signaling: Complexity at the surface. *Journal of Cell Science, 119*, 395–402.

Carlson, B. A., Moustafa, M. E., Sengupta, A., Schweizer, U., Shrimali, R., Rao, M., et al. (2007). Selective restoration of the selenoprotein population in a mouse hepatocyte selenoproteinless background with different mutant selenocysteine tRNAs lacking Um34. *Journal of Biological Chemistry, 282*, 32591–32602.

Carlson, B. A., Xu, X. M., Gladyshev, V. N., & Hatfield, D. L. (2005). Selective rescue of selenoprotein expression in mice lacking a highly specialized methyl group in selenocysteine tRNA. *Journal of Biological Chemistry, 280*, 5542–5548.

Cho, H. Y., Jedlicka, A. E., Reddy, S. P., Kensler, T. W., Yamamoto, M., Zhang, L. Y., et al. (2002). Role of NRF2 in protection against hyperoxic lung injury in mice. *American Journal of Respiratory Cell and Molecular Biology, 26*, 175–182.

Chowdhry, S., Zhang, Y., McMahon, M., Sutherland, C., Cuadrado, A., & Hayes, J. D. (2012). Nrf2 is controlled by two distinct beta-TrCP recognition motifs in its Neh6 domain, one of which can be modulated by GSK-3 activity. *Oncogene*, http://dx.doi.org/10.1038/onc.2012.388.

Chu, F. F., Esworthy, R. S., Chu, P. G., Longmate, J. A., Huycke, M. M., Wilczynski, S., et al. (2004). Bacteria-induced intestinal cancer in mice with disrupted Gpx1 and Gpx2 genes. *Cancer Research, 64*, 962–968.

Cohen, P., & Frame, S. (2001). The renaissance of GSK3. *Nature Reviews. Molecular Cell Biology, 2*, 769–776.

Cullinan, S. B., Gordan, J. D., Jin, J., Harper, J. W., & Diehl, J. A. (2004). The Keap1-BTB protein is an adaptor that bridges Nrf2 to a Cul3-based E3 ligase: Oxidative stress sensing by a Cul3-Keap1 ligase. *Molecular and Cellular Biology, 24*, 8477–8486.

Die, L., Yan, P., Jun Jiang, Z., Min Hua, T., Cai, W., & Xing, L. (2012). Glycogen synthase kinase-3 beta inhibitor suppresses Porphyromonas gingivalis lipopolysaccharide-induced CD40 expression by inhibiting nuclear factor-kappa B activation in mouse osteoblasts. *Molecular Immunology, 52*, 38–49.

Ding, V. W., Chen, R. H., & McCormick, F. (2000). Differential regulation of glycogen synthase kinase 3beta by insulin and Wnt signaling. *Journal of Biological Chemistry, 275*, 32475–32481.

Fang, W., Han, A., Bi, X., Xiong, B., & Yang, W. (2010). Tumor inhibition by sodium selenite is associated with activation of c-Jun NH2-terminal kinase 1 and suppression of beta-catenin signaling. *International Journal of Cancer, 127*, 32–42.

Felix, K., Gerstmeier, S., Kyriakopoulos, A., Howard, O. M., Dong, H. F., Eckhaus, M., et al. (2004). Selenium deficiency abrogates inflammation-dependent plasma cell tumors in mice. *Cancer Research, 64*, 2910–2917.

Ferrer-Sueta, G., Manta, B., Botti, H., Radi, R., Trujillo, M., & Denicola, A. (2011). Factors affecting protein thiol reactivity and specificity in peroxide reduction. *Chemical Research in Toxicology, 24*, 434–450.

Flohé, L. (1989). The selenoprotein glutathione peroxidase. In D. Dolphin, R. Poulson, & O. Aramovic (Eds.), *Glutathione: Chemical, biochemical and medical aspects—Part A* (pp. 643–731). New York: John Wiley & Sons Inc.

Flohé, L., Günzler, W. A., & Schock, H. H. (1973). Glutathione peroxidase: A selenoenzyme. *FEBS Letters, 32*, 132–134.

Florian, S., Wingler, K., Schmehl, K., Jacobasch, G., Kreuzer, O. J., Meyerhof, W., et al. (2001). Cellular and subcellular localization of gastrointestinal glutathione peroxidase in normal and malignant human intestinal tissue. *Free Radical Research, 35*, 655–663.

Forman, H. J., Maiorino, M., & Ursini, F. (2010). Signaling functions of reactive oxygen species. *Biochemistry, 49*, 835–842.

Fourquet, S., Guerois, R., Biard, D., & Toledano, M. B. (2010). Activation of NRF2 by nitrosative agents and H2O2 involves KEAP1 disulfide formation. *Journal of Biological Chemistry, 285*, 8463–8471.

Friling, R. S., Bensimon, A., Tichauer, Y., & Daniel, V. (1990). Xenobiotic-inducible expression of murine glutathione S-transferase Ya subunit gene is controlled by an electrophile-responsive element. *Proceedings of the National Academy of Sciences of the United States of America, 87*, 6258–6262.

Funato, Y., Michiue, T., Asashima, M., & Miki, H. (2006). The thioredoxin-related redox-regulating protein nucleoredoxin inhibits Wnt-beta-catenin signalling through dishevelled. *Nature Cell Biology, 8*, 501–508.

Hernandez, F., Langa, E., Cuadros, R., Avila, J., & Villanueva, N. (2010). Regulation of GSK3 isoforms by phosphatases PP1 and PP2A. *Molecular and Cellular Biochemistry, 344*, 211–215.

Hintze, K. J., Wald, K. A., Zeng, H., Jeffery, E. H., & Finley, J. W. (2003). Thioredoxin reductase in human hepatoma cells is transcriptionally regulated by sulforaphane and other electrophiles via an antioxidant response element. *Journal of Nutrition, 133*, 2721–2727.

Irmak, M. B., Ince, G., Ozturk, M., & Cetin-Atalay, R. (2003). Acquired tolerance of hepatocellular carcinoma cells to selenium deficiency: A selective survival mechanism? *Cancer Research, 63*, 6707–6715.

Ishii, T., Itoh, K., Ruiz, E., Leake, D. S., Unoki, H., Yamamoto, M., et al. (2004). Role of Nrf2 in the regulation of CD36 and stress protein expression in murine macrophages: Activation by oxidatively modified LDL and 4-hydroxynonenal. *Circulation Research, 94*, 609–616.

Jain, A. K., & Jaiswal, A. K. (2007). GSK-3beta acts upstream of Fyn kinase in regulation of nuclear export and degradation of NF-E2 related factor 2. *Journal of Biological Chemistry, 282*, 16502–16510.

Kajla, S., Mondol, A. S., Nagasawa, A., Zhang, Y., Kato, M., Matsuno, K., et al. (2012). A crucial role for Nox 1 in redox-dependent regulation of Wnt-beta-catenin signaling. *The FASEB Journal, 26*, 2049–2059.

Kipp, A., Banning, A., & Brigelius-Flohé, R. (2007). Activation of the glutathione peroxidase 2 (GPx2) promoter by beta-catenin. *Biological Chemistry, 388*, 1027–1033.

Kipp, A., Banning, A., van Schothorst, E. M., Méplan, C., Schomburg, L., Evelo, C., et al. (2009). Four selenoproteins, protein biosynthesis, and Wnt signalling are particularly sensitive to limited selenium intake in mouse colon. *Molecular Nutrition & Food Research, 53*, 1561–1572.

Kipp, A. P., Banning, A., van Schothorst, E. M., Meplan, C., Coort, S. L., Evelo, C. T., et al. (2012). Marginal selenium deficiency down-regulates inflammation-related genes in splenic leukocytes of the mouse. *The Journal of Nutritional Biochemistry, 23*, 1170–1177.

Kipp, A. P., Müller, M. F., Göken, E. M., Deubel, S., & Brigelius-Flohé, R. (2012). The selenoproteins GPx2, TrxR2 and TrxR3 are regulated by Wnt signalling in the intestinal epithelium. *Biochimica et Biophysica Acta, 1820*, 1588–1596.

Kobayashi, A., Kang, M.-I., Watai, Y., Tong, K. I., Shibata, T., Uchida, K., et al. (2006). Oxidative and electrophilic stresses activate Nrf2 through inhibition of ubiquitination activity of Keap1. *Molecular and Cellular Biology, 26*, 221–229.

Korswagen, H. C. (2006). Regulation of the Wnt/beta-catenin pathway by redox signaling. *Developmental Cell, 10*, 687–688.

Krehl, S., Loewinger, M., Florian, S., Kipp, A., Banning, A., Wessjohann, L. A., et al. (2012). Glutathione peroxidase-2 and selenium decreased inflammation and tumors in a mouse model of inflammation-associated carcinogenesis whereas sulforaphane effects differed with selenium supply. *Carcinogenesis, 33*, 620–628.

Lawrence, R. A., & Burk, R. F. (1976). Glutathione peroxidase activity in selenium-deficient rat liver. *Biochemical and Biophysical Research Communications, 71*, 952–958.

Lawrence, R. A., Parkhill, L. K., & Burk, R. F. (1978). Hepatic cytosolic non selenium-dependent glutathione peroxidase activity: Its nature and the effect of selenium deficiency. *Journal of Nutrition*, *108*, 981–987.

Li, V. S., Ng, S. S., Boersema, P. J., Low, T. Y., Karthaus, W. R., Gerlach, J. P., et al. (2012). Wnt signaling through inhibition of beta-catenin degradation in an intact axin1 complex. *Cell*, *149*, 1245–1256.

Locy, M. L., Rogers, L. K., Prigge, J. R., Schmidt, E. E., Arnér, E. S., & Tipple, T. E. (2012). Thioredoxin reductase inhibition elicits Nrf2-mediated responses in clara cells: Implications for oxidant-induced lung injury. *Antioxidants & Redox Signaling*, *17*, 1407.

Logan, C. Y., & Nusse, R. (2004). The Wnt signaling pathway in development and disease. *Annual Review of Cell and Developmental Biology*, *20*, 781–810.

Luo, H., Yang, Y., Huang, F., Li, F., Jiang, Q., Shi, K., et al. (2012). Selenite induces apoptosis in colorectal cancer cells via AKT-mediated inhibition of beta-catenin survival axis. *Cancer Letters*, *315*, 78–85.

Malbon, C. C., & Wang, H. Y. (2006). Dishevelled: A mobile scaffold catalyzing development. *Current Topics in Developmental Biology*, *72*, 153–166.

Masukawa, T., Nishimura, T., & Iwata, H. (1984). Differential changes of glutathione S-transferase activity by dietary selenium. *Biochemical Pharmacology*, *33*, 2635–2639.

McMahon, M., Itoh, K., Yamamoto, M., & Hayes, J. D. (2003). Keap1-dependent proteasomal degradation of transcription factor Nrf2 contributes to the negative regulation of antioxidant response element-driven gene expression. *Journal of Biological Chemistry*, *278*, 21592–21600.

McManus, E. J., Sakamoto, K., Armit, L. J., Ronaldson, L., Shpiro, N., Marquez, R., et al. (2005). Role that phosphorylation of GSK3 plays in insulin and Wnt signalling defined by knockin analysis. *EMBO Journal*, *24*, 1571–1583.

Medina, D., & Shepherd, F. (1980). Selenium-mediated inhibition of mouse mammary tumorigenesis. *Cancer Letters*, *8*, 241–245.

Medina, D., & Shepherd, F. (1981). Selenium-mediated inhibition of 7,12-dimethylbenz[a] anthracene-induced mouse mammary tumorigenesis. *Carcinogenesis*, *2*, 451–455.

Metcalfe, C., & Bienz, M. (2011). Inhibition of GSK3 by Wnt signalling-two contrasting models. *Journal of Cell Science*, *124*, 3537–3544.

Miki, H., & Funato, Y. (2012). Regulation of intracellular signalling through cysteine oxidation by reactive oxygen species. *Journal of Biochemistry*, *151*, 255–261.

Müller, M., Banning, A., Brigelius-Flohé, R., & Kipp, A. (2010). Nrf2 target genes are induced under marginal selenium-deficiency. *Genes & Nutrition*, *5*, 297–307.

Naito, A. T., Akazawa, H., Takano, H., Minamino, T., Nagai, T., Aburatani, H., et al. (2005). Phosphatidylinositol 3-kinase-Akt pathway plays a critical role in early cardiomyogenesis by regulating canonical Wnt signaling. *Circulation Research*, *97*, 144–151.

Narayanan, B. A., Narayanan, N. K., Desai, D., Pittman, B., & Reddy, B. S. (2004). Effects of a combination of docosahexaenoic acid and 1,4-phenylene bis(methylene) selenocyanate on cyclooxygenase 2, inducible nitric oxide synthase and beta-catenin pathways in colon cancer cells. *Carcinogenesis*, *25*, 2443–2449.

Ng, S. S., Mahmoudi, T., Danenberg, E., Bejaoui, I., de Lau, W., Korswagen, H. C., et al. (2009). Phosphatidylinositol 3-kinase signaling does not activate the wnt cascade. *Journal of Biological Chemistry*, *284*, 35308–35313.

Novoselov, S. V., Calvisi, D. F., Labunskyy, V. M., Factor, V. M., Carlson, B. A., Fomenko, D. E., et al. (2005). Selenoprotein deficiency and high levels of selenium compounds can effectively inhibit hepatocarcinogenesis in transgenic mice. *Oncogene*, *24*, 8003–8011.

Rachakonda, G., Xiong, Y., Sekhar, K. R., Stamer, S. L., Liebler, D. C., & Freeman, M. L. (2008). Covalent modification at Cys151 dissociates the electrophile sensor Keap1 from the ubiquitin ligase CUL3. *Chemical Research in Toxicology*, *21*, 705–710.

Rada, P., Rojo, A. I., Chowdhry, S., McMahon, M., Hayes, J. D., & Cuadrado, A. (2011). SCF/{beta}-TrCP promotes glycogen synthase kinase 3-dependent degradation of the Nrf2 transcription factor in a Keap1-independent manner. *Molecular and Cellular Biology, 31*, 1121–1133.

Rada, P., Rojo, A. I., Evrard-Todeschi, N., Innamorato, N. G., Cotte, A., Jaworski, T., et al. (2012). Structural and functional characterization of Nrf2 degradation by the glycogen synthase kinase 3/beta-TrCP axis. *Molecular and Cellular Biology, 32*, 3486–3499.

Rao, C. V., Cooma, I., Rodriguez, J. G., Simi, B., El-Bayoumy, K., & Reddy, B. S. (2000). Chemoprevention of familial adenomatous polyposis development in the APC(min) mouse model by 1,4-phenylene bis(methylene)selenocyanate. *Carcinogenesis, 21*, 617–621.

Reiter, R., & Wendel, A. (1983). Selenium and drug metabolism—I. Multiple modulations of mouse liver enzymes. *Biochemical Pharmacology, 32*, 3063–3067.

Reiter, R., & Wendel, A. (1985). Selenium and drug metabolism—III. Relation of glutathione-peroxidase and other hepatic enzyme modulations to dietary supplements. *Biochemical Pharmacology, 34*, 2287–2290.

Ritskes-Hoitinga, M. (2004). Nutrition of laboratory mice. In H. Hedrich (Ed.), *The laboratory mouse* (pp. 463–479). San Diego: Elsevier Academic Press.

Rochat, A., Fernandez, A., Vandromme, M., Moles, J. P., Bouschet, T., Carnac, G., et al. (2004). Insulin and wnt1 pathways cooperate to induce reserve cell activation in differentiation and myotube hypertrophy. *Molecular and Cellular Biology, 15*, 4544–4555.

Rojo, A. I., Rada, P., Egea, J., Rosa, A. O., Lopez, M. G., & Cuadrado, A. (2008). Functional interference between glycogen synthase kinase-3 beta and the transcription factor Nrf2 in protection against kainate-induced hippocampal cell death. *Molecular and Cellular Neuroscience, 39*, 125–132.

Rotruck, J. T., Pope, A. L., Ganther, H. E., Swanson, A. B., Hafeman, D. G., & Hoekstra, W. G. (1973). Selenium: Biochemical role as a component of glutathione peroxidase. *Science, 179*, 588–590.

Rushmore, T. H., & Pickett, C. B. (1990). Transcriptional regulation of the rat glutathione S-transferase Ya subunit gene. Characterization of a xenobiotic-responsive element controlling inducible expression by phenolic antioxidants. *Journal of Biological Chemistry, 265*, 14648–14653.

Saifo, M. S., Rempinski, D. R., Jr., Rustum, Y. M., & Azrak, R. G. (2010). Targeting the oncogenic protein beta-catenin to enhance chemotherapy outcome against solid human cancers. *Molecular Cancer, 9*, 310.

Sakurai, A., Nishimoto, M., Himeno, S., Imura, N., Tsujimoto, M., Kunimoto, M., et al. (2005). Transcriptional regulation of thioredoxin reductase 1 expression by cadmium in vascular endothelial cells: Role of NF-E2-related factor-2. *Journal of Cellular Physiology, 203*, 529–537.

Salazar, M., Rojo, A. I., Velasco, D., de Sagarra, R. M., & Cuadrado, A. (2006). Glycogen synthase kinase-3beta inhibits the xenobiotic and antioxidant cell response by direct phosphorylation and nuclear exclusion of the transcription factor Nrf2. *Journal of Biological Chemistry, 281*, 14841–14851.

Seiler, A., Schneider, M., Forster, H., Roth, S., Wirth, E. K., Culmsee, C., et al. (2008). Glutathione peroxidase 4 senses and translates oxidative stress into 12/15-lipoxygenase dependent- and AIF-mediated cell death. *Cell Metabolism, 8*, 237–248.

Sekhar, K. R., Rachakonda, G., & Freeman, M. L. (2010). Cysteine-based regulation of the CUL3 adaptor protein Keap1. *Toxicology and Applied Pharmacology, 244*, 21–26.

Sengupta, A., Carlson, B. A., Weaver, J. A., Novoselov, S. V., Fomenko, D. E., Gladyshev, V. N., et al. (2008). A functional link between housekeeping selenoproteins and phase II enzymes. *Biochemical Journal, 413*, 151–161.

Shamberger, R. J., & Frost, D. V. (1969). Possible protective effect of selenium against human cancer. *Canadian Medical Association Journal, 100*, 682.

Spallholz, J. E. (1994). On the nature of selenium toxicity and carcinostatic activity. *Free Radical Biology & Medicine, 17*, 45–64.

Sunde, R. A. (2012). Selenoproteins: Hierarchy, requirements, and biomarkers. In D. L. Hatfield, M. J. Berry, & V. N. Gladyshev (Eds.), *Selenium: Its molecular biology and role in human health* (pp. 137–152, 3rd ed.). New York: Springer Science + Business Media, LLC.

Surh, Y. J., Kundu, J. K., & Na, H. K. (2008). Nrf2 as a master redox switch in turning on the cellular signaling involved in the induction of cytoprotective genes by some chemopreventive phytochemicals. *Planta Medica, 74*, 1526–1539.

Suvorova, E. S., Lucas, O., Weisend, C. M., Rollins, M. F., Merrill, G. F., Capecchi, M. R., et al. (2009). Cytoprotective Nrf2 pathway is induced in chronically txnrd 1-deficient hepatocytes. *PLoS One, 4*, e6158.

Taelman, V. F., Dobrowolski, R., Plouhinec, J. L., Fuentealba, L. C., Vorwald, P. P., Gumper, I., et al. (2010). Wnt signaling requires sequestration of glycogen synthase kinase 3 inside multivesicular endosomes. *Cell, 143*, 1136–1148.

Taguchi, K., Motohashi, H., & Yamamoto, M. (2011). Molecular mechanisms of the Keap1-Nrf2 pathway in stress response and cancer evolution. *Genes to Cells, 16*, 123–140.

Takaya, K., Suzuki, T., Motohashi, H., Onodera, K., Satomi, S., Kensler, T. W., et al. (2012). Validation of the multiple sensor mechanism of the Keap1-Nrf2 system. *Free Radical Biology & Medicine, 53*, 817–827.

Tamura, T., & Stadtman, T. C. (1996). A new selenoprotein from human lung adenocarcinoma cells: Purification, properties, and thioredoxin reductase activity. *Proceedings of the National Academy of Sciences of the United States of America, 93*, 1006–1011.

van Es, J. H., Barker, N., & Clevers, H. (2003). You Wnt some, you lose some: Oncogenes in the Wnt signaling pathway. *Current Opinion in Genetics and Development, 13*, 28–33.

Wakabayashi, N., Slocum, S. L., Skoko, J. J., Shin, S., & Kensler, T. W. (2010). When NRF2 talks, who's listening? *Antioxidants & Redox Signaling, 13*, 1649–1663.

Wild, A. C., Moinova, H. R., & Mulcahy, R. T. (1999). Regulation of gamma-glutamylcysteine synthetase subunit gene expression by the transcription factor Nrf2. *Journal of Biological Chemistry, 274*, 33627–33636.

Wingler, K., Böcher, M., Flohé, L., Kollmus, H., & Brigelius-Flohé, R. (1999). mRNA stability and selenocysteine insertion sequence efficiency rank gastrointestinal glutathione peroxidase high in the hierarchy of selenoproteins. *European Journal of Biochemistry, 259*, 149–157.

Yang, H., Jia, X., Chen, X., Yang, C. S., & Li, N. (2012). Time-selective chemoprevention of vitamin E and selenium on esophageal carcinogenesis in rats: The possible role of nuclear factor kappaB signaling pathway. *International Journal of Cancer, 131*, 1517–1527.

Yoo, M. H., Carlson, B. A., Tsuji, P. A., Tobe, R., Naranjo-Suarez, S., Lee, B. J., et al. (2012). Selenoproteins harboring a split personality in both preventing and promoting cancer. In D. L. Hatfield, M. J. Berry, & V. N. Gladyshev (Eds.), *Selenium: Its molecular biology and role in human health* (pp. 325–333, 3rd ed.). New York: Springer Science + Business Media, LLC.

Zeng, X., Huang, H., Tamai, K., Zhang, X., Harada, Y., Yokota, C., et al. (2008). Initiation of Wnt signaling: Control of Wnt coreceptor Lrp6 phosphorylation/activation via frizzled, dishevelled and axin functions. *Development, 135*, 367–375.

Zhang, D. D., Lo, S.-C., Cross, J. V., Templeton, D. J., & Hannink, M. (2004). Keap1 is a redox-regulated substrate adaptor protein for a Cul3-dependent ubiquitin ligase complex. *Molecular and Cellular Biology*, *24*, 10941–10953.

Zhang, W., Yan, S., Liu, M., Zhang, G., Yang, S., He, S., et al. (2010). beta-Catenin/TCF pathway plays a vital role in selenium induced-growth inhibition and apoptosis in esophageal squamous cell carcinoma (ESCC) cells. *Cancer Letters*, *296*, 113–122.

CHAPTER FIVE

Selenoprotein W as Biomarker for the Efficacy of Selenium Compounds to Act as Source for Selenoprotein Biosynthesis

Anna Patricia Kipp, Janna Frombach, Stefanie Deubel, Regina Brigelius-Flohé[1]

Department Biochemistry of Micronutrients, German Institute of Human Nutrition Potsdam-Rehbruecke, Nuthetal, Germany

[1]Corresponding author: e-mail address: flohe@dife.de

Contents

1. Introduction 88
2. Experimental 90
   2.1 Cell culture 90
   2.2 Biomarkers for the selenium status 91
   2.3 Statistics 94
3. Results 94
   3.1 Toxicity of individual selenium compounds 94
   3.2 Estimation of biomarkers 96
4. Discussion 102
   4.1 Metabolism of selenium compounds 104
   4.2 Efficacy of selenium compounds to increase selenoprotein level and activity 105
   4.3 Efficacy of selenium compounds to increase protein levels of selenoproteins 106
   4.4 Efficacy of selenium compounds to stabilize selenoprotein mRNA 106
5. Conclusions 108
Acknowledgments 109
References 109

Abstract

Selenium is an essential trace element and, like all elements, present in many different compounds with unequivocal functions. This fact is only sporadically mentioned when recommended intake or supplementation is indicated just as "selenium." In mammals, selenium is an integral part of selenoproteins as selenocysteine. Selenocysteine is formed from serine at the respective $tRNA^{(ser)sec}$, a reaction that requires selenophosphate formed from selenide and ATP. Thus, only compounds that can be metabolized into selenide can serve as sources for selenoprotein biosynthesis. We

Methods in Enzymology, Volume 527
ISSN 0076-6879
http://dx.doi.org/10.1016/B978-0-12-405882-8.00005-2

therefore tested the ability of selenium compounds such as sodium selenite, methylseleninic acid (MeSeA), Se-methyl selenocysteine, and selenomethionine to increase the activity, protein, or mRNA levels of commonly used biomarkers of the selenium status, glutathione peroxidase-1 (GPx1) and thioredoxin reductase, and of putatively new biomarkers, selenoprotein W1 (SepW1), selenoprotein H, and selenoprotein 15 in three different cell lines. Selenite and MeSeA were most efficient in increasing all markers tested, whereas the other compounds had only marginal effects. Effects were higher in the noncancerous young adult mouse colon cells than in the cancer cell lines HepG2 and HT-29. At the protein level, SepW1 responded as well as GPx1 and at the mRNA level, even better. Thus, the outcome of selenium treatment strongly depends on the chemical form, the cell type, and the biomarker used for testing efficacy.

1. INTRODUCTION

Based on the early epidemiological (Schrauzer, White, & Schneider, 1977; Shamberger & Frost, 1969) and subsequent clinical studies (Clark et al., 1996) reviewed by Rayman (2012), selenium has been propagated as suitable to prevent carcinogenesis. This general statement, however, does not consider that selenium is an element like, for example, oxygen, carbon, or nitrogen, and as such is present in a large variety of compounds with a broad spectrum of different functions. The predominant biologically active form of selenium in mammals is selenocysteine, which is an integral part of selenoproteins. For incorporation of selenocysteine into selenoproteins, selenium has to be converted into selenide from which monoselenophosphate is formed to generate selenocysteyl-tRNA$^{(Ser)Sec}$ (Carlson et al., 2004; Turanov et al., 2011). Thus, only selenium compounds convertible into selenide can serve as selenium source for selenoproteins.

Not all selenoproteins respond uniformly to selenium supply. Some selenoproteins disappear quickly when selenium becomes limiting, whereas others remain synthesized until deficiency becomes more severe. This phenomenon is called the hierarchy of selenoproteins (Behne, Hilmert, Scheid, Gessner, & Elger, 1988; Berry, 2005) (for a recent review, see Sunde & Raines, 2011) in which low-ranking selenoproteins disappear faster than high-ranking ones. One of the lowest ranking selenoproteins is glutathione peroxidase-1 (GPx1) (Barnes, Evenson, Raines, & Sunde, 2009; Hill, Lyons, & Burk, 1992; Wingler, Böcher, Flohé, Kollmus, & Brigelius-Flohé, 1999), whereas GPx2 (Wingler et al., 1999) and GPx4 (Bermano, Arthur, & Hesketh, 1996; Lei, Evenson, Thompson, & Sunde, 1995) belong to the highest ranking ones. Apart from the decrease in the incorporation of

SeCys into the growing peptide chain, a different degree of degradation of the respective mRNA in selenium deficiency contributes to the hierarchy (Bermano et al., 1996).

Selenium-dependent mRNA stability of low-ranking selenoproteins led to attempts to identify sensitive selenoproteins also by measuring mRNA levels. Indeed, by analyzing the whole selenoproteome in mice fed a diet of selenium made adequate by selenomethionine (SeMet) and a moderately selenium-deficient diet, the RNA of four selenoproteins was found to significantly respond to moderate selenium deficiency in the colon (Kipp et al., 2009) and leukocytes (Kipp et al., 2012) from which selenoprotein W1 (SepW1) mRNA revealed to be most sensitive. Feeding to rats a different amount of selenium as sodium selenite highly affected the stability of the mRNA of SepW1, selenoprotein H (SelH), selenoprotein K, GPx3 and GPx1 in liver, kidney, and muscle, suggesting mRNA levels of these proteins as suitable markers of the selenium status also in other tissues. However, none of the selenoprotein RNAs required higher selenium intake than GPx1 to reach maximum levels(Barnes et al., 2009) making GPx1 RNA the most sensitive biomarker in these tissues, at least in the rat.

A crucial influence of the chemical form of selenium on the outcome of feeding studies as well as clinical trials is increasingly recognized. Selenium compounds so far tested in clinical trials in humans are selenite, SeMet, and selenium-enriched yeast mainly containing SeMet but also different concentrations of selenite or selenomethylselenocysteine (Kotrebai, Birringer, Tyson, Block, & Uden, 2000). The latter compound, for example, is the major compound in highly selenium-enriched garlic (Kotrebai et al., 2000). In animal experiments apart from SeMet and selenite, monomethylated forms were used and turned out to be more efficient in chemoprevention (reviewed in Brigelius-Flohé, 2008). It may well be that different response to different selenium compounds and the lack of knowledge of which selenoproteins are affected have contributed to the inconsistent outcome of studies undertaken to prove the anticancer effect of selenium (Rayman, Infante, & Sargent, 2008).

We here tested the efficacy of four different selenium compounds, sodium selenite, methylseleninic acid (MeSeA), Se-methyl selenocysteine (SeMeSeCys), and SeMet (Scheme 5.1) to upregulate classical markers of the selenium status, namely GPx and thioredoxin reductase (TrxR) activity in three different cell lines, HepG2, HT-29, and young adult mouse colon (YAMC) cells. In contrast to the cancer cell lines HepG2 and HT-29, YAMC cells are primary mouse colon cells which were further used to

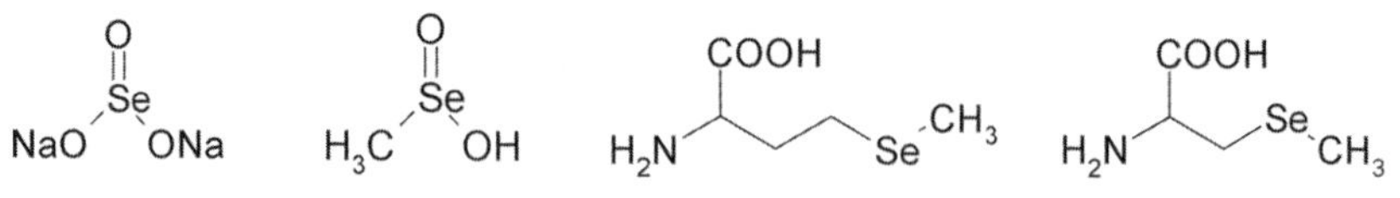

Selenite MeSeA SeMet SeMeSeCys

Scheme 5.1 Structures of selenium compounds used in this study.

analyze protein as well as mRNA levels of selenium-sensitive GPx1, SepW1, and selenoprotein 15 (Sep15).

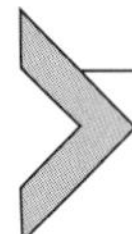

2. EXPERIMENTAL

2.1. Cell culture

HepG2 (human liver carcinoma cells; ATCC HB8065) and HT-29 (human colon adenocarcinoma cells; German Collection of Microorganisms and Cell Culture, ACC 299) cells were grown in a 5% CO_2 atmosphere in respective media, pH 7.1, containing 10% heat-inactivated fetal calf serum (FCS; Biochrom, Berlin, Germany), 100 U/ml penicillin and 100 μg/ml streptomycin (all Gibco, Karlsruhe, Germany), at 37 °C. Medium for HepG2 cells was RPMI1640 (Gibco) supplemented with 2 m*M* L-alanyl-L-glutamine (Glutamax, Gibco). Medium for HT-29 cells was DMEM (high glucose, Gibco) with 1% nonessential amino acids. HepG2 and HT-29 cells were seeded at 1 Mio cells/T75 cm^2. YAMC cells are conditionally immortalizable cells isolated from H-$2K^b$-tsA58 mice (Whitehead, VanEeden, Noble, Ataliotis, & Jat, 1993) and were a kind gift from Robert H. Whitehead (Vanderbilt University Centre, Nashville, USA). For maintenance, cells were seeded at 6 Mio cells/T175-cm^2 flasks in RPMI1640 with 10% FCS, 2 m*M* L-alanyl-L-glutamine, 10 μ*M* α-thioglycerol (Sigma-Aldrich, Munich, Germany), 0.1 μ*M* hydrocortisone (Sigma), and 5 U/ml γ-interferon (γ-IFN, Peprotech, Hamburg, Germany) at 33 °C. 100 μg/ml Normocin (InvivoGen, Toulouse, France) was added to the culture medium to prevent mycoplasma infection. At the permissive temperature of 33 °C, YAMC cells express the functional SV40 large T-antigen and proliferate. For experiments, cells were grown at the nonpermissive temperature of 37 °C without γ-IFN to prevent SV40 large T-antigen expression. Under these conditions, they stop proliferation and gain characteristics of primary intestinal epithelial cells (Whitehead et al., 1993). In addition, they become unable to form tumor xenografts (Whitehead & Joseph, 1994).

2.1.1 Treatment with selenium compounds

The batch of FCS used for control (Biochrom) contained 1.57 μg Se/L resulting in about 2 n*M* selenium of unknown structure and bioavailability in the final culture media. Sodium selenite was obtained from Merck (Darmstadt, Germany); SeMet from Aeros Organics (Geel, Belgium); SeMeSeCys and MeSeA acid from Sigma-Aldrich. Structures are shown in Scheme 5.1. Cells were seeded in their complete culture medium at the cell numbers and volumes indicated under respective experimental procedures (see above) and grown for 72 h with the selenium compounds and concentrations as indicated in figure legends.

The toxicity of selenium compounds was estimated by quantification of MTT reduction at the end of selenium exposure as described (Banning et al., 2008).

2.1.2 Sample preparation

At the end of incubation times w/o selenium compounds, cells were harvested and homogenized in 150 μl Tris buffer (100 m*M* Tris, 300 m*M* KCl, pH 7.6 with 0.1% Triton X-100 (Serva, Heidelberg, Germany)). Lysis was obtained by sonification with 10 strokes at 70% energy and 0.5-s cycle time (Hielscher Ultrasonix UP50H, Teltow, Germany).

2.2. Biomarkers for the selenium status

2.2.1 GPx activity

GPx activity in principle was measured with a glutathione reductase (GR)-coupled assay as described (Brigelius-Flohé, Lötzer, Maurer, Schultz, & Leist, 1995) and discussed (Brigelius-Flohé, Wingler, & Müller, 2002) (Eqs. 5.1 and 5.2) with modifications for estimation in 96-well plates if sample numbers were large.

$$\mathrm{ROOH}+2\mathrm{GSH}\xrightarrow{\mathrm{GPx}}\mathrm{GSSG}+\mathrm{ROH}+\mathrm{H_2O} \tag{5.1}$$

$$\mathrm{GSSH}+\mathrm{NADPH}+\mathrm{H}^+\xrightarrow{\mathrm{GR}}2\mathrm{GSH}+\mathrm{NADP}^+ \tag{5.2}$$

Assay performance: Addition of 2.5 μl sample, 15 μl Tris buffer (see above), 213.75 μl GPx reaction buffer (100 m*M* Tris–HCl, 1 m*M* Na-azide, 5 m*M* EDTA, pH 7.6), 2.5 μl 10% Triton X-100 solution, 2.5 μl NADPH solution (20 m*M* NADPH in 0.1% (v/v) $NaHCO_3$), 2.5 μl GSH solution (300 m*M* GSH in 0.2 m*M* HCl to avoid GSH autoxidation), 1.25 μl GR solution (23 U/ml GR-buffer, i.e., 3.2 *M* $(NH_4)_2SO_4$, pH 6.0) was made into each well. As blank, 2.5 μl Tris buffer was added instead of sample. A GPx solution of known activity was carried along as positive control. With the exception of sample and homogenization buffer, all solutions can be mixed in advance and used as test mixture. However, care should be taken that GSH and NADPH are prepared freshly every day. After 10-min preincubation at 37 °C, the reaction was started by the addition of 10 μl H_2O_2 to reach a final concentration of 50 μ*M*. Thus, the final volume was 250 μl/well. H_2O_2 was used to measure GPx1 or total GPx activity, and for GPx4 a final concentration of 50 μ*M* phosphatidylcholine hydroperoxide was used as substrate. For simultaneous addition into each well, a 1.225-m*M* H_2O_2 solution was filled into the dispenser of the microplate absorbance reader (e.g., Biotech Instruments, Bad Friedrichshall, Germany, as used for the data obtained here). Decrease of absorption as measured at 340 nm for 2 min. During this time, a clear linear decrease should be observed, otherwise the sample has to be diluted or the volume enhanced, respectively. The absorbance reader corrects the various filling levels to a light path of 1 cm. In this method, activity can be calculated by using Lambert–Beer's law. One unit generally is defined as consumption of 1 μmol NADPH/min and expressed as mU/mg protein.

2.2.2 TrxR activity

TrxR activity is estimated via the reduction of 5,5′-dithiobis(2-nitrobenzoic acid) (DTNB) to 5′-thionitrobenzoate (TNB) by NADPH which can be followed by the increase in absorption at 412 nm (Eq. 5.3) (Gromer & Gross, 2002).

$$\mathrm{DTNB} + \mathrm{NADPH} + \mathrm{H}^+ \xrightarrow{\mathrm{TrxR}} 2\mathrm{TNB} + \mathrm{NADP}^+ \qquad (5.3)$$

Also, in this case, 96-well plates can be used. The final volume in each well is 250 μl. 5 μl of sample was mixed with 20 μl Tris buffer (see above) and 185 μl reaction mixture (100 m*M* KP_i, 2 m*M* EDTA, pH 7.4) and 15 μl DTNB (50 m*M* in DMSO) in a 96-well plate. As blank, 25 μl Tris buffer was used. DTNB solution is stored at −20 °C in the dark. Reaction temperature is 25 °C. The reaction was started by the addition of 25 μl of a freshly prepared 20 m*M* NADPH solution in the reaction mixture by the instrumental dispenser. Each sample was measured a second time without NADPH, as DTNB can be unspecifically reduced by thiol groups present in the sample. A putatively observed increased absorption was subtracted from the absorption of the respective sample. Absorption was followed for 2 min at

412 nm. Due to the correction for the different filling levels, the Lambert–Beer's law is used for calculation with $\varepsilon_{TNB412nm} = 13.6\ mM^{-1}\ cm^{-1}$. As two molecules TNB are produced by one molecule NADPH, the resulting activity is divided by 2 and expressed as mU/mg protein.

2.2.3 SDS-PAGE and Western blotting

For Western blotting, HepG2 and HT-29 cells were seeded at 500,000 cells/well into 6-well-plates and incubated with respective selenium compounds for 72 h. From YAMC cells 1 Mio cells were seeded into a 175-cm^2 flask.

Cells were lyzed for 15 min on ice in 150 μl (HepG2 and HT-29) or 250 μl (YAMC) RIPA buffer (50 m*M* Tris, 150 m*M* NaCl, 2 m*M* EGTA, 0.1% SDS, 0.5% sodium deoxycholate, 1% Nonidet P-40, pH 7.4). Aliquots (50 μg protein/lane) were subjected to SDS-PAGE and blots performed as described (Banning, Deubel, Kluth, Zhou, & Brigelius-Flohé, 2005). GPx1 was detected with a rabbit–antihuman GPx1 antiserum (ab22604, Abcam, Cambridge, UK), dilution 1:8000, SepW1 with a rabbit-antihuman/mouse SepW1 antiserum (Rockland 600-401-A29S), dilution 1:500, and Sep15 with a rabbit–antihuman/mouse monoclonal antibody (ab 124840, Abcam), dilution 1:1000. As secondary antibody peroxidase-conjugated goat-antirabbit-IgG (Chemicon, Hofheim, Germany), dilution 1:50,000 was used. Rabbit-anti-β-actin served as internal control (ab8227, Abcam, dilution 1:5000). Signals were detected in a Fuji LAS3000-CCD System (Raytest, Straubenhardt, Germany) with Supersignal West Dura (Perbio, Bonn, Germany) as substrate.

2.2.4 Quantitative real-time PCR

RNA was isolated using the guanidinium thiocyanate–phenol–chloroform extraction (Chomczynski & Sacchi, 1987) provided as TRIzol Kit by Invitrogen (Karlsruhe, Germany). Cell pellets were suspended in 800 μl cold TRIzol and the procedure followed as described in the manufacturer's protocol. Genomic DNA was digested with 10 U RQ1 RNase-free DNase (Promega, Mannheim, Germany), and RNA was cleaned up with phenol–chloroform extraction. Concentration and purity of RNA were confirmed in a Nano Drop ND-1000 (Peqlab Biotechnology, Erlangen, Germany). Reverse transcription was performed with 3 μg RNA, 150 fmol oligo (dT) 15 primers, and 180 U Moloney murine leukemia virus reverse transcriptase (Promega) in a total volume of 45 μl. Real time PCRs (Mx 3005P^{TM} qPCR System, Stratagene, Amsterdam, Netherlands) were performed in triplicate with 1 μl 10× diluted cDNA in 25 μl reaction mixtures using SYBR Green 1 (Molecular Probes Eugene, OR, USA) as fluorescent reporter. Standard curves were monitored to quantify PCR products. cDNA-specific primers (Table 5.1) were designed with Perl Primer v1.1.14 (Marshall, 2004). β-Actin was used as reference gene.

2.3. Statistics

Data are shown as means + SD of three experiments with the exception of Western blots measured in triplicate. Statistical differences were tested by one-way ANOVA or Student's *t*-test as appropriate and are indicated in figure legends (GraphPad Prism® version 5.0, San Diego, CA, USA) (Figs. 5.1–5.5). A p-value <0.05 was considered significant.

3. RESULTS

Based on the sensitive response of SepW1, GPx1, SelH, and SelM to selenium supply in the colon of mice (Kipp et al., 2009), cultured cells were selected to get an idea of how these proteins respond to different selenium compounds. HT-29 and HepG2 cells are human colon and liver cancer cells, respectively, in which a putatively similar response as in the mouse colon can be verified. YAMC cells are colon cells of mice with characteristics of primary cells (see Section 2.1). Thus, a difference or similarity in mouse and human cells as well as in cancer and noncancer cells was expected to become detectable.

3.1. Toxicity of individual selenium compounds

We first tested toxicity of the compounds by the MTT test after incubating the three cell lines with increasing concentrations of the individual selenium compounds for 72 h. Reduction of vitality to less than 80% was taken as the first sign of toxicity and IC_{80} values calculated (Table 5.2). YAMC cells reacted most sensitively; HepG2 cells were more resistant to the compounds. The highest toxicity was revealed in selenite whereas organic compounds (SeMet and SeMeSeCys) were better accepted. Nevertheless, concentrations chosen for experiments were below IC_{80} levels with the exception of the use of 1000 n*M* MeSeA in YAMC cells. The outcome, however, was not significantly influenced (see Figs. 5.1–5.3).

Table 5.1 Primer sequences (5′ → 3′)

Gene	Account number	Primer sequence	Position	Exon	Product (bp)
Gpx1	NM_008160.5	fwd TACACCGAGATGAACGATCTG	196–277	1	101
		rev ATTCTTGCCATTCTCCTGGT	217–297	1 and 2	
Selh	NM_001037279.1	fwd CCTTATTCCACCAACGCGCCA	85–105	1	154
		rev GCGTCAGCTCGTACAATGCTC	219–239	1 and 2	
Sep15	NM_053102.2	fwd GTTTCAAGCGGCGTCTGCTC	363–383	1	159
		rev TGCTTCTTCCTGACAGCACCC	501–522	1 and 2	
Sepw1	NM_009156.2	fwd ATGCCTGGACATTTGTGGCGA	187–207	3 and 4	152
		rev GCAGCTTTGATGGCGGTCAC	320–339	5	
Reference gene					
β-Actin	NM_007393.3	fwd CACTGCCGCATCCTCTTCCT	763–783	4	128 bp
		rev GATTCCATACCCAAGAAGGAAGGC	868–891	4 and 5	

Table 5.2 IC_{80} values for the indicated Se compounds in three different cell lines as measured by the MTT test

	IC_{80} (μ*M*)		
Se compound	**HepG2**	**HT-29**	**YAMC**
Selenite	2.3	2.5	0.8
MeSeA	33.3	3.4	0.6
SeMet	165.8	148.5	99.5
SeMeSeCys	Not toxic up to 200 μ*M*	49.8	32.0

3.2. Estimation of biomarkers

3.2.1 GPx activity

Dose-dependent effects of SeMeSeCys, SeMet, MeSeA, and selenite on GPx activity were tested in HepG2, HT-29, and YAMC cells (Fig. 5.1). In HepG2 cells, maximal activity was reached with 50 n*M* MeSeA and selenite and was not further increased with higher concentrations (compare to untreated controls indicated by "–"). SeMeSeCys and SeMet showed a clear concentration-dependency and appeared not to have reached maximal activity with the highest dosage. HT-29 cells responded similarly, although with a generally lower fold change. The exception was the response to SeMet, for which only the highest dosage was effective. In YAMC cells, again the lowest concentration of selenite was sufficient for a maximal increase whereas MeSeA, in contrast to all the other lines, needed at least 200 n*M* to reach the plateau. Enhanced activity with SeMet and SeMeSeCys was only observed with the highest dosages.

In sum, in all cell lines, selenite and MeSeA were more effective than selenium-containing amino acids. The plateau of activation was reached at lower concentrations but with a lower fold change in cancer cells than in the noncancerous cell line YAMC. The efficiency of the selenium compounds to increase total GPx activity can be ranked as follows:

$$\text{Selenite} \sim \text{MeSeA} > \text{SeMeSeCys} > \text{SeMet}$$

3.2.2 TrxR activity

In all cells, stimulation of TrxR activity was generally less than that of total GPx (Fig. 5.2). In HepG2 (Fig. 5.2A) cells, again selenite reached the plateau (1.8-fold increase) with the lowest concentration tested, whereas the other

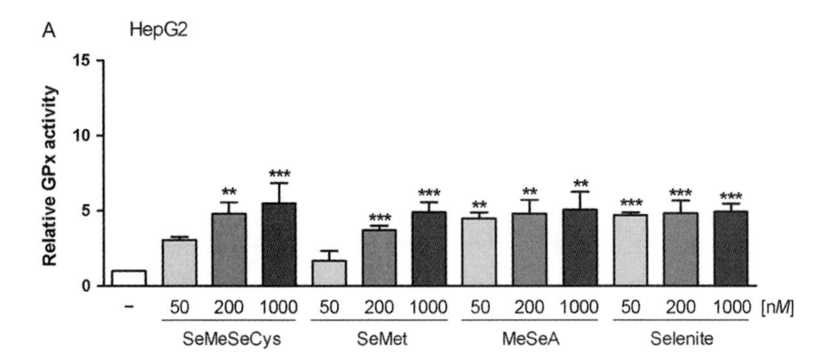

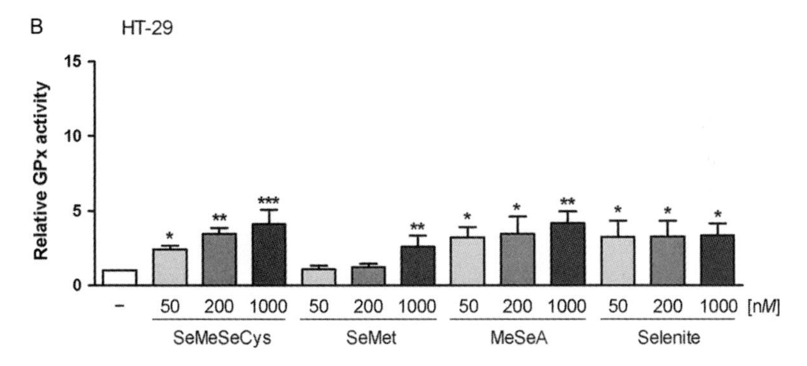

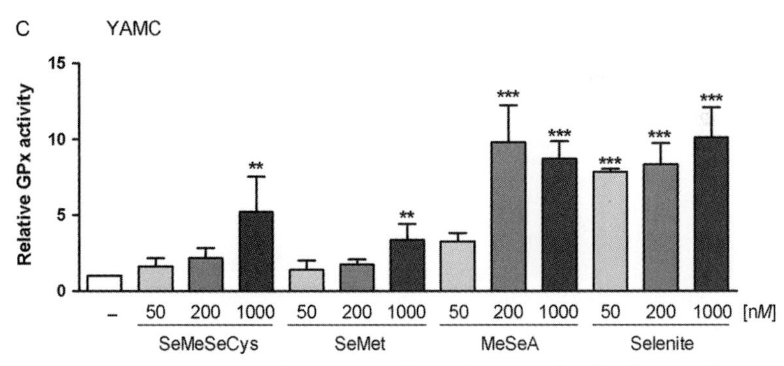

Figure 5.1 GPx activity in HepG2 (A), HT-29 (B), and YAMC (C) cells after incubation with SeMeSeCys, SeMet, MeSeA, or selenite. Cells were incubated for 72 h in medium without (–) or with indicated concentrations of the different selenium compounds. GPx activity was measured as described in "Experimental Protocols" (Section 2.2.1) and calculated as fold change relative to untreated controls (–) set to 1. Values are means + SD of three independent experiments measured in triplicate. $^{*}p < 0.05$, $^{**}p < 0.01$, and $^{***}p < 0.001$ tested with one-way ANOVA with Bonferroni's post test versus untreated control.

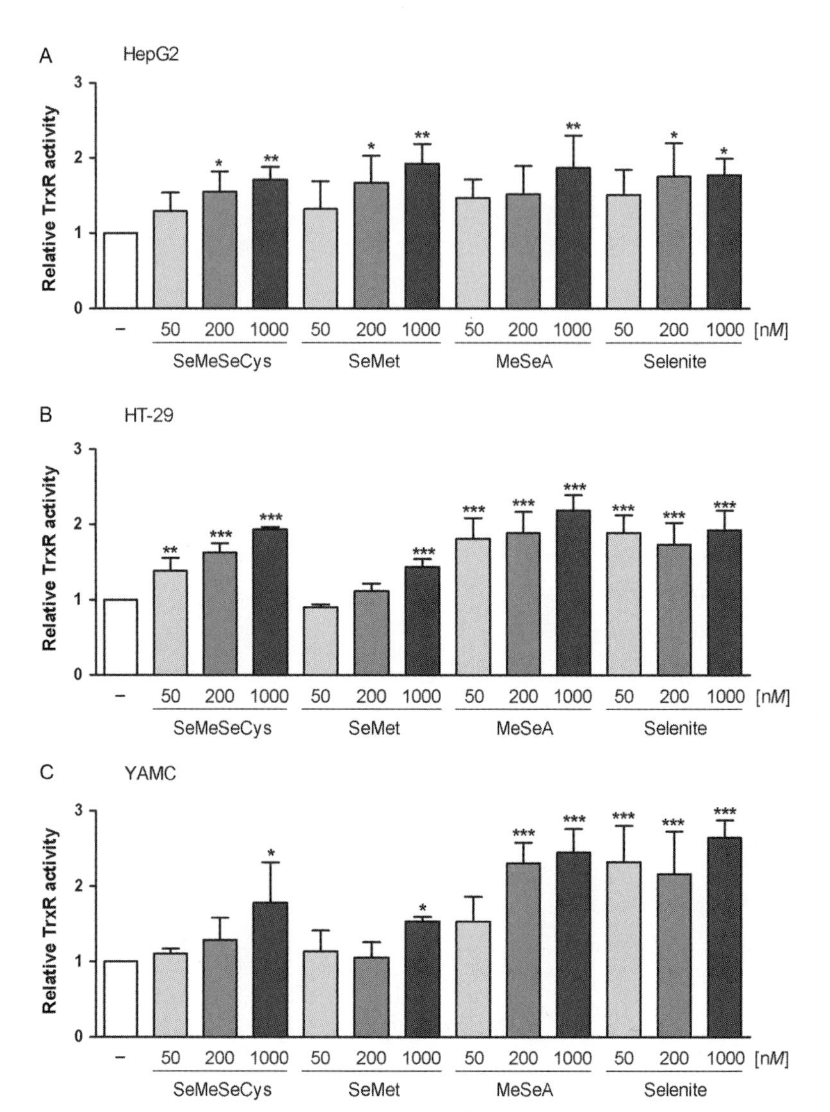

Figure 5.2 Thioredoxin reductase (TrxR) activity in HepG2 (A), HT-29 (B), and YAMC (C) cells. Cells were incubated for 72 h in medium without or with indicated concentrations of the different selenium compounds. TrxR activity was measured as described in "Experimental Protocols" (Section 2.2.2) and calculated as fold change relative to untreated controls (–) set to 1. Values are means + SD ($n=3$) measured in triplicate. $^{*}p<0.05$, $^{**}p<0.01$, and $^{***}p<0.001$ tested with one-way ANOVA with Bonferroni's post test versus untreated control.

compounds caused about 1.8-fold increase only with the highest concentration (1000 n*M*). In HT-29 (Fig. 5.2B) and YAMC (Fig. 5.2C) cells, TrxR activity responded to selenium compounds as did GPx activities. Thus, the ranking of selenium compounds regarding their efficiency to stimulate selenoprotein activity was the same for GPx and TrxR activity:

$$\text{Selenite} \sim \text{MeSeA} > \text{SeMeSeCys} > \text{SeMet}$$

3.2.3 Protein levels of selenoproteins

As the response of selenoenzyme activation was most distinct in YAMC cells, protein and RNA levels were additionally analyzed in these cells. Protein levels of GPx1 (Fig. 5.3A) correlated well with activities of total GPx (compare to Fig. 5.1C). The slightly higher activity at low concentrations of SeMeSeCys and SeMet may indicate that some of the total activity was derived from the higher ranking GPx4. Also TrxR is relatively high ranking and, thus, its activity is less influenced by lower concentrations of selenium compounds. We therefore focussed on low-ranking selenoproteins as GPx1, SepW1, and Sep15. SepW1 activity could not be estimated due to the lack of available assays. Protein expression, however, was possible to be analyzed and most effectively enhanced with selenite at the lowest dosage (Fig. 5.3B). From MeSeA, 1000 n*M* were required to cause a significant effect, which, however, might not yet result in the maximum protein level. SeMeSeCys and SeMet did not affect SepW1 protein expression.

Sep15 protein was already enhanced by the 50 n*M* MeSeA and selenite, but in contrast to GPx1 and SepW1 further upregulated with increasing selenite concentration. SeMeSeCys and SeMet did not significantly affect Sep15 protein levels (Fig. 5.3C).

As the plateau of SepW1 and GPx1 protein was reached with 50 n*M* selenite, additional lower concentrations were tested. Protein levels of GPx1 (Fig. 5.4A) and SepW1 (Fig. 5.4B) started to increase from 5 n*M* on and reached the plateau with 50 n*M* as in Fig. 5.3A and B, respectively.

Taken together, at protein levels of low-ranking selenoproteins, selenite and MeSeA had a larger effect than organic forms with a surprisingly low effect of SeMet.

3.2.4 mRNA levels of selenoproteins

As protein levels responded to lower concentrations of selenium compounds, concentrations below 50 n*M* were used for RNA analyses. GPx1 mRNA (Fig. 5.5A) was slightly increased with higher concentrations of

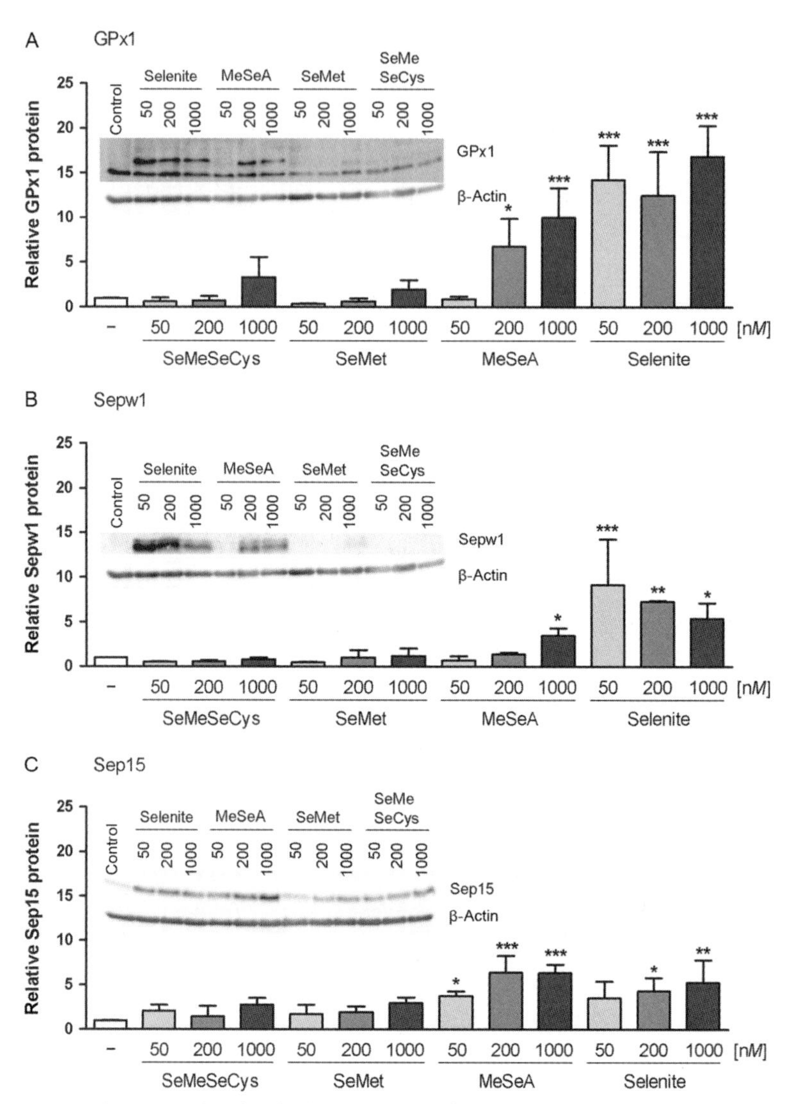

Figure 5.3 Changes in levels of selenoproteins after incubation of YAMC cells with different selenium compounds. Cells were incubated for 72 h without or with indicated Se compounds and concentrations. Levels of GPx1 (A), SepW1 (B), and Sep15 (C) were analyzed by Western blots, quantified by densitometry, normalized to β-actin and expressed relative to untreated cells set to 1. Respective representative blots are shown as inserts. Bands of GPx1 (A), SepW (B), and β-actin (A, B) were stained on the same blot since molecular weights are different enough (21, 9.6, and 40 kDa) to be separated by the used procedure. For details see "Experimental Protocols" (Section 2.2.2). Values are means + SD ($n=3$). $^{*}p<0.05$, $^{**}p<0.01$, and $^{***}p<0.001$ tested with one-way ANOVA and Bonferroni's post test versus untreated control.

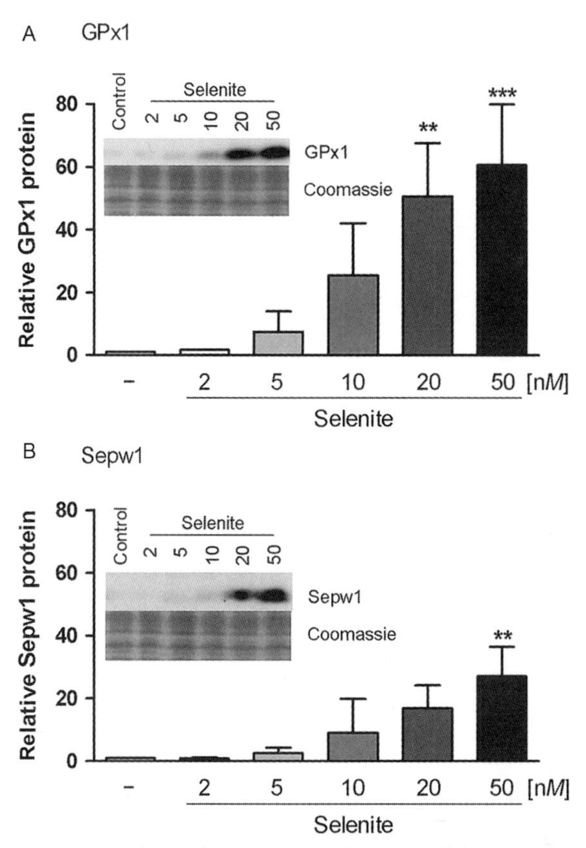

Figure 5.4 Concentration-dependent increase of GPx1 and SepW1 protein by selenite in YAMC cells. Cells were incubated for 72 h without or with indicated concentrations of selenite. Levels of GPx1 (A) and SepW1 (B) were analyzed by Western Blots, quantified by densitometry, normalized to the Coomassie staining of the gel, and expressed relative to untreated cells set to 1. Respective representative blots are shown as inserts. Values are means + SD ($n = 3$). $^{**}p < 0.01$ and $^{***}p < 0.001$ tested with one-way ANOVA and Bonferroni's post test versus untreated control.

selenite; however, the increase was significant only when compared directly to the control (*t*-test). Again, MeSeA and selenite increased GPx1 mRNA with the higher dosages only. SeMeSeCys and SeMet did not affect GPx1 mRNA, which remained as low as in cells incubated in the absence of selenium.

In contrast, SepW1 mRNA was significantly stabilized by selenite from a concentration of 20 n*M* on and reached the plateau at 50 n*M* (Fig. 5.5B). Also MeSeA stabilized SepW1 RNA, however, only with concentrations of 100 n*M* or more. SeMeSeCys and SeMet were not able to enhance SepW1 mRNA levels.

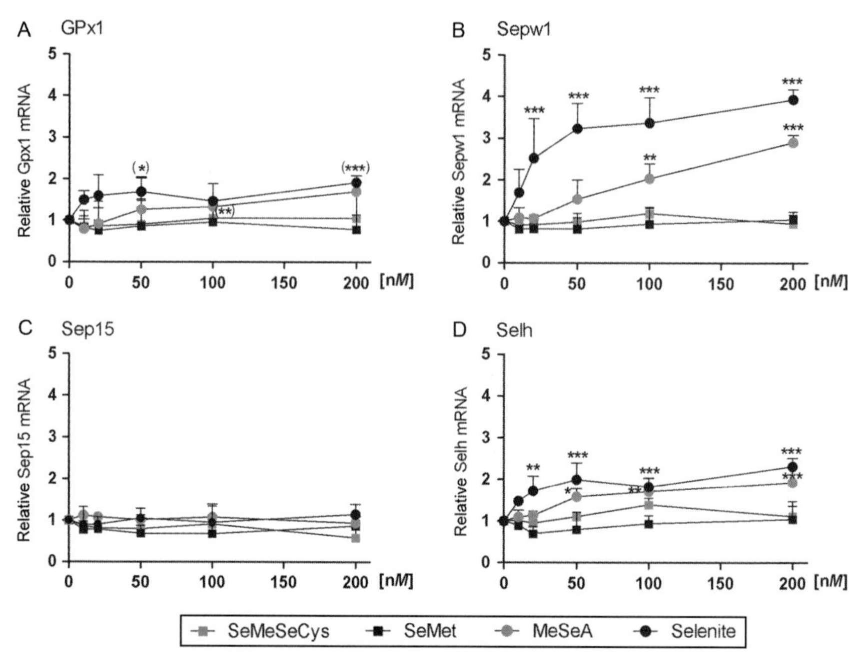

Figure 5.5 Changes in levels of selenoprotein mRNA by different selenium compounds in YAMC cells. YAMC cells were incubated for 72 h without (–) or with indicated Se compounds and concentrations. mRNA of GPx1 (A), SepW1 (B), Sep15 (C), and SelH (D) was analyzed by qPCR as described in "Experimental Protocols" (Section 2.2.4), normalized to β-actin, and changes calculated relative to untreated controls set to 1. Values are means + SD ($n = 3$) measured in triplicate. $^{*}p < 0.05$, $^{**}p < 0.01$, and $^{***}p < 0.001$ tested with one-way ANOVA and Bonferroni's post test. Asterisks denote significance calculated with unpaired *t*-test versus untreated control.

Sep15 mRNA was not changed by any of the selenium compounds (Fig. 5.4C), indicating that Sep15 mRNA does not respond to selenium levels, at least not in YAMC cells, whereas in the colon of mice it was slightly decreased in moderate selenium deficiency (Kipp et al., 2009).

As SelH mRNA levels were also significantly decreased in the colon of moderate Se-deficient mice (Kipp et al., 2009), we tested it in YAMC cells as well (Fig. 5.5D). Clearly, MeSeA and selenite were able to enhance mRNA levels of SelH, but not SeMeSeCys and SeMet.

4. DISCUSSION

We tested four different selenium compounds for their capacity to serve as a source for selenoprotein biosynthesis, especially for selenoprotein

W, as a possible additional biomarker of the selenium status. In general, selenite and MeSeA were most potent in all cells and for all selenoproteins tested, whereas SeMeSeCys and SeMet were less efficient if at all. Differences in efficacy can easily be explained by differences in their metabolism to selenide (H_2Se), which is a prerequisite for incorporation into selenocysteine and thus selenoproteins. The only direct source for selenide

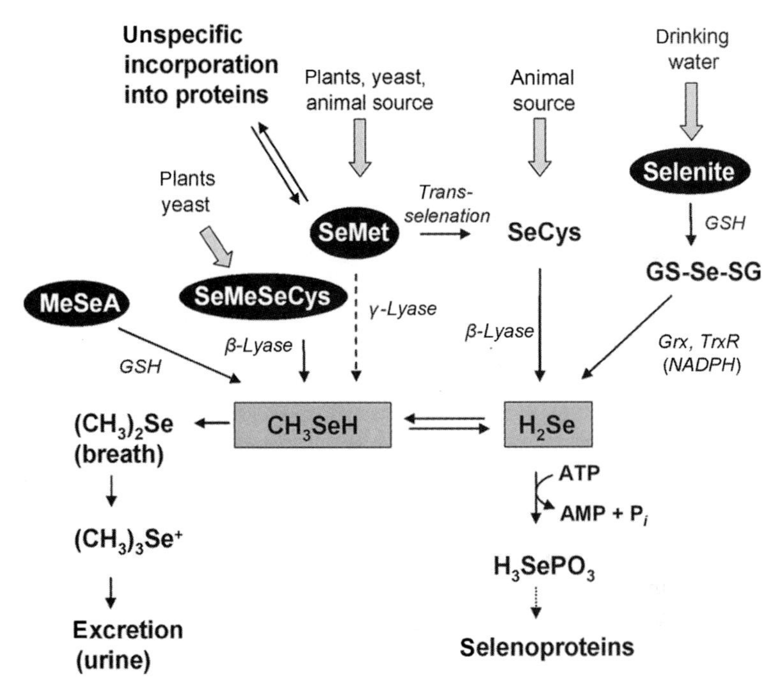

Scheme 5.2 Metabolic pathways of selenium compounds used in this study. All Se compounds derived from nutritional sources are transformed to selenide (H_2Se). Selenite supplied with drinking water is reduced by 4 mol GSH to the intermediate selenodiglutathione (GS-Se-SG) which is reduced to selenide by glutaredoxin (Grx), the Trx/TrxR system or GR. Selenocysteine liberated from selenoproteins can be transformed to selenide by β-lyase directly. SeMet provided by plant and animal sources is split into methylselenol (CH_3—SeH) by a γ-lyase. Alternatively SeMet is converted into SeCys via the *trans*-selenation pathway in analogy to the *trans*-sulfuration pathway. In addition, SeMet is incorporated instead of Met into growing peptide chains since the SeMet-specific tRNA does not differentiate between Met and SeMet. SeMeSeCys primarily present in plants is transformed into methylselenol by β-lyase. MeSeA is reduced to methylselenol in a GSH-dependent reaction (MeSeA → MeSeOH → MeSe-SG → MeSeH) with the last step performed by NADPH-linked GR. Methylselenol is demethylated to selenide, which is phosphorylated and utilized for selenoprotein synthesis. Excess methylselenol is further methylated via *S*-adenosylmethionine for excretion. Further details and references see text.

production is SeCys delivered from proteins after proteolysis. It is transformed into H_2Se and alanine by selenocysteine β-lyase (Esaki, Karai, Nakamura, Tanaka, & Soda, 1985). Major transformation pathways for the selenium compounds used here are shown in Scheme 5.2.

4.1. Metabolism of selenium compounds

Selenite mainly delivered via drinking water can be reduced to selenide in a sequence requiring GSH, glutaredoxin, the Trx/TrxR system, and GR (Björnstedt, Kumar, & Holmgren, 1992; Spallholz, 1994; Wallenberg, Olm, Hebert, Bjornstedt, & Fernandes, 2010). MeSeA is a synthetic compound widely used to study the mechanisms exerted by selenium in cancer prevention. In addition, it is also an oxidation product of the major urinary metabolite selenosugar (Ogra, Hatano, Ohmichi, & Suzuki, 2003). In selenium deficiency, selenium from selenosugar can be reused through MeSeA (Suzuki, Somekawa, & Suzuki, 2006) which is reduced to methylselenol (CH_3—SeH) and subsequently demethylated to $H_2Se + CH_3$—OH (Ip, Thompson, Zhu, & Ganther, 2000; Suzuki, Kurasaki, Ogawa, & Suzuki, 2006; Suzuki, Kurasaki, & Suzuki, 2007). Thus, the formation of selenide from both compounds is achieved by reductive steps, appears to be straight forward, and can explain their high efficiency as a source for selenoprotein biosynthesis.

In contrast, the fate of SeMet varies. Primarily it is incorporated into proteins instead of methionine as the respective $tRNA^{Met}$ does not differentiate between SeMet and Met (McConnell & Hoffman, 1972). Thus, unspecific incorporation explains why SeMet usually efficiently increases selenium contents in cultured cells but not necessarily also, for example, GPx activity (Brigelius-Flohé et al., 1995; Leist et al., 1999). This also has convincingly been shown in a randomized, placebo-controlled intervention study, where 200, 400, or 600 μg/d selenium in the form of sodium selenite, high-selenium yeast, or SeMet was applied to healthy selenium-adequate volunteers and plasma selenium concentration, plasma GPx activity, and plasma selenoprotein P (SePP) content were measured as biomarkers (Burk, Norsworthy, Hill, Motley, & Byrne, 2006). Selenite was without any effect, whereas SeMet and high-selenium yeast increased plasma selenium but not enzymes. The utilization of SeMet for selenoprotein biosynthesis requires transselenation into SeCys which enters the biosynthetic pathway after transformation by β-lyase (Kurokawa et al., 2011; Suzuki et al., 2007). Alternatively, SeMet is split by a pyridoxal phosphate (PLP)-dependent γ-lyase into

CH_3—SeH, ammonia, and α-ketobutyrate (Pinto, Lee, Sinha, MacEwan, & Cooper, 2011). H_2Se is formed by subsequent demethylation of CH_3—SeH.

SeMeSeCys is the predominant selenium compound in selenium-accumulating plants, such as onions, garlic, cruciferous vegetables, or Brazil nuts (Arnault & Auger, 2006; Dumont, Vanhaecke, & Cornelis, 2006). It is also a substrate for a PLP-dependent β-lyase generating CH_3—SeH, pyruvate, and ammonia followed by H_2Se formation (Pinto et al., 2011). All CH_3—SeH derived from the methylated selenium compounds used here can be demethylated into H_2Se by methylselenol demethylase for selenoprotein biosynthesis (Suzuki et al., 2007). In case of excess, CH_3—SeH is further methylated for elimination via tidal air as CH_3—Se—CH_3 or in the urine as $(CH_3)_3Se^+$ (Foster, Kraus, & Ganther, 1986). Excess H_2Se is excreted in the urine as selenosugar (Kobayashi et al., 2002; Suzuki, Ohta, & Suzuki, 2006). Thus, the metabolism of SeMet and SeMeSeCys is more complex and allows more deviations from the biosynthesis pathway, which explains their lower capability to serve as a source of selenium for selenoproteins.

4.2. Efficacy of selenium compounds to increase selenoprotein level and activity

From 10 different selenocompounds tested for their ability to increase SePP and GPx biosynthesis in HepG2 cells (Hoefig, Renko, Köhrle, Birringer, & Schomburg, 2010), selenite and SeMeSeCys turned out to be most effective in enhancing SePP protein as well as GPx activity and to have the lowest toxicity. MeSeA was less efficient. Compounds were categorized into three classes with decreasing efficiency to enhance SePP:

$$\begin{aligned}&(\text{Selenite, L-SeCys, DL-SeCys, GSSeSG, SeMeSeCys})\\&\quad > (\text{selenate, L-SeMet, DL-SeMet, MeSeA}) \gg \text{ebselen}\end{aligned}$$

Ranking was similar when compounds were tested for stimulation of GPx activity.

Our study more or less confirms these results. A very low toxicity of SeMeSeCys was observed in HepG2 cells but was much higher in the intestinal HT-29 cancer cell line and in primary-like intestinal YAMC cells (Table 5.2). Maximal stimulation of GPx activity was not reached even with 1000 n*M* SeMeSeCys (Fig. 5.1A) or with the 100 n*M* concentration used in (Hoefig et al., 2010). However, in our case, SeMeSeCys was less efficient than selenite and MeSeA. MeSeA and selenite enhanced GPx activity to maximal levels with 50 n*M* in all three cell lines, whereas SeMet needed at least 200 n*M* to exert any effect. This might explain the failure of SeMet to stimulate GPx activity in the study by Hoefig et al. (2010), where the highest concentration used was 100 n*M*. The situation was similar in HT-29 cells. In YAMC cells, selenite and MeSeA definitely were most effective (Fig. 5.1C).

This clearly shows that toxicity as well as efficiency to enhance the activity of specific selenoproteins can be different in cells from different tissues, and that cancer and noncancer cells appear to respond differently. Even the same selenoprotein can react differently in different cells. Whereas TrxR activity was increased less than twofold in HepG2 cells with all Se-compounds, it increased twofold in HT-29 cells with all compounds but SeMet, and it was increased more than twofold with MeSeA and selenite in YAMC cells and was lowest with SeMet (Fig. 5.2). Thus, suitability of the choice of function and protein level of a selenoprotein as biomarker for the selenium status depend on the form and concentration of selenium applied and the cell type analyzed.

4.3. Efficacy of selenium compounds to increase protein levels of selenoproteins

Comparison of protein expression in YAMC cells revealed that GPx1 protein started to increase with 5 n*M* selenite (Fig. 5.4) and reached the plateau with 50 n*M*, whereas from MeSeA at least 200 n*M* were needed (Fig. 5.3A). This correlates well with GPx activity (Fig. 5.1C). SepW1 protein reacted to selenite like GPx1 but only responded to concentrations of MeSeA which come close to toxicity (Fig. 5.3B). Unfortunately, there is no activity assay for SepW1, and therefore, protein levels cannot be correlated with function. The same holds true for Sep15, which responded only to selenite and MeSeA, too, however, with a very low fold change (Fig. 5.3C). Protein levels have to be analyzed by Western blotting which is not easy to quantify if changes are small. Reliable results were only obtained with the more efficient compounds selenite and MeSeA. The lack of an effect of SeMet might be surprising, but under the condition used here it is a regular observation. In animal experiments, SeMet is undoubtedly able to enhance GPx activity and protein when compared to Se-deficient controls. However, selenium compounds usually are fed for at least 4 weeks, whereas cells are treated for 72 h.

4.4. Efficacy of selenium compounds to stabilize selenoprotein mRNA

Traditionally used biomarkers for the selenium status are plasma selenium, plasma GPx activity reflecting the activity of GPx3, or erythrocyte GPx activity which results from GPx1. More recently, protein levels of plasma SePP turned out to be more appropriate (Xia, Hill, Byrne, Xu, & Burk, 2005). SePP needed a higher SeMet intake to reach a plateau than that required for maximal GPx activity and needed longer time (Xia et al., 2010). As plasma selenium does not necessarily reflect the expression level or activity of specific selenoproteins, attempts were undertaken to use molecular biomarkers that respond to the selenium status. The dramatic change in mRNA levels of some hierarchically low-ranking selenoproteins makes their mRNA a promising candidate for additional biomarkers. Related research was started in Roger Sunde's group (Evenson, Wheeler, Blake, & Sunde, 2004). They found a comparable selenoprotein mRNA expression in blood and major tissues of rats with the highest decrease of GPx1 mRNA in Se deficiency. Thus, white blood cells may be an easily available source for testing selenoprotein mRNA as biomarkers for the selenium status. We, indeed, found a set of selenoprotein mRNAs highly regulated by the selenium status in leukocytes of mice with GPx1 and SepW1 showing the largest effect (Kipp et al., 2012).

In the present study, SepW1 mRNA most drastically increased with selenite and MeSeA, whereas again SeMeSeCys and SeMet were without any effect. This means that RNA levels of SepW1 were dramatically decreased in selenium deficiency, and that transcription is required before protein biosynthesis and incorporation of selenocysteine can start. Increase of GPx1 and SelH mRNA was also observed only with selenite and MeSeA, but the fold change was much less than that of SepW1 mRNA. Sep15 mRNA did not change at all, indicating that it was not degraded in selenium-deficient cells and that the increase in protein levels (Fig. 5.3C) resulted from translation of still present mRNA. So far, at first glance, the need to produce selenide to stabilize mRNA of selenoproteins is not easy to explain. It was, however, shown recently that selenium-responsive factors are required for stabilization. The eukaryotic initiation factor 4a3 (eIF4a3) is induced in selenium deficiency and interacts with a subset of selenoprotein mRNAs and prevents formation of the selenoprotein synthesis complex (Budiman et al., 2009). GPx1 but not GPx4 belongs to this subset. Another factor which discriminates between selenoprotein-specific SECIS (selenocysteine insertion sequence) is nucleolin, which in contrast to eIF4a3, is a positive regulator of Gpx4 but not GPx1 (Miniard, Middleton, Budiman, Gerber, & Driscoll, 2010). Thus, specific selenium-sensitive factors appear to contribute to the ranking of at least GPxs in the hierarchy of selenoproteins.

The distinct response of SepW1 to selenium availability makes it a sensitive marker for selenium supply. mRNA levels of SepW1 significantly decreased in selenium deficiency whenever tested. Examples are liver of pigs (Zhou et al., 2009), various tissues of mice (Barger et al., 2012; Kipp et al., 2012, 2009; Mallonee, Crowdus, Barger, Dawson, & Power, 2011) and rats (Barnes et al., 2009; Bosse et al., 2010), as well as muscles of chicken (Zhang et al., 2011). Consequently, it was included in the suggested panel of biomarkers suitable to asses the selenium status in rats at least (Sunde, 2010).

5. CONCLUSIONS

We here demonstrate a drastic difference in various selenium compounds and their ability to increase selenoprotein biosynthesis. Most effective were the inorganic forms selenite and MeSeA; however, they were the most toxic forms. The low dosages needed to maximally stimulate GPx1 and SepW1 synthesis may balance toxicity. The effects of SeMeSeCys were only convincing on GPx1 activity and protein level but needed high concentrations, which are compensated by a relatively low toxicity. SeMet revealed only a poor efficacy in cultured cells, which is in contrast to animal and human studies in which higher concentrations and longer exposure times are used. The unspecific incorporation of SeMet into proteins may change their structure and function. In addition, SeMet in proteins can easily be oxidized to methionine sulfoxide with unknown consequences. Regarding selenite, GPx1 and SepW1 protein levels were equally suitable as biomarkers for sufficient selenium supply. At the mRNA level, SepW1 appears to be much more reliable as, although its plateau is reached at the same concentration as GPx1, it responds to selenium repletion with a much larger fold change that makes an analysis much easier. The difficulties in translating findings made with selenoprotein transcripts in cultured cells or animal studies to human requirements are extensively discussed by Reszka, Jablonska, Gromadzinska, and Wasowicz (2012) who come to the conclusion that further studies on large groups of healthy individuals are necessary. SepW1 mRNA should be included into such studies if blood parameters are investigated, as leukocytes are easily available also from humans. SepW1 mRNA appears to be a serious alternative biomarker if suitable antibodies to estimate protein levels are not available.

ACKNOWLEDGMENTS

The kind gift of YAMC cells from Robert H. Whitehead (Vanderbilt University Centre, Nashville, TN, USA) is highly appreciated.

REFERENCES

Arnault, I., & Auger, J. (2006). Seleno-compounds in garlic and onion. *Journal of Chromatography. A, 1112*, 23–30.

Banning, A., Deubel, S., Kluth, D., Zhou, Z., & Brigelius-Flohé, R. (2005). The GI-GPx gene is a target for Nrf2. *Molecular and Cellular Biology, 25*, 4914–4923.

Banning, A., Kipp, A., Schmitmeier, S., Loewinger, M., Florian, S., Krehl, S., et al. (2008). Glutathione peroxidase 2 inhibits cyclooxygenase-2-mediated migration and invasion of HT-29 adenocarcinoma cells but supports their growth as tumors in nude mice. *Cancer Research, 68*, 9746–9753.

Barger, J. L., Kayo, T., Pugh, T. D., Vann, J. A., Power, R., Dawson, K., et al. (2012). Gene expression profiling reveals differential effects of sodium selenite, selenomethionine, and yeast-derived selenium in the mouse. *Genes & Nutrition*, 7, 155–165.

Barnes, K. M., Evenson, J. K., Raines, A. M., & Sunde, R. A. (2009). Transcript analysis of the selenoproteome indicates that dietary selenium requirements of rats based on selenium-regulated selenoprotein mRNA levels are uniformly less than those based on glutathione peroxidase activity. *The Journal of Nutrition, 139*, 199–206.

Behne, D., Hilmert, H., Scheid, S., Gessner, H., & Elger, W. (1988). Evidence for specific selenium target tissues and new biologically important selenoproteins. *Biochimica et Biophysica Acta, 966*, 12–21.

Bermano, G., Arthur, J. R., & Hesketh, J. E. (1996). Selective control of cytosolic glutathione peroxidase and phospholipid hydroperoxide glutathione peroxidase mRNA stability by selenium supply. *FEBS Letters, 387*, 157–160.

Berry, M. J. (2005). Insights into the hierarchy of selenium incorporation. *Nature Genetics, 37*, 1162–1163.

Björnstedt, M., Kumar, S., & Holmgren, A. (1992). Selenodiglutathione is a highly efficient oxidant of reduced thioredoxin and a substrate for mammalian thioredoxin reductase. *The Journal of Biological Chemistry, 267*, 8030–8034.

Bosse, A. C., Pallauf, J., Hommel, B., Sturm, M., Fischer, S., Wolf, N. M., et al. (2010). Impact of selenite and selenate on differentially expressed genes in rat liver examined by microarray analysis. *Bioscience Reports, 30*, 293–306.

Brigelius-Flohé, R. (2008). Selenium compounds and selenoproteins in cancer. *Chemistry and Biodiversity, 5*, 389–395.

Brigelius-Flohé, R., Lötzer, K., Maurer, S., Schultz, M., & Leist, M. (1995). Utilization of selenium from different chemical entities for selenoprotein biosynthesis by mammalian cell lines. *Biofactors, 5*, 125–131.

Brigelius-Flohé, R., Wingler, K., & Müller, C. (2002). Estimation of individual types of glutathione peroxidases. *Methods in Enzymology, 347*, 101–112.

Budiman, M. E., Bubenik, J. L., Miniard, A. C., Middleton, L. M., Gerber, C. A., Cash, A., et al. (2009). Eukaryotic initiation factor 4a3 is a selenium-regulated RNA-binding protein that selectively inhibits selenocysteine incorporation. *Molecular Cell, 35*, 479–489.

Burk, R. F., Norsworthy, B. K., Hill, K. E., Motley, A. K., & Byrne, D. W. (2006). Effects of chemical form of selenium on plasma biomarkers in a high-dose human supplementation trial. *Cancer Epidemiology, Biomarkers & Prevention, 15*, 804–810.

Carlson, B. A., Xu, X. M., Kryukov, G. V., Rao, M., Berry, M. J., Gladyshev, V. N., et al. (2004). Identification and characterization of phosphoseryl-tRNA[Ser]Sec kinase. *Proceedings of the National Academy of Science of the United States of America, 101*, 12848–12853.

Chomczynski, P., & Sacchi, N. (1987). Single-step method of RNA isolation by acid guanidinium thiocyanate-phenol-chloroform extraction. *Analytical Biochemistry, 162*, 156–159.
Clark, L. C., Combs, G. F., Jr., Turnbull, B. W., Slate, E. H., Chalker, D. K., Chow, J., et al. (1996). Effects of selenium supplementation for cancer prevention in patients with carcinoma of the skin. A randomized controlled trial. Nutritional Prevention of Cancer Study Group. *Journal of the American Medical Association, 276*, 1957–1963.
Dumont, E., Vanhaecke, F., & Cornelis, R. (2006). Selenium speciation from food source to metabolites: A critical review. *Analytical and Bioanalytical Chemistry, 385*, 1304–1323.
Esaki, N., Karai, N., Nakamura, T., Tanaka, H., & Soda, K. (1985). Mechanism of reactions catalyzed by selenocysteine beta-lyase. *Archives of Biochemistry and Biophysics, 238*, 418–423.
Evenson, J. K., Wheeler, A. D., Blake, S. M., & Sunde, R. A. (2004). Selenoprotein mRNA is expressed in blood at levels comparable to major tissues in rats. *The Journal of Nutrition, 134*, 2640–2645.
Foster, S. J., Kraus, R. J., & Ganther, H. E. (1986). The metabolism of selenomethionine, Se-methylselenocysteine, their selenonium derivatives, and trimethylselenonium in the rat. *Archives of Biochemistry and Biophysics, 251*, 77–86.
Gromer, S., & Gross, J. H. (2002). Methylseleninate is a substrate rather than an inhibitor of mammalian thioredoxin reductase. Implications for the antitumor effects of selenium. *The Journal of Biological Chemistry, 277*, 9701–9706.
Hill, K. E., Lyons, P. R., & Burk, R. F. (1992). Differential regulation of rat liver selenoprotein mRNAs in selenium deficiency. *Biochemical and Biophysical Research Communications, 185*, 260–263.
Hoefig, C. S., Renko, K., Köhrle, J., Birringer, M., & Schomburg, L. (2010). Comparison of different selenocompounds with respect to nutritional value vs. toxicity using liver cells in culture. *The Journal of Nutritional Biochemistry, 22*, 945–955.
Ip, C., Thompson, H. J., Zhu, Z., & Ganther, H. E. (2000). In vitro and in vivo studies of methylseleninic acid: Evidence that a monomethylated selenium metabolite is critical for cancer chemoprevention. *Cancer Research, 60*, 2882–2886.
Kipp, A. P., Banning, A., van Schothorst, E. M., Meplan, C., Coort, S. L., Evelo, C. T., et al. (2012). Marginal selenium deficiency down-regulates inflammation-related genes in splenic leukocytes of the mouse. *The Journal of Nutritional Biochemistry, 23*, 1170–1177.
Kipp, A., Banning, A., van Schothorst, E. M., Méplan, C., Schomburg, L., Evelo, C., et al. (2009). Four selenoproteins, protein biosynthesis, and Wnt signalling are particularly sensitive to limited selenium intake in mouse colon. *Molecular Nutrition & Food Research, 53*, 1561–1572.
Kobayashi, Y., Ogra, Y., Ishiwata, K., Takayama, H., Aimi, N., & Suzuki, K. T. (2002). Selenosugars are key and urinary metabolites for selenium excretion within the required to low-toxic range. *Proceedings of the National Academy of Science of the United States of America, 99*, 15932–15936.
Kotrebai, M., Birringer, M., Tyson, J. F., Block, E., & Uden, P. C. (2000). Selenium speciation in enriched and natural samples by HPLC-ICP-MS and HPLC-ESI-MS with perfluorinated carboxylic acid ion-pairing agents. *Analyst, 125*, 71–78.
Kurokawa, S., Takehashi, M., Tanaka, H., Mihara, H., Kurihara, T., Tanaka, S., et al. (2011). Mammalian selenocysteine lyase is involved in selenoprotein biosynthesis. *Journal of Nutritional Science and Vitaminology, 57*, 298–305.
Lei, X. G., Evenson, J. K., Thompson, K. M., & Sunde, R. A. (1995). Glutathione peroxidase and phospholipid hydroperoxide glutathione peroxidase are differentially regulated in rats by dietary selenium. *The Journal of Nutrition, 125*, 1438–1446.
Leist, M., Maurer, S., Schultz, M., Elsner, A., Gawlik, D., & Brigelius-Flohé, R. (1999). Cytoprotection against lipid hydroperoxides correlates with increased glutathione peroxidase activities, but not selenium uptake from different selenocompounds. *Biological Trace Element Research, 68*, 159–174.

Mallonee, D. H., Crowdus, C. A., Barger, J. L., Dawson, K. A., & Power, R. F. (2011). Use of stringent selection parameters for the identification of possible selenium-responsive marker genes in mouse liver and gastrocnemius. *Biological Trace Element Research, 143*, 992–1006.

Marshall, O. J. (2004). PerlPrimer: Cross-platform, graphical primer design for standard, bisulphite and real-time PCR. *Bioinformatics, 20*, 2471–2472.

McConnell, K. P., & Hoffman, J. L. (1972). Methionine-selenomethionine parallels in rat liver polypeptide chain synthesis. *FEBS Letters, 24*, 60–62.

Miniard, A. C., Middleton, L. M., Budiman, M. E., Gerber, C. A., & Driscoll, D. M. (2010). Nucleolin binds to a subset of selenoprotein mRNAs and regulates their expression. *Nucleic Acids Research, 38*, 4807–4820.

Ogra, Y., Hatano, T., Ohmichi, M., & Suzuki, K. T. (2003). Oxidative production of monomethylated selenium from the major urinary selenometabolite, selenosugar. *Journal of Analytical Atomic Spectrometry, 18*, 1252–1255.

Pinto, J. T., Lee, J. I., Sinha, R., MacEwan, M. E., & Cooper, A. J. (2011). Chemopreventive mechanisms of alpha-keto acid metabolites of naturally occurring organoselenium compounds. *Amino Acids, 41*, 29–41.

Rayman, M. P. (2012). Selenium and human health. *The Lancet, 379*, 1256–1268.

Rayman, M. P., Infante, H. G., & Sargent, M. (2008). Food-chain selenium and human health: Spotlight on speciation. *The British Journal of Nutrition, 100*, 238–253.

Reszka, E., Jablonska, E., Gromadzinska, J., & Wasowicz, W. (2012). Relevance of selenoprotein transcripts for selenium status in humans. *Genes & Nutrition, 7*, 127–137.

Schrauzer, G. N., White, D. A., & Schneider, C. J. (1977). Cancer mortality correlation studies—III: Statistical associations with dietary selenium intakes. *Bioinorganic Chemistry, 7*, 23–31.

Shamberger, R. J., & Frost, D. V. (1969). Possible protective effect of selenium against human cancer. *Canadian Medical Association Journal, 100*, 682.

Spallholz, J. E. (1994). On the nature of selenium toxicity and carcinostatic activity. *Free Radical Biology & Medicine, 17*, 45–64.

Sunde, R. A. (2010). Molecular biomarker panels for assessment of selenium status in rats. *Experimental Biology and Medicine (Maywood, N.J.), 235*, 1046–1052.

Sunde, R. A., & Raines, A. M. (2011). Selenium regulation of the selenoprotein and nonselenoprotein transcriptomes in rodents. *Advances in Nutrition (Bethesda, MD.), 2*, 138–150.

Suzuki, K. T., Kurasaki, K., Ogawa, S., & Suzuki, N. (2006). Metabolic transformation of methylseleninic acid through key selenium intermediate selenide. *Toxicology and Applied Pharmacology, 215*, 189–197.

Suzuki, K. T., Kurasaki, K., & Suzuki, N. (2007). Selenocysteine beta-lyase and methylselenol demethylase in the metabolism of Se-methylated selenocompounds into selenide. *Biochimica et Biophysica Acta, 1770*, 1053–1061.

Suzuki, K. T., Ohta, Y., & Suzuki, N. (2006). Availability and metabolism of 77Se-methylseleninic acid compared simultaneously with those of three related selenocompounds. *Toxicology and Applied Pharmacology, 217*, 51–62.

Suzuki, K. T., Somekawa, L., & Suzuki, N. (2006). Distribution and reuse of 76Se-selenosugar in selenium-deficient rats. *Toxicology and Applied Pharmacology, 216*, 303–308.

Turanov, A. A., Xu, X. M., Carlson, B. A., Yoo, M. H., Gladyshev, V. N., & Hatfield, D. L. (2011). Biosynthesis of selenocysteine, the 21st amino acid in the genetic code, and a novel pathway for cysteine biosynthesis. *Advances in Nutrition, 2*, 122–128.

Wallenberg, M., Olm, E., Hebert, C., Bjornstedt, M., & Fernandes, A. P. (2010). Selenium compounds are substrates for glutaredoxins: A novel pathway for selenium metabolism and a potential mechanism for selenium-mediated cytotoxicity. *The Biochemical Journal, 429*, 85–93.

Whitehead, R. H., & Joseph, J. L. (1994). Derivation of conditionally immortalized cell lines containing the Min mutation from the normal colonic mucosa and other tissues of an "Immortomouse"/Min hybrid. *Epithelial Cell Biology, 3*, 119–125.
Whitehead, R. H., VanEeden, P. E., Noble, M. D., Ataliotis, P., & Jat, P. S. (1993). Establishment of conditionally immortalized epithelial cell lines from both colon and small intestine of adult H-2Kb-tsA58 transgenic mice. *Proceedings of the National Academy of Science of the United States of America, 90*, 587–591.
Wingler, K., Böcher, M., Flohé, L., Kollmus, H., & Brigelius-Flohé, R. (1999). mRNA stability and selenocysteine insertion sequence efficiency rank gastrointestinal glutathione peroxidase high in the hierarchy of selenoproteins. *European Journal of Biochemistry, 259*, 149–157.
Xia, Y., Hill, K. E., Byrne, D. W., Xu, J., & Burk, R. F. (2005). Effectiveness of selenium supplements in a low-selenium area of China. *The American Journal of Clinical Nutrition, 81*, 829–834.
Xia, Y., Hill, K. E., Li, P., Xu, J., Zhou, D., Motley, A. K., et al. (2010). Optimization of selenoprotein P and other plasma selenium biomarkers for the assessment of the selenium nutritional requirement: A placebo-controlled, double-blind study of selenomethionine supplementation in selenium-deficient Chinese subjects. *The American Journal of Clinical Nutrition, 92*, 525–531.
Zhang, J. L., Li, J. L., Huang, X. D., Bo, S., Rihua, W., Li, S., et al. (2011). Dietary selenium regulation of transcript abundance of selenoprotein N and selenoprotein W in chicken muscle tissues. *Biometals, 25*, 297–307.
Zhou, J. C., Zhao, H., Li, J. G., Xia, X. J., Wang, K. N., Zhang, Y. J., et al. (2009). Selenoprotein gene expression in thyroid and pituitary of young pigs is not affected by dietary selenium deficiency or excess. *The Journal of Nutrition, 139*, 1061–1066.

CHAPTER SIX

Peroxiredoxins and Sulfiredoxin at the Crossroads of the NO and H_2O_2 Signaling Pathways

Kahina Abbas, Sylvie Riquier, Jean-Claude Drapier[1]

Institut de Chimie des Substances Naturelles, Centre National de la Recherche Scientifique, Gif-sur-Yvette, France

[1]Corresponding author: e-mail address: jean-claude.drapier@cnrs.fr

Contents

1. Introduction 114
2. The Effect of NO on the Level of 2-Cys-Prx Overoxidation 116
   2.1 Cell culture 116
   2.2 General methods 117
   2.3 Two-dimensional electrophoresis analysis 118
   2.4 MS analyses 119
   2.5 Immunoblotting 120
3. Detection of Srx 122
4. Comments 123
Acknowledgments 125
References 125

Abstract

Peroxiredoxins (Prxs) are a family of peroxidases that maintain thiol homeostasis by catalyzing the reduction of organic hydroperoxides, H_2O_2, and peroxynitrite. Eukaryotic 2-Cys-Prxs, also referred to as typical Prxs, can be inactivated by oxidation of the catalytic cysteine to sulfinic acid, which may regulate the intracellular messenger function of H_2O_2. A small redox protein, sulfiredoxin (Srx), has been shown to reduce sulfinylated 2-Cys-Prxs and thus to regenerate active 2-Cys-Prxs. We previously reported that cytokine-induced nitric oxide (NO) intervenes in this pathway by decreasing the level of 2-Cys overoxidation and by upregulating Srx through the activation of the transcription factor nuclear factor erythroid 2-related factor (Nrf2). Here, we describe the methods used to monitor the interplay between NO and H_2O_2 in the regulation of the Prx/Srx system in immunostimulated macrophages, which produce both reactive oxygen species and NO.

Methods in Enzymology, Volume 527
ISSN 0076-6879
http://dx.doi.org/10.1016/B978-0-12-405882-8.00006-4

1. INTRODUCTION

The role of nitric oxide (NO) and reactive oxygen species (ROS) in cellular signaling has attracted considerable interest over the past two decades. Rapidly, it became evident that such reactive molecules could react with each other to yield higher oxides like peroxynitrite, the product of the diffusion-limited reaction of NO with superoxide (O_2^{-}). Even though NO can react directly with H_2O_2 *in vitro* (Noronha-Dutra, Epperlein, & Woolf, 1993), it has long been reported that NO can protect cells against H_2O_2-mediated cell death by indirect interconnection (Wink et al., 2001). In this chapter, we present the methods that enabled us to describe an NO-dependent antioxidant pathway mediated by the peroxiredoxin (Prx)–sulfiredoxin (Srx) system.

Prxs are abundant thiol peroxidases that catalyze the reduction of peroxides, and recent kinetic studies have revealed that many are highly efficient in reducing H_2O_2 (Peskin et al., 2007). The fact that some of them are also active as peroxynitrite reductases (Trujillo, Ferrer-Sueta, Thomson, Flohe, & Radi, 2007) was a spur to investigate whether NO could upregulate Prxs and exert feedback control on peroxynitrite concentration. In mammals, there are six Prx isoforms: four 2-Cys-Prxs referred to as typical (Prx1–4), the atypical 2-Cys-Prx5, and the 1-Cys-Prx6 (Rhee, Woo, Kil, & Bae, 2012). Prx1 was found to be upregulated by NO in LPS-stimulated Kupffer cells (Immenschuh, Tan, & Ramadori, 1999), and we showed that upon immunostimulation, murine macrophages display high expression of Prx1, Prx5, and Prx6. Analysis of differentially expressed genes in cells from inducible NO synthase $(iNOS)^{-/-}$ mice showed that Prx1 and Prx6-but not Prx5-regulation is totally dependent on iNOS-derived NO (Diet et al., 2007). Upon catalysis, 2-Cys-Prxs are active as homodimers with the peroxidatic cysteine (S_P) oxidized to a sulfenic acid by the peroxide reacting with the resolving cysteine (S_R) on another monomer of Prx to form an intermolecular disulfide bridge (Hall, Karplus, & Poole, 2009). The thioredoxin (Trx)/Trx reductase system reduces oxidized 2-Cys-Prx to complete the catalytic cycle (Fig. 6.1, left). In eukaryotes and cyanobacteria, which constitute an exception among prokaryotes (Pascual, Mata-Cabana, Florencio, Lindahl, & Cejudo, 2010), sulfenic acid can react with a second molecule of peroxide to form a sulfinic acid in what is generally called hyperoxidation or overoxidation, which leads to inactivation of peroxidase activity. These findings have provided a basis for the floodgate model, which suggests that peroxide-mediated inactivation of Prx allows H_2O_2-mediated signaling in eukaryotes

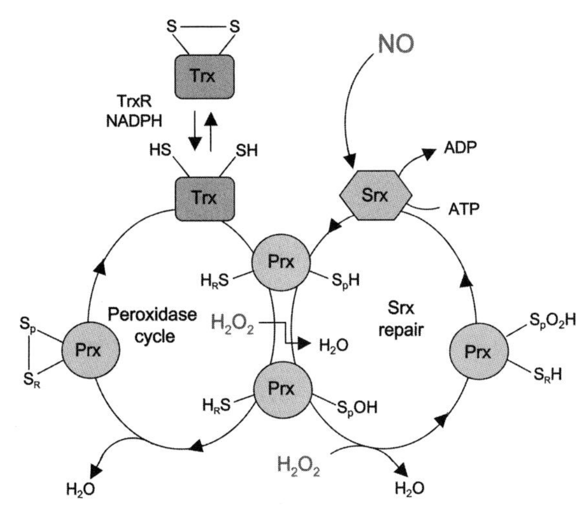

Figure 6.1 Two-Cys-Prx catalytic cycles. Peroxidase cycle (left). Peroxides oxidize Prx-Sp into a sulfenate, Prx-SpOH. This sulfenate rapidly generates a disulfide with the Cys-S_RH of another Prx molecule to form a covalent dimer. This dimer can easily be reduced by the Trx/TrxR system. As a general rule, in eukaryotes Prx-SpOH can react with another H_2O_2 molecule to form a sulfinate S_pO_2H (right), resulting in peroxidase inactivation which can be repaired by Srx. In murine macrophages, iNOS-derived NO is an inducer of Srx expression. *Adapted from Abbas, Breton, and Drapier (2008).* (For color version of this figure, the reader is referred to the online version of this chapter.)

(Wood, Poole, & Karplus, 2003). It has also been proposed that sulfinylation is not only a way to inactivate Prx but also a regulatory switch converting Prxs from peroxidases to chaperones (Jang et al., 2004).

Two-Cys-Prxs are sulfinylated in a reaction herein after referred to as overoxidation that cannot be reduced by the Trx/TrxR system but can be slowly regenerated by a sulfinyl reductase called sulfiredoxin (Srx), which catalyzes an ATP-dependent reduction of the sulfinic acid group of 2-Cys-Prx (Biteau, Labarre, & Toledano, 2003; Jeong, Bae, Toledano, & Rhee, 2012). Srx homologs are restricted to eukaryotes with *Caenorhabditis elegans* as an exception (Jonsson & Lowther, 2007) and to cyanobacteria (Boileau et al., 2011). Therefore, the ratio between the reduced form of Prx (active) and its overoxidized form (inactive) varies, depending on several parameters, and particularly upstream local H_2O_2 levels and downstream Srx expression and activity, which allow regeneration (Fig. 6.1, right). Recently, it was reported that the oscillation between the reduced form and the overoxidized form of Prx is rhythmic and independent of transcription and translation (O'Neill et al., 2011). This metabolic clock is conserved across all three phylogenetic domains and may have arisen around the rise of oxygen on Earth, 2.5 billion years ago. The rationale of this cyclic inactivation/reactivation of 2-Cys-Prx is still poorly understood, but it has been proposed that ROS control systems may have driven modern circadian rhythms (Edgar et al., 2012).

Macrophages produce a plethora of reactive molecules, and upon adequate stimulation can express Nox2, iNOS, and cystathionase (Oh et al., 2006), which produce NO, ROS, and hydrogen sulfide (H_2S), respectively. They are therefore a good cellular model for studying the crossroads between these redox signaling molecules. NO markedly affects the process of oxidation/reduction of 2-Cys-Prx in two main ways. First, NO protects 2-Cys-Prxs from overoxidation by peroxides (Diet et al., 2007; Fang, Nakamura, Cho, Gu, & Lipton, 2007), very likely by reacting with the peroxidatic cysteine or the resolving cysteine (Wu et al., 2010). S-nitrosation of Prx1 by Cys-NO or iNOS has been reported in several cell types including RAW264.7 macrophages, epithelial cells, and endothelial cells (Forrester et al., 2009; Lam et al., 2010; Martinez-Ruiz & Lamas, 2004). S-nitrosation of Prx2 in neurons exposed to NO or stimulated for NOS1 expression has also been described (Fang et al., 2007). In addition, we found that NO upregulates Srx at the transcriptional level in immunostimulated macrophages and therefore potentially mitigates H_2O_2 accumulation (Abbas et al., 2011; Diet et al., 2007).

Srx is induced by electrophiles and chemopreventive compounds (Soriano et al., 2008), synaptic activity in rat neurons (Papadia et al., 2008), and endogenously produced redox species like iNOS-derived NO (Abbas et al., 2011) or ethanol-induced ROS (Bae et al., 2011). As a consequence, expression of Srx allows faster regeneration of 2-Cys-Prx. Srx expression is dependent on the activation of the transcription factors AP-1 and Nrf2 (Soriano et al., 2008). *In vivo* loss of function studies showed that Srx protects mice against LPS-induced endotoxic shock (Planson et al., 2011) and against oxidative damage in the liver after chronic ethanol feeding (Bae et al., 2011). *In vitro* and *in vivo* studies showed that macrophages from Srx null mice exhibit higher ROS production (Abbas et al., 2011; Planson et al., 2011)

2. THE EFFECT OF NO ON THE LEVEL OF 2-Cys-Prx OVEROXIDATION

2.1. Cell culture

Macrophages (RAW 264.7) are grown in DMEM supplemented with 5% FBS and 100 μg/mL gentamicin. Bone marrow cells are obtained by flushing femurs of C57BL/6 mice with Hanks medium and differentiated into macrophages in RPMI 1640 supplemented with 10% fetal bovine serum, 10% L929 cell-conditioned medium, and 1% penicillin/streptomycin. Mouse bone marrow-derived macrophages (BMMs) are cultured in RPMI 1640 medium supplemented with 10% L929 cell-conditioned medium as previously described (Weischenfeldt & Porse, 2008). BMMs are treated with 100 units/mL INF-γ (R&D Systems, Abingdon, UK) and 500 ng/mL LPS (Alexis, TLR4 grade) for 16 h. The iNOS inhibitor 1400W (Sigma) is added to evaluate the possibility of an iNOS-independent contribution of immunostimulation. Some experiments are performed in parallel with BMM from iNOS$^{-/-}$ mice to emphasize the intrinsic contribution of NO or NO-derived species. Nitrite, the stable end product of NO, is quantified in the conditioned medium by using the Griess reagent as reported (Drapier & Hibbs, 1988). Moreover, cells are also exposed for 16 h to an exogenous source of authentic NO, diethyltriamine NONOate (DETA-NO from Cayman Chemicals) a slow releasing NO donor that can release NO spontaneously under physiological conditions. Decomposition rate of DETA-NO is determined by loss of the chromophore at 252 nm. For 10-cm diameter dishes, BMM are washed with 3 mL of PBS and then placed in fresh RPMI 1640 with 100 μM H_2O_2 for 20 min. Cells are washed two times with 3 mL cold PBS and lysed on ice for 10 min in 400 μl of a buffer containing 25 mM Tris, pH 7.4, 50 mM NaCl, 2 mM EDTA, and 0.5% (v/v) Triton X-100 added with the Protease Inhibitor Cocktail (Sigma P8340). Cells are scraped and lysates are centrifuged at 4 °C for 10 min at $12{,}000 \times g$. The protein content of supernatant is determined spectrophotometrically at 562 nm by using the bicinchoninic acid assay.

2.2. General methods

The level of overoxidized 2-Cys-Prx is assessed by immunoblotting using an anti-Prx-$SO_{2/3}$ antibody. This antibody detects the sulfinylated as well as the sufonylated forms of 2-Cys-Prxs, but it is generally assumed that only sulfinylated Prxs are formed in a cellular context. At least in macrophages, overoxidized 2-Cys-Prxs are routinely detected in the absence of stimulation, which is consistent with the basal oxidative metabolism of this cell type. To see whether NO-producing cells display a lower level of sulfinylated Prx, cells can be challenged by H_2O_2 or by ROS inducers like *tert*-butylhydroquinone (*t*-BHQ) given as a bolus. It may be more relevant to perform experiments that more closely reflect physiological conditions.

Thus, addition of glucose oxidase (GOX) in a glucose-rich culture medium exposes cells to low steady-state levels of H_2O_2. In macrophages, Nox2 expression can be activated by exposure to pharmacological activators of PKC such as phorbol myristate acetate (PMA). Superoxide anion is rapidly dismutated to H_2O_2 either spontaneously or by superoxide dismutases. Abatement by NO of 2-Cys-Prx overoxidation in response to GOX and PMA is illustrated in Fig. 6.2.

2.3. Two-dimensional electrophoresis analysis

It may be difficult to separate Prxs by one-dimensional (1D) electrophoresis, particularly as Prx1 and Prx2 are very similar and often comigrate on 1D PAGE. Therefore, separation of Prxs on two-dimensional (2D) gels is a more demanding but rewarding approach. Identification of mammalian Prxs by 2D PAGE and mass spectrometry was pioneered by Rabilloud et al. (2002) and Rhee and coworkers (Yang et al., 2002).

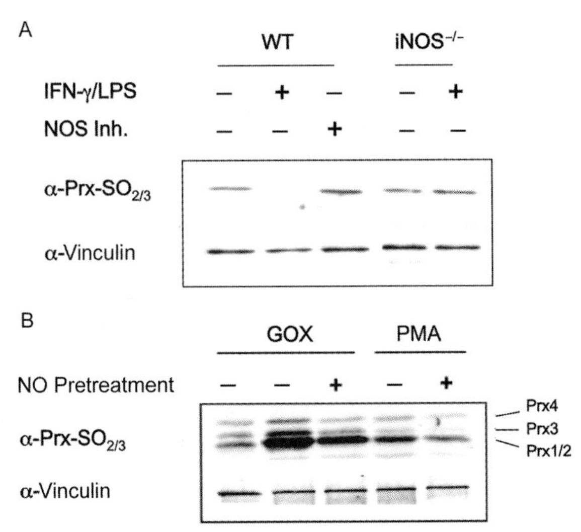

Figure 6.2 NO decreases basal and H_2O_2-triggered overoxidation of 2-Cys-Prx. (A) WT and iNOS$^{-/-}$ BMMs were stimulated with 100 units/mL IFN-γ and 500 ng/mL LPS (I/L) in the presence or not of a mix of 25 μ*M* 1400W and 100 μ*M* *S*-ethylisothiourea, two NOS inhibitors (Inh.). (B) Effect of NO on the H_2O_2-triggered overoxidation of Prxs. BMM were preexposed to 500 μ*M* DETA-NO for 18 h. Cell monolayers were then exhaustively washed and exposed for 3 h to 100 milliunits of glucose oxidase (GOX) or to 500 n*M* PMA. Two-Cys-Prxs were then assessed for overoxidation by immunoblotting with an anti-Prx-$SO_{2/3}$ antibody. Vinculin was used as a loading control. *Part of this research was originally published in Diet et al. (2007) © the American Society for Biochemistry and Molecular Biology.*

For the first dimension, 250 μg of protein after precipitation by 2D cleanup (Bio-Rad) is diluted in 150 μl of a rehydration buffer containing 7 *M* urea, 2 *M* thiourea, 4% CHAPS (w/v), 1% immobilized pH gradient (IPG) buffer pH 3–10 (v/v), and 3 μl of 0.25% bromophenol blue (w/v). Samples are loaded on 11-cm IPG strips with a nonlinear gradient of pH 3–10 for 2 h at room temperature. Isoelectric focusing is performed at 20 °C with a Protean isoelectric focusing Cell (Bio-Rad): rehydration for 9 h at 50 V, and isoelectric focusing for 30 min at 200 V constant, 1 h at 1000 V constant and 5 h at 8000 V constant, reaching 40,000 Vh at the end of the run. Strips are equilibrated for 15 min in 5 mL of the equilibration solution (6 *M* urea, 375 m*M* Tris–HCl pH 8.8, 20% glycerol, 2% SDS (w/v) containing 20 mg/mL DTT), then for another 15 min in 5 mL of the equilibration solution containing 25 μg/mL iodoacetamide, in the dark. Separation in the second dimension is performed on SDS-PAGE (12% acrylamide) using a Criterion Cell (Bio-Rad) after 2-h migration/gel. Proteins are stained using the SimplyBlue kit (Invitrogen).

2.4. MS analyses

2.4.1 Peptide preparation

Protein spots protein resolved by SDS-PAGE are manually excised and washed three times in 50 μL of 25 m*M* ammonium bicarbonate pH 8.3 and dehydrated in 50 μL of acetonitrile. The pieces of gel are dried in a Speed Vac for 15 min. The gel pieces are rehydrated in 10 μL of 50 ng/mL trypsin (Promega Gold) in 50 m*M* ammonium bicarbonate, and incubated on ice for 30 min. After removing excess of trypsin, 30 μL of 25 m*M* ammonium bicarbonate pH 8.3 is added, and protein digestion is performed overnight at 37 °C while mild shaking. The peptides are extracted and the gel is washed with 30 μl of 70% acetonitrile, 1% TFA in water for 15 min. After centrifugation, supernatants containing extracted peptides are pooled and dried under vacuum.

2.4.2 MALDI MS analyses

Peptide mixtures produced by trypsin digestion are analyzed by MALDI-TOF MS using a 4800 MALDI-TOF–TOF (Applied Biosystems, AB/Sciex) mass spectrometer. Proteolyzed samples are dissolved in 3 μL of 0.1% TFA (v/v) and 0.5 μL of eluate is spotted on a stainless steel MALDI target plate and covered with 0.5 μL of 5 mg/mL α-cyano-4-hydroxycinnamic acid in 70% acetonitrile (v/v), 0.1% TFA (v/v). MALDI-TOF spectra are acquired in the positive and reflector modes, and external calibration is performed using an equal mix of peptide Cal mix 1 and Cal mix2 standards (Applied Biosystems). Monoisotopic peptide mass values extracted by Data Explorer (Applied Biosystems) are used for protein identification by searching against the UniProtKB/Swiss-Prot database using the online Mascot search engine (www.matrixscience.com). The searches are conducted using the following settings: trypsin as digestion enzyme, one missed cleavage allowed, 30 ppm tolerance, carbamidomethyl as fixed modification, and methionine oxidation as variable modification. As can be seen in Fig. 6.3A, the four 2-cys-Prx of RAW264.7 macrophages are easily separated, especially Prx1 and Prx2 whose isoelectric points are quite different. After SimplyBlue staining (Invitrogen), Prxs are identified by MALDI-MS analyses. Focusing on Prx1, analysis of differentiated expressed spots allows detection of the sulfinylated form at a more acidic position of the gel (Fig. 6.3A, lower).

2.5. Immunoblotting

Prxs and their modified forms can also be detected conveniently and accurately quantified by two-color Western blot using the LI-COR Odyssey Imaging System. This method is less time-consuming and more cost effective than mass spectrometry analyses. Moreover, IR fluorescence provides a wide linear range for quantitation, and low background fluorescence of membranes enables high sensitivity.

The proteins are resolved by SDS-PAGE and transferred onto Immobilon-FL PVDF membrane at 250 mA/gel for 90 min in a Trans-Blot cell (Bio-Rad). Transfer buffer: 12.5 m*M* Tris–HCl, 96 m*M* Glycine, 20% ethanol (v/v). Membranes are blocked in 5% (w/v) skim milk in $TBST_{20}$ (150 m*M* NaCl, 50 m*M* Tris–HCl pH 8, and 0.05% Tween 20). In our studies, we routinely achieve two-color detection of the different forms of Prxs by multiplexing IRDye 800CW and IRDye 680CW conjugates. The membranes are incubated overnight at 4 °C with a primary antibody against Prx-$SO_{2/3}$ (Abcam ab16830) at 1:5000–1:10,000 dilution in $TBST_{20}$. After washing in $TBST_{20}$, membranes are incubated with IRDye 680CW or IRDye 800CW secondary antibody conjugates (1:10,000, respectively). From this step forward samples should be protected from light. In Fig. 6.3B (upper), 2-Cys-Prxs are detected by an antibody which reacts with the four typical 2-Cys-Prxs (Abcam ab1765), 1/2000. It is worth noting that we have observed that Prx1 is not very sensitive to this antibody and we recommend reincubating the membrane with an anti-Prx1 antibody (Abcam ab15571) at the dilution 1:10,000 to obtain satisfactory results. Detection

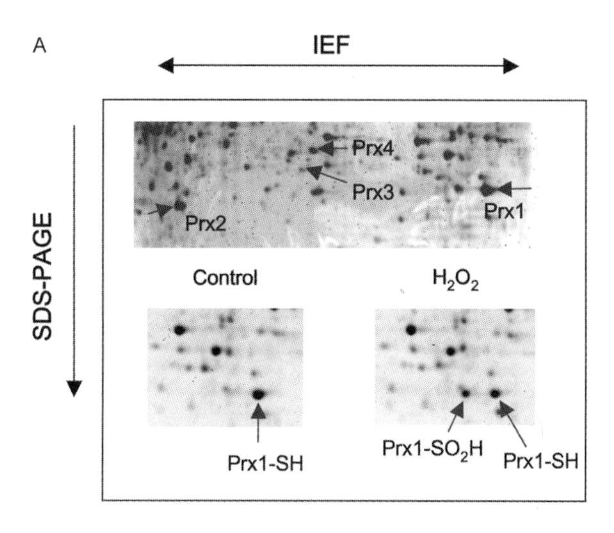

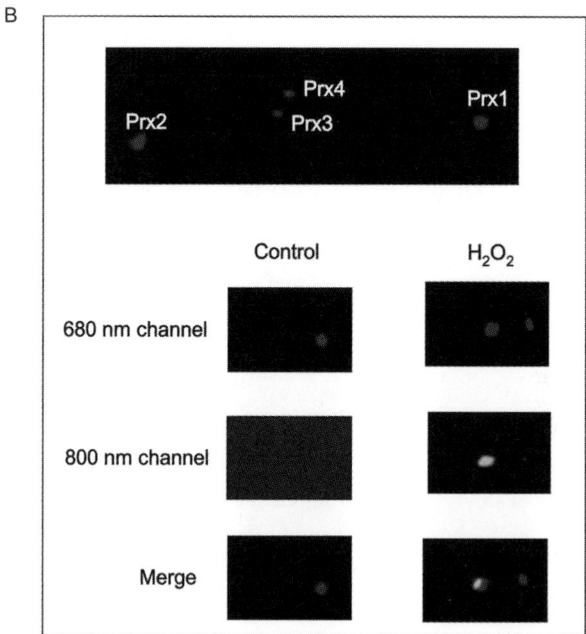

Figure 6.3 Modification of Prx1 in macrophages upon exposure to H_2O_2. (A) Upper: cell extracts from RAW264.7 cultured in DMEM were analyzed by 2D electrophoresis. The images shown focus on the Prx spot containing gel. Proteins were visualized by SimplyBlue staining. The 2-Cys-Prx spots were excised and identified by MALDI-TOF. Lower: extracts obtained from cells cultured in DMEM under control conditions or treated with 100 μM H_2O_2. Normal Prx1 spot and the more acidic additional Prx1 spot were identified by MALDI-TOF. (B) Upper: cell extracts from RAW264.7 cultured in DMEM were analyzed by 2D electrophoresis following 2D electrophoresis. The gel was transferred to an Immobilon-FL PVDF membrane for fluorescent Western blot detection. The 2D membrane was incubated with an antibody raised against all 2-Cys-Prxs and with an anti-Prx1 antibody (otherwise the Prx1 signal may be weak) and, after washing, with an

is achieved by incubation with a fluorescent secondary antibody coupled to IRDye 680CW (red) at the dilution 1:10,000. Images are scanned with an Odyssey scanner (Li-Cor Biosciences). Figure 6.3B (lower) gives an example of multiplex fluorescent Western blot which allows detection of both reduced and overoxidized Prx1. IRDye 800CW- or IRDye 680CW-labeled secondary antibodies detect overoxidized 2-Cys-Prx1 (green) and global (reduced as well as overoxidized) 2-Cys-Prx1 (red), respectively. Merge colors are shown as yellow. In gel detection of proteins using fluorescence, antibodies coupled to fluorescent dyes have been described (Kettenhofen, Wang, Gladwin, & Hogg, 2008). This procedure is effective but consumes large amounts of antibodies and should be preferred for detection of proteins that poorly transfer to membranes. Otherwise, for detection of proteins in cell lysates, immunoblotting appears to be more sensitive and more cost effective.

3. DETECTION OF Srx

Srx is an inducible enzyme and is generally not an abundant protein. Before SDS-PAGE analysis, it is wise to concentrate the samples before loading them on the gel. Because Srx is a small protein (17 kD), small pore size membranes are suitable. Moreover, some commercially available anti-Srx antibodies display low sensitivity and moderate specificity. Thus, selected primary antibodies, careful transfer and blocking, and highly sensitive chemiluminescence-based assays are recommended for detection.

Primary macrophages and the RAW264.7 cell line are cultured as described earlier. Protein extracts (40 μg) are added with reducing sample buffer 1 × with SDS (2% w/v), final concentration (sample buffer 5 × concentrated: 2% SDS (w/v), 250 m*M* Tris–HCl pH 6.8, 50% glycerol, 450 m*M* DTT). Lysates are analyzed by SDS-PAGE in 16% polyacrylamide gel under reducing conditions. After the electrophoretic run (25 m*M* Tris, 192 m*M* glycine, 0.1% SDS, pH 8.3) and protein immobilization for 90 min at 100 mA in 12.5 m*M* Tris, 96 m*M* glycine, 20% ethanol (Novex XCell SureLoc, Invitrogen), Immobilon-FL PVDF membranes (Millipore) are

IRDye 680CW-conjugated secondary antibody. Lower: two-color Western blot detection was performed using secondary antibodies conjugated to IRDyes 680CW and 800CW to allow visualization of Prx1 overoxidation. Prx1-SH was detected at 680 nm (red) and Prx1-SO_2 (green) at 800 nm. Yellow indicates overlap of signals and confirms protein modification. (For interpretation of the references to color in this figure legend, the reader is referred to the online version of this chapter.)

blocked for 1 h at room temperature with 5% (w/v) skim milk in $TBST_{20}$ (150 m*M* NaCl, 50 m*M* Tris–HCl pH 8, and 0.05% Tween 20) and incubated overnight at 4 °C with anti-Srx antibody (sc-166566, Santa Cruz) used at a concentration of 1:600 in $TBST_{20}$. β-Act-in or vinculin is used as a loading control. The membrane is washed with $TBST_{20}$ and incubated with HRP-conjugated secondary antibodies (DAKO Cytomation). The blot is imaged using the Chemidoc XRS system (Bio-Rad).

Results in Fig. 6.4 compare the level of Srx protein in macrophages from WT and Srx (upper), iNOS (middle), and Nrf2 (lower) deficient mice upon exposure or not to exogenous or endogenous NO. Data reveal robust expression of Srx protein and activation of the transcription factor Nrf2 following exposure to NO.

4. COMMENTS

In this chapter, we have focused on the intersection between NO and H_2O_2 signaling pathways. 2D gel electrophoresis followed by two-color Western blot is a useful and accurate tool for rapid visualization of several forms of redox posttranslational modifications. Examples include normal and sulfinylated Prxs (this chapter), S-nitrosylated proteins detected by a biotin switch-derived method often referred to as "fluorescence or dye switch" (Kettenhofen et al., 2008; Tello et al., 2009), and sulfhydrated and S-nitrosylated proteins, as recently described for the p65 subunit of NF-κB in H_2S-producing cells (Sen et al., 2012). As an NO-inducible enzyme, Srx is also pivotal in this interplay. Intriguingly, tumor cells, in contrast to normal cells, seem to express high levels of Srx, which is associated with tumorigenicity and metastasis (Wei et al., 2011). In this regard, increase in Srx level was shown to enhance proliferation rate of cancer cell lines and to play a role in cell motility (Bowers, Manevich, Townsend, & Tew, 2012; Lei, Townsend, & Tew, 2008). These data suggest that Srx is involved in reactions other than Prx regeneration or that the balance between reduced and overoxidized 2-Cys-Prxs has farther-reaching consequences than once expected. Moreover, as an Nrf2-regulated gene, Srx can also be induced by environmental pollutants like cigarette smoke and by dietary-derived electrophiles like cruciferous isothiocyanates (Soriano et al., 2008). Therefore, it may be worth considering that Srx expression will be sought in many sorts of physiological and pathological preclinical settings.

There is ample evidence that NO and ROS pathways operate in virtually every mammalian cell. As regard H_2O_2, recent progress in the development

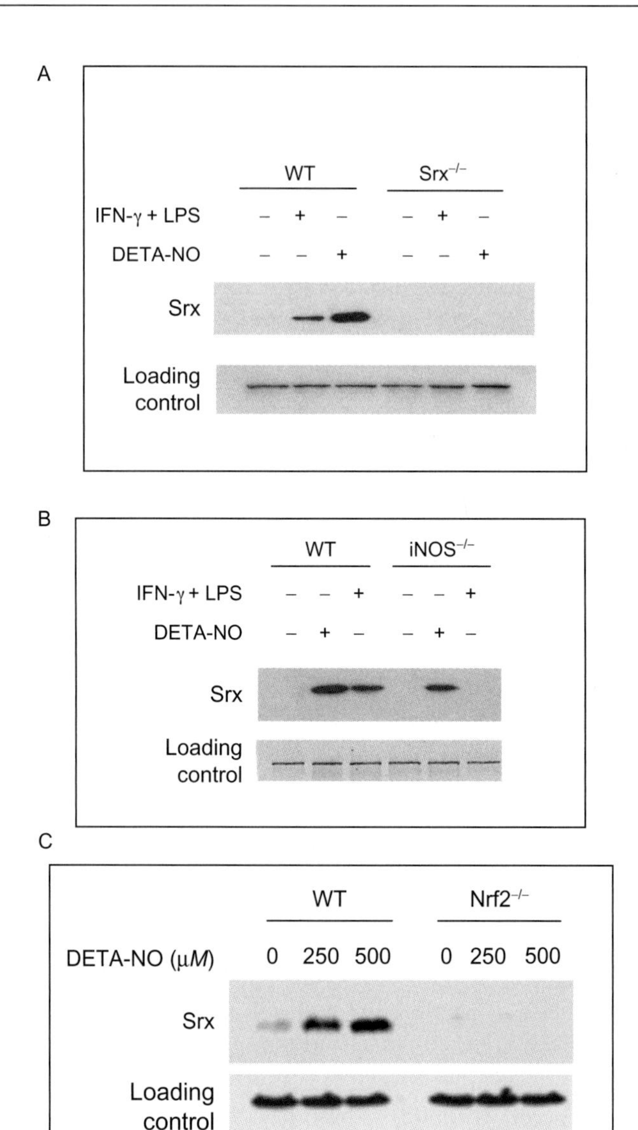

Figure 6.4 Srx protein expression in immunostimulated macrophages. (A) BMMs from WT and $Srx^{-/-}$ mice were stimulated with 50 units/mL IFN-γ and 100 ng/mL LPS or exposed to 500 μM DETA-NO. Total absence of the band corresponding to Srx in extracts from $Srx^{-/-}$ cells shows antibody specificity. (B) BMMs from WT and $iNOS^{-/-}$ mice were treated as in A. Cell lysates (40 μg protein) were collected and Srx protein expression was analyzed by immunoblotting using an anti-Srx antibody. Vinculin was used as a loading control. (C) BMMs from WT and $Nrf2^{-/-}$ mice were exposed for 16 h to DETA-NO. Cell lysates were analyzed as in B. *Partially reproduced with permission from Abbas et al. (2011).*

of sensitive and specific ratiometric fluorescent probes as well as live imaging (Belousov et al., 2006; Dickinson, Huynh, & Chang, 2010; Morgan, Sobotta, & Dick, 2011) has uncovered its presence in many cells or tissues, and revealed unexpected roles, for instance, in leukocyte migration and during wound healing (Niethammer, Grabher, Look, & Mitchison, 2009; Yoo, Starnes, Deng, & Huttenlocher, 2011). Accordingly, there is still much to learn about how Srx and companion 2-Cys-Prxs regulate cell signaling. The data presented here were mainly obtained with macrophages, but looking ahead, future work may want to evaluate the effects of NO synthesis on the Prx–Srx system in other cell types or tissues, notably those known to produce ROS in normal or pathological settings including cardiovascular signaling (Prosser, Ward, & Lederer, 2011; Sartoretto, Kalwa, Pluth, Lippard, & Michel, 2011), lung diseases like chronic obstructive pulmonary disease (Bowler, Barnes, & Crapo, 2004; Comhair & Erzurum, 2002), liver injury like cirrhosis or fibrosis (Sanchez-Valle, Chavez-Tapia, Uribe, & Mendez-Sanchez, 2012; Szabo et al., 2010), and cerebral stroke (Moskowitz, Lo, & Iadecola, 2010).

ACKNOWLEDGMENTS

The authors acknowledge the skills and insight provided by past and present collaborators including Cécile Bouton, Jacques Breton, Alexandre Diet, and Michel B. Toledano. This project was supported by the ANR Grant NOPEROX.

REFERENCES

Abbas, K., Breton, J., & Drapier, J. C. (2008). The interplay between nitric oxide and peroxiredoxins. *Immunobiology*, *213*, 815–822.

Abbas, K., Breton, J., Planson, A. G., Bouton, C., Bignon, J., Seguin, C., et al. (2011). Nitric oxide activates an Nrf2/sulfiredoxin antioxidant pathway in macrophages. *Free Radical Biology & Medicine*, *51*, 107–114.

Bae, S. H., Sung, S. H., Cho, E. J., Lee, S. K., Lee, H. E., Woo, H. A., et al. (2011). Concerted action of sulfiredoxin and peroxiredoxin I protects against alcohol-induced oxidative injury in mouse liver. *Hepatology*, *53*, 945–953.

Belousov, V. V., Fradkov, A. F., Lukyanov, K. A., Staroverov, D. B., Shakhbazov, K. S., Terskikh, A. V., et al. (2006). Genetically encoded fluorescent indicator for intracellular hydrogen peroxide. *Nature Methods*, *3*, 281–286.

Biteau, B., Labarre, J., & Toledano, M. B. (2003). ATP-dependent reduction of cysteine-sulphinic acid by *S. cerevisiae* sulphiredoxin. *Nature*, *425*, 980–984.

Boileau, C., Eme, L., Brochier-Armanet, C., Janicki, A., Zhang, C. C., & Latifi, A. (2011). A eukaryotic-like sulfiredoxin involved in oxidative stress responses and in the reduction of the sulfinic form of 2-Cys peroxiredoxin in the cyanobacterium Anabaena PCC 7120. *The New Phytologist*, *191*, 1108–1118.

Bowers, R. R., Manevich, Y., Townsend, D. M., & Tew, K. D. (2012). Sulfiredoxin redox-sensitive interaction with S100A4 and non-muscle myosin IIA regulates cancer cell motility. *Biochemistry*, *51*, 7740–7754.

Bowler, R. P., Barnes, P. J., & Crapo, J. D. (2004). The role of oxidative stress in chronic obstructive pulmonary disease. *COPD, 1*, 255–277.

Comhair, S. A., & Erzurum, S. C. (2002). Antioxidant responses to oxidant-mediated lung diseases. *American Journal of Physiology. Lung Cellular and Molecular Physiology, 283*, L246–L255.

Dickinson, B. C., Huynh, C., & Chang, C. J. (2010). A palette of fluorescent probes with varying emission colors for imaging hydrogen peroxide signaling in living cells. *Journal of the American Chemical Society, 132*, 5906–5915.

Diet, A., Abbas, K., Bouton, C., Guillon, B., Tomasello, F., Fourquet, S., et al. (2007). Regulation of peroxiredoxins by nitric oxide in immunostimulated macrophages. *The Journal of Biological Chemistry, 282*, 36199–36205.

Drapier, J. C., & Hibbs, J. B., Jr. (1988). Differentiation of murine macrophages to express nonspecific cytotoxicity for tumor cells results in L-arginine-dependent inhibition of mitochondrial iron-sulfur enzymes in the macrophage effector cells. *Journal of Immunology, 140*, 2829–2838.

Edgar, R. S., Green, E. W., Zhao, Y., van Ooijen, G., Olmedo, M., Qin, X., et al. (2012). Peroxiredoxins are conserved markers of circadian rhythms. *Nature, 485*, 459–464.

Fang, J., Nakamura, T., Cho, D. H., Gu, Z., & Lipton, S. A. (2007). S-nitrosylation of peroxiredoxin 2 promotes oxidative stress-induced neuronal cell death in Parkinson's disease. *Proceedings of the National Academy of Sciences of the United States of America, 104*, 18742–18747.

Forrester, M. T., Thompson, J. W., Foster, M. W., Nogueira, L., Moseley, M. A., & Stamler, J. S. (2009). Proteomic analysis of S-nitrosylation and denitrosylation by resin-assisted capture. *Nature Biotechnology, 27*, 557–559.

Hall, A., Karplus, P. A., & Poole, L. B. (2009). Typical 2-Cys peroxiredoxins—Structures, mechanisms and functions. *The FEBS Journal, 276*, 2469–2477.

Immenschuh, S., Tan, M., & Ramadori, G. (1999). Nitric oxide mediates the lipopolysaccharide dependent upregulation of the heme oxygenase-1 gene expression in cultured rat Kupffer cells. *Journal of Hepatology, 30*, 61–69.

Jang, H. H., Lee, K. O., Chi, Y. H., Jung, B. G., Park, S. K., Park, J. H., et al. (2004). Two enzymes in one; two yeast peroxiredoxins display oxidative stress-dependent switching from a peroxidase to a molecular chaperone function. *Cell, 117*, 625–635.

Jeong, W., Bae, S. H., Toledano, M. B., & Rhee, S. G. (2012). Role of sulfiredoxin as a regulator of peroxiredoxin function and regulation of its expression. *Free Radical Biology & Medicine, 53*, 447–456.

Jonsson, T. J., & Lowther, W. T. (2007). The peroxiredoxin repair proteins. *Sub-Cellular Biochemistry, 44*, 115–141.

Kettenhofen, N. J., Wang, X., Gladwin, M. T., & Hogg, N. (2008). In-gel detection of S-nitrosated proteins using fluorescence methods. *Methods in Enzymology, 441*, 53–71.

Lam, Y. W., Yuan, Y., Isaac, J., Babu, C. V., Meller, J., & Ho, S. M. (2010). Comprehensive identification and modified-site mapping of S-nitrosylated targets in prostate epithelial cells. *PLoS One, 5*, e9075.

Lei, K., Townsend, D. M., & Tew, K. D. (2008). Protein cysteine sulfinic acid reductase (sulfiredoxin) as a regulator of cell proliferation and drug response. *Oncogene, 27*, 4877–4887.

Martinez-Ruiz, A., & Lamas, S. (2004). Detection and proteomic identification of S-nitrosylated proteins in endothelial cells. *Archives of Biochemistry and Biophysics, 423*, 192–199.

Morgan, B., Sobotta, M. C., & Dick, T. P. (2011). Measuring E(GSH) and H2O2 with roGFP2-based redox probes. *Free Radical Biology & Medicine, 51*, 1943–1951.

Moskowitz, M. A., Lo, E. H., & Iadecola, C. (2010). The science of stroke: Mechanisms in search of treatments. *Neuron, 67*, 181–198.

Niethammer, P., Grabher, C., Look, A. T., & Mitchison, T. J. (2009). A tissue-scale gradient of hydrogen peroxide mediates rapid wound detection in zebrafish. *Nature*, *459*, 996–999.

Noronha-Dutra, A. A., Epperlein, M. M., & Woolf, N. (1993). Reaction of nitric oxide with hydrogen peroxide to produce potentially cytotoxic singlet oxygen as a model for nitric oxide-mediated killing. *FEBS Letters*, *321*, 59–62.

Oh, G. S., Pae, H. O., Lee, B. S., Kim, B. N., Kim, J. M., Kim, H. R., et al. (2006). Hydrogen sulfide inhibits nitric oxide production and nuclear factor-kappaB via heme oxygenase-1 expression in RAW264.7 macrophages stimulated with lipopolysaccharide. *Free Radical Biology & Medicine*, *41*, 106–119.

O'Neill, J. S., van Ooijen, G., Dixon, L. E., Troein, C., Corellou, F., Bouget, F. Y., et al. (2011). Circadian rhythms persist without transcription in a eukaryote. *Nature*, *469*, 554–558.

Papadia, S., Soriano, F. X., Leveille, F., Martel, M. A., Dakin, K. A., Hansen, H. H., et al. (2008). Synaptic NMDA receptor activity boosts intrinsic antioxidant defenses. *Nature Neuroscience*, *11*, 476–487.

Pascual, M. B., Mata-Cabana, A., Florencio, F. J., Lindahl, M., & Cejudo, F. J. (2010). Overoxidation of 2-Cys peroxiredoxin in prokaryotes: Cyanobacterial 2-Cys peroxiredoxins sensitive to oxidative stress. *The Journal of Biological Chemistry*, *285*, 34485–34492.

Peskin, A. V., Low, F. M., Paton, L. N., Maghzal, G. J., Hampton, M. B., & Winterbourn, C. C. (2007). The high reactivity of peroxiredoxin 2 with H(2)O(2) is not reflected in its reaction with other oxidants and thiol reagents. *The Journal of Biological Chemistry*, *282*, 11885–11892.

Planson, A. G., Palais, G., Abbas, K., Gerard, M., Couvelard, L., Delaunay, A., et al. (2011). Sulfiredoxin protects mice from lipopolysaccharide-induced endotoxic shock. *Antioxidants & Redox Signaling*, *14*, 2071–2080.

Prosser, B. L., Ward, C. W., & Lederer, W. J. (2011). X-ROS signaling: Rapid mechano-chemo transduction in heart. *Science*, *333*, 1440–1445.

Rabilloud, T., Heller, M., Gasnier, F., Luche, S., Rey, C., Aebersold, R., et al. (2002). Proteomics analysis of cellular response to oxidative stress. Evidence for in vivo overoxidation of peroxiredoxins at their active site. *The Journal of Biological Chemistry*, *277*, 19396–19401.

Rhee, S. G., Woo, H. A., Kil, I. S., & Bae, S. H. (2012). Peroxiredoxin functions as a peroxidase and a regulator and sensor of local peroxides. *The Journal of Biological Chemistry*, *287*, 4403–4410.

Sanchez-Valle, V., Chavez-Tapia, N. C., Uribe, M., & Mendez-Sanchez, N. (2012). Role of oxidative stress and molecular changes in liver fibrosis: A review. *Current Medicinal Chemistry*, *19*, 4850–4860.

Sartoretto, J. L., Kalwa, H., Pluth, M. D., Lippard, S. J., & Michel, T. (2011). Hydrogen peroxide differentially modulates cardiac myocyte nitric oxide synthesis. *Proceedings of the National Academy of Sciences of the United States of America*, *108*, 15792–15797.

Sen, N., Paul, B. D., Gadalla, M. M., Mustafa, A. K., Sen, T., Xu, R., et al. (2012). Hydrogen sulfide-linked sulfhydration of NF-kappaB mediates its antiapoptotic actions. *Molecular Cell*, *45*, 13–24.

Soriano, F. X., Leveille, F., Papadia, S., Higgins, L. G., Varley, J., Baxter, P., et al. (2008). Induction of sulfiredoxin expression and reduction of peroxiredoxin hyperoxidation by the neuroprotective Nrf2 activator 3H-1,2-dithiole-3-thione. *Journal of Neurochemistry*, *107*, 533–543.

Szabo, G., Wands, J. R., Eken, A., Osna, N. A., Weinman, S. A., Machida, K., et al. (2010). Alcohol and hepatitis C virus—Interactions in immune dysfunctions and liver damage. *Alcoholism, Clinical and Experimental Research*, *34*, 1675–1686.

Tello, D., Tarin, C., Ahicart, P., Breton-Romero, R., Lamas, S., & Martinez-Ruiz, A. (2009). A "fluorescence switch" technique increases the sensitivity of proteomic detection and identification of S-nitrosylated proteins. *Proteomics*, *9*, 5359–5370.

Trujillo, M., Ferrer-Sueta, G., Thomson, L., Flohe, L., & Radi, R. (2007). Kinetics of peroxiredoxins and their role in the decomposition of peroxynitrite. *Sub-Cellular Biochemistry*, *44*, 83–113.

Wei, Q., Jiang, H., Xiao, Z., Baker, A., Young, M. R., Veenstra, T. D., et al. (2011). Sulfiredoxin-Peroxiredoxin IV axis promotes human lung cancer progression through modulation of specific phosphokinase signaling. *Proceedings of the National Academy of Sciences of the United States of America*, *108*, 7004–7009.

Weischenfeldt, J., & Porse, B. (2008). Bone marrow-derived macrophages (BMM): Isolation and applications. *CSH Protocols*, pdb prot5080.

Wink, D. A., Miranda, K. M., Espey, M. G., Pluta, R. M., Hewett, S. J., Colton, C., et al. (2001). Mechanisms of the antioxidant effects of nitric oxide. *Antioxidants & Redox Signaling*, *3*, 203–213.

Wood, Z. A., Poole, L. B., & Karplus, P. A. (2003). Peroxiredoxin evolution and the regulation of hydrogen peroxide signaling. *Science*, *300*, 650–653.

Wu, C., Liu, T., Chen, W., Oka, S., Fu, C., Jain, M. R., et al. (2010). Redox regulatory mechanism of transnitrosylation by thioredoxin. *Molecular & Cellular Proteomics*, *9*, 2262–2275.

Yang, K. S., Kang, S. W., Woo, H. A., Hwang, S. C., Chae, H. Z., Kim, K., et al. (2002). Inactivation of human peroxiredoxin I during catalysis as the result of the oxidation of the catalytic site cysteine to cysteine-sulfinic acid. *The Journal of Biological Chemistry*, *277*, 38029–38036.

Yoo, S. K., Starnes, T. W., Deng, Q., & Huttenlocher, A. (2011). Lyn is a redox sensor that mediates leukocyte wound attraction in vivo. *Nature*, *480*, 109–112.

CHAPTER SEVEN

Glutathione and γ-Glutamylcysteine in Hydrogen Peroxide Detoxification

Ruben Quintana-Cabrera, Juan P. Bolaños[1]

Institute of Functional Biology and Genomics (IBFG), Department of Biochemistry and Molecular Biology, University of Salamanca-CSIC, Salamanca, Spain

[1]Corresponding author: e-mail address: jbolanos@usal.es

Contents

1. Introduction 130
2. Materials 131
3. Previous Considerations 132
   3.1 Buffer conditions 132
   3.2 Substrates and enzymes 132
4. Procedure with Purified Enzymes and Substrates 133
   4.1 Preparation of reactants 134
   4.2 Reaction procedure and reading 135
5. Analysis in Biological Samples 137
   5.1 Analysis of γGC antioxidant activity 137
   5.2 Samples collection and H_2O_2 determination 139
6. Further Applications: H_2O_2 Produced by NOS 140
7. Conclusions 140
Acknowledgments 141
References 141

Abstract

Hydrogen peroxide (H_2O_2) is an important regulator of cell redox status and signaling pathways. However, if produced in excess, it can trigger oxidative damage, which can be counteracted by the antioxidant systems. Amongst these, the glutathione (GSH) precursor, γ-glutamylcysteine (γGC), has recently been shown to detoxify H_2O_2 in a glutathione peroxidase-1 (GPx1)-dependent fashion. To analyze how both γGC and GSH reduce H_2O_2, we have taken advantage of a colorimetric assay that allows simple and reliable quantification of H_2O_2 in the micromolar range. Whereas most assays rely on coupled enzymatic reactions, this method determines the formation of a ferric thiocyanate derivative after direct Fe^{2+} oxidation by H_2O_2. Here, we detail the procedure and considerations to determine H_2O_2 reduction by both γGC and GSH, either from cell samples or *in vitro* reactions with purified enzymes from GSH metabolism.

Methods in Enzymology, Volume 527
ISSN 0076-6879
http://dx.doi.org/10.1016/B978-0-12-405882-8.00007-6

1. INTRODUCTION

Mitochondrial O_2 consumption during oxidative phosphorylation is a vital process required for energy production and cell survival (Saraste, 1999). Coupled to this process, reactive oxygen species are formed due to the incomplete reduction of O_2 to superoxide anion ($O_2^{\cdot -}$) (Fernandez-Fernandez, Almeida, & Bolanos, 2012). Hydrogen peroxide (H_2O_2), formed by $O_2^{\cdot -}$ dismutation, is considered a major player in the cell redox reactions. In excess, H_2O_2 can oxidize macromolecules likely compromising their function (Avery, 2011); however, in low—physiological—concentrations, it acts as an important intracellular messenger (D'Autreaux & Toledano, 2007). Cells require, therefore, efficient enzymatic systems to tightly regulate their H_2O_2 levels.

Glutathione peroxidase (GPx) accounts for a large proportion of cellular H_2O_2 reduction (Toppo, Flohe, Ursini, Vanin, & Maiorino, 2009), which takes place at the expense of the oxidation of glutathione (GSH), the main antioxidant thiol derivative (Dringen, 2000). However, several authors have suggested the ability of γ-glutamylcysteine (γGC)—the immediate biosynthetic GSH precursor—to exert antioxidant functions (Faulkner, Veeravalli, Gon, Georgiou, & Beckwith, 2008; Grant, MacIver, & Dawes, 1997; Ristoff et al., 2002), although its ability to reduce H_2O_2 by acting as a GPx1 cofactor was only recently demonstrated (Quintana-Cabrera et al., 2012). Tissue levels of γGC are normally very low, unless its conversion to GSH is impaired (Faulkner et al., 2008) such as that occurring during glutathione synthetase (GSS) deficiency (Njalsson, 2005; Ristoff et al., 2002). Analyzing the ability of tissues to use γGC as a GPx1 cofactor in H_2O_2 detoxification is, therefore, of interest to gain a better understanding of the thiol antioxidant system status under different pathophysiological circumstances.

To quantify H_2O_2, traditional methods are commonly based in *in vitro* colorimetric, luminometric, or fluorimetric assays (Desagher, Glowinski, & Premont, 1996; Dringen, Kussmaul, & Hamprecht, 1998; Rapoport, Hanukoglu, & Sklan, 1994; Rhee, Chang, Jeong, & Kang, 2010). To counteract some of their limitations, in the past few years, several techniques have been developed for *in vitro* and *in vivo* H_2O_2 determinations. Emerging genetically encoded redox sensor proteins have gained special attention, allowing real-time dynamic measurements of H_2O_2 levels in different cell compartments (Belousov et al., 2006; Cannon & Remington, 2008; Roma et al., 2012).

These techniques are commonly based on a ratiometric quantification of the sensor-derived fluorescence after selective oxidation by H_2O_2 (Rhee et al., 2010; Roma et al., 2012). Despite the advantages, they have minor limitations such as the influence of pH, delayed response time, or high reducing midpoint potential (Cannon & Remington, 2008; Roma et al., 2012). One should also be aware of the fact that the expression of high levels of green fluorescent protein-based redox sensors can increase the intracellular levels of H_2O_2 (Rhee et al., 2010).

In view of the normally low γGC concentrations, the analysis of its antioxidant activity in a physiological context may not be an easy task. Therefore, genetic tools may be required to increase γGC concentration in the cell to measure subsequent changes in H_2O_2 levels. In addition, an approach using purified enzymes complements the kinetic analysis of H_2O_2 reduction by γGC or GSH. Accordingly, we have adapted a protocol first published by Hildebrandt, Roots, Tjoe, and Heinemeyer (1978) for H_2O_2 quantification in either cell extracts or *in vitro*-catalyzed reactions. The proposed colorimetric assay allows a fast—just 10 min incubation period required—and reliable quantification of peroxide concentrations. In kinetic analysis, reactions can be finished at the desired time points by simply adding concentrated HCl. H_2O_2 detection is based on Fe^{2+} oxidation and subsequent formation of a ferric thiocyanate derivative, registering its absorbance at 492 nm. In sum, here we detail a procedure to quantify H_2O_2 reduction by γGC or GSH in the presence of purified GPx1. We also refer further hallmarks and considerations for the analysis of the role of γGC and GSH in H_2O_2 detoxification in a more physiological, cellular context. Finally, the application of the method for the quantification of H_2O_2 in other enzymatic reactions, such as H_2O_2 derived from brain nitric oxide synthase (NOS) activity, is also proposed.

2. MATERIALS

Reduced γGC [(des-Gly)-glutathione, ammonium salt] can be purchased from Bachem (Ref.: G4305; Bachem, Germany), reduced GSH (Ref.: G4251), NADPH (Ref.: N7505), GPx1 from bovine erythrocytes (EC 1.11.1.9; Ref.: G6137), and glutathione reductase (GSR) from baker's yeast (EC 1.6.4.2; Ref.: G3664), from Sigma. All other chemicals can be obtained at analytical grade from Sigma, namely 30% (w/v) H_2O_2, ammonium ferrous sulfate hexahydrate [$(NH_4)_2Fe(SO_4)_2$], potassium thiocyanate (KSCN), and sodium phosphate ($NaHPO_4$). Genetic downregulation, by small interference RNA (siRNA), of GSS, glutamate–cysteine ligase (GCL), catalytic subunit, GSR, and GPx1, has been previously described for both human and rat (Quintana-Cabrera et al., 2012). siRNAs can be purchased from Dharmacon (Abgene, Thermo Fisher, Epsom, UK) and 10-kDa molecular mass cutoff devices from Amicon (Millipore). Pharmacological GCL inhibitor L-buthionine sulfoximine (L-BSO; L-buthionine(*S*,*R*)-sulfoximine; Ref.: B2640) and acivicin (Ref.: A2295) can be purchased from Sigma. In reactions involving NOS, catalase from bovine liver (Ref.: C9322) and all other reactants were obtained at analytical grade from Sigma. Absorbance at 492 nm was recorded using a Fluoroskan Ascent FL (Thermo Scientific, Rockford, IL, USA) fluorimeter in 96-well plates, which can be purchased from Nunc (Roskilde, Denmark).

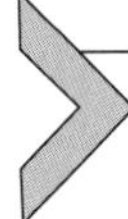

3. PREVIOUS CONSIDERATIONS

3.1. Buffer conditions

To measure H_2O_2 reduction by γGC or GSH, we have used a phosphate buffered saline (PBS; 136 m*M* NaCl, 2.7 m*M* KCl, 7.8 m*M* Na_2HPO_4, 1.7 m*M* KH_2PO_4, pH 7.4) solution. However, other nonoxidizing buffer conditions can be used in determinations with different redox couples, depending on the specific requirements. The buffer should avoid containing any ferric anion binding or chelating compound, such as desferrioxamine, diethylenetriaminepentaacetic acid, or ethylenediaminetetraacetic acid (EDTA) (Rhee et al., 2010); however, if required, the latest could be used at concentrations up to 0.5 m*M*. These considerations are of special relevance when using noncommercial purified enzymes. NADPH(H^+) should be added in reactions including GSR in order to regenerate GSH from its oxidized form (GSSG). In view of its rapid degradation at room temperature, fresh NADPH(H^+) solutions should be kept on ice and added to the buffer extemporarily. Buffer pH should be checked to avoid any possible loss of reducing capacity of NADPH(H^+) at pH 7. If necessary, NADPH(H^+) degradation can be easily determined by measuring the absorbance ratios A_{260}/A_{340} and A_{260}/A_{290}, corresponding to values of ~2.3 and >10, respectively, for pure preparations (Wu, Wu, & Knight, 1986).

3.2. Substrates and enzymes

γGC and GSH are commercially available both in reduced or oxidized forms. We have performed reactions with reduced γGC or GSH in the presence of GPx. However, oxidized derivatives can also be used to determine GSR activity in coupled reactions. Working solutions of γGC or GSH can be kept up to 7 days at 2–8 °C for temporary storage; however, we highly recommend preparing fresh solutions to prevent spontaneous oxidation. For long-term storage, reconstitute solid γGC or GSH in PBS and freeze in aliquots at −80 °C. To avoid misleading results due to the presence of salts in formulations, a molecular weight of 268.90 g/mol for γGC and 307.32 g/mol for GSH should be considered for the correct extrapolation of results.

GPx (Awasthi, Dao, Lal, & Srivastava, 1979; Maddipati & Marnett, 1987; Mills, 1959) or GSR (Carlberg & Mannervik, 1977; Mavis & Stellwagen, 1968) can be purified when analysis of specific activities from tissue or cell extracts is required. Otherwise, commercial GPx1 from bovine erythrocytes is suitable for the assay. Once reconstituted in PBS, aliquots of the enzyme solutions can be frozen at −20 °C. Addition of 1 mg/ml of IgG or 1 m*M* dithiothreitol (DTT) may help to stabilize the enzyme in diluted solutions and protect sulfhydryl groups, respectively. Note that one enzyme unit (U) catalyzes the oxidation of 1 μmol/min of GSH by H_2O_2 at pH 7.0 and 25 °C. pH and temperature strongly modulate GPx activity, reaching its maximum effect at pH 8.8 and 50 °C, when GPx1 activity is 10 times higher than that at pH 7. However, these optimal conditions for GPx1 may result in thiol and other reactants instability, as well as nonenzymatic oxidations. The GSR used in our assay is commercially available as ammonium sulfate solution; if this was not available, reconstitute lyophilized enzyme in potassium phosphate buffer (100 m*M* KH_2PO_4, 1 m*M* EDTA, 0.1 m*M* DTT, 38 mg/ml trehalose buffer, pH 7.5) and store aliquots of the reconstituted enzyme at −20 °C; one GSR enzyme unit catalyzes the reduction of 1 μmol/min of GSSG at pH 7.6 and 25 °C.

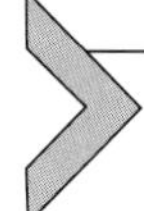

4. PROCEDURE WITH PURIFIED ENZYMES AND SUBSTRATES

Determination of H_2O_2 reduction by γGC or GSH is based on a protocol described by Hildebrandt et al. (1978) adapted for a microtiter plate level. This method allows the direct quantification of the peroxide by measuring ferric thiocyanate absorbance at 492 nm, after Fe^{2+} oxidation by H_2O_2 and subsequent reaction with KSCN. Unless otherwise indicated, incubations should be performed in the presence of GPx1 from bovine erythrocytes—the absence of the enzyme results in complete loss of H_2O_2 reduction by γGC or GSH (Quintana-Cabrera et al., 2012).

4.1. Preparation of reactants

The following steps describe the preparation of reactants, as calculated in excess for a single determination in a 96-well dish. For longer preparations, scale up amounts and volumes proportionally.

1. Prepare at least 20 ml PBS adjusted to pH 7.4. NADPH(H^+) should be added to a final concentration of 0.2 m*M* in the reaction buffer if GSR is included in the assay.
2. Dilute γGC or GSH in 5 ml PBS to obtain a twofold (2 ×) solution of up to 5 m*M*; keep the solutions in ice until the start of the reactions.
3. Prepare 4 ml of a 400 μ*M* H_2O_2 (4 ×) master solution from 30% (w/v) commercial stock (~8.82 m*M*; MW = 34.01 g/mol). Perform a H_2O_2 standard curve ranging from 0 to 120 μ*M* final concentrations, as shown in Table 7.1, calculated to make triplicates. Thereafter, prepare 5 ml of 200 μ*M* H_2O_2 by diluting 1:2 the master solution; this should be used as the working solution in the reactions including the antioxidants.
4. For a 96-well dish, reconstitute GPx1 and dilute it to 5×10^{-3} U/ml (2 ×) in 1.6 ml PBS for the H_2O_2 concentration curve (R1) and in 5 ml γGC or GSH solution (R2), prepared in step 2.
5. In reactions including GSR, add the enzyme to obtain 5 U/ml (2 ×) in the solutions also containing GPx1.
6. Make 2.5 ml aqueous solutions of both 25.6 m*M* $(NH_4)_2Fe(SO_4)_2$ and 1.44 *M* KSCN. These solutions should be mixed at equal amounts (R3) just before the readings, giving concentrations of 3.2 m*M* $(NH_4)_2Fe(SO_4)_2$ and 180 m*M* KSCN in the final reaction buffer.

Table 7.1 Preparation of H_2O_2 standard curve

H_2O_2 concentration curve

400 μ*M* H_2O_2 (μl)	PBS or reaction buffer (μl)	H_2O_2 in final reaction volume (μ*M*)
0	200	0
10	190	10
20	180	20
40	160	40
60	140	60
80	120	80
100	100	100
120	80	120

7. Prepare at least 5 ml of concentrated (∼11.8 *M*) HCl to stop reactions at the desired time points.

4.2. Reaction procedure and reading

Reactions should be performed at room temperature by using a multichannel pipette to simultaneously start the reactions in triplicate. Perform reactions under a chemical hood to avoid possible inhalation of HCl. Proceed with the following steps to measure H_2O_2-dependent formation of ferric thiocyanate (Fig. 7.1B).

1. For the calibration curve, add 50 μl/well of each H_2O_2 concentration (Table 7.1) plus 50 μl of the reaction buffer containing the enzyme/s (R1).
2. Perform triplicates of 50 μl/well of γGC or GSH at a fixed concentration (R2) for each time point to be analyzed (e.g., 0–5 min). Note that the enzymes are already included in these solutions.
3. Add 50 μl/well of concentrated HCl to both H_2O_2 standards and time "0" wells.
4. Start reactions by adding 50 μl of 200 μ*M* H_2O_2 to sample wells. We recommend starting the reactions for each time point separated by a fixed

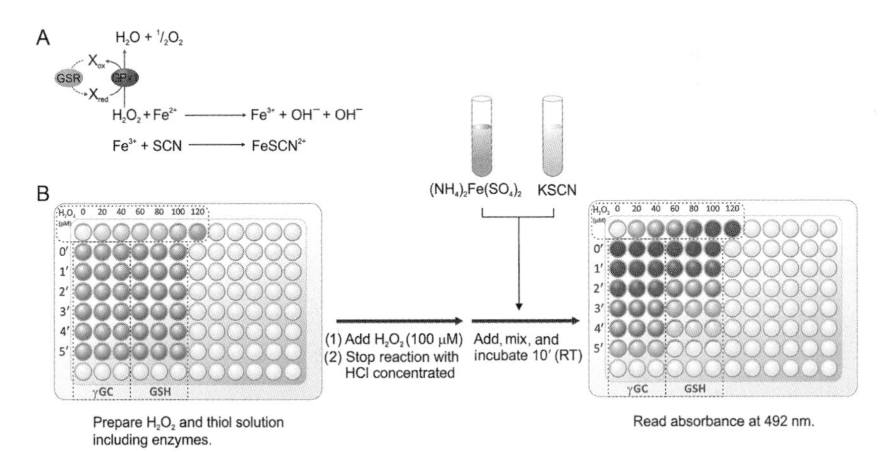

Figure 7.1 (A) H_2O_2 quantification is based on the formation of ferric thiocyanate ($FeSCN^{2+}$) in the presence of KSCN after Fe^{2+} oxidation by the peroxide. Gray lines represent the reduction of H_2O_2 by GPx1 at the expense of the thiolic compounds; X, GSH, or γGC. GSR only regenerates GSH (dotted line). (B) Schematic schedule of the assay to analyze the reduction of 100 μ*M* H_2O_2 by a fixed concentration of γGC or GSH. Triplicates for each time point ranging from 1 to 5 min are represented. (See Color Insert.)

interval (i.e., 10 s); this delay should be considered to stop the reactions at the precise time points required.

5. Finish the reactions by adding 50 μl/well of concentrated HCl.
6. Mix equal volumes of $(NH_4)_2Fe(SO_4)_2$ and KSCN (R3) and add 50 μl to both samples and H_2O_2 standards. A colored ferric thiocyanate derivative will be formed immediately after addition to wells. To obtain stable derivatives, incubate for 10 min at room temperature avoiding light exposure.
7. Read absorbance at 492 nm on a multiwell reader spectrophotometer.

H_2O_2 concentration is calculated by extrapolating the read absorbance to the peroxide standard curve. As shown in Fig. 7.2A, the increase in the absorbance (ΔA) ranged from 0.09 ± 0.008 to 0.57 ± 0.028, and it was linear ($R^2 = 0.9986$) from 0 to 120 μ*M* H_2O_2 final concentration in reaction buffer. The peroxide levels are stable during the analysis, both in the presence or absence of γGC or GSH alone (Fig. 7.2B, left panel). Accordingly, the addition of GPx1 is essential for H_2O_2 clearance by γGC or GSH (Fig. 7.2B, left panel), hence indicating the absence of spontaneous reduction by the antioxidants in the proposed assay conditions. Addition of GSR does not change H_2O_2 concentrations, even in the presence of thiols (Fig. 7.2B, right panel); this enzyme is used to regenerate GSH—not γGC—from its oxidized form, thus increasing the rate of H_2O_2 reduction

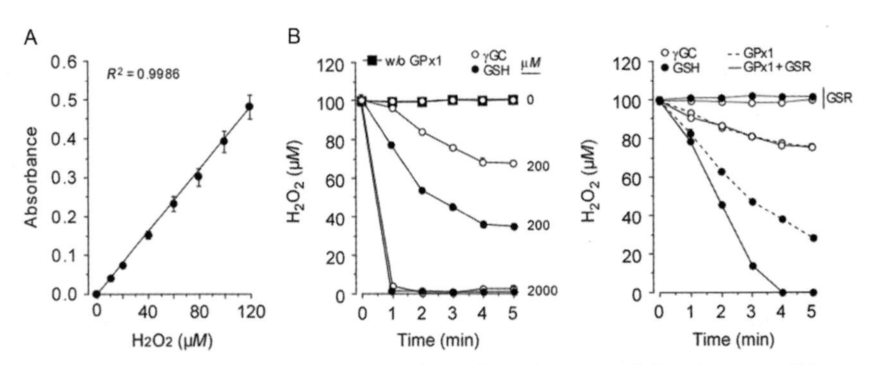

Figure 7.2 (A) H_2O_2 concentration curve shows linear increase of absorbance at 492 nm in the range of 0–120 μ*M* final concentrations used in the assay; R^2, linear coefficient. (B) Concentration-dependent reduction of 100 μ*M* H_2O_2 by γGC and GSH in the absence or presence of GPx1; when thiols and/or GPx1 are not included, H_2O_2 concentration remains constant in the assay conditions (left panel). Addition of GSR increases the rate of H_2O_2 reduction by 200 μ*M* GSH but not that by γGC. The only presence of the enzyme has no effect on H_2O_2 levels (right panel). All data are expressed as mean values ± SEM of four independent experiments.

(Fig. 7.2B, right panel). It is of note that, at 2 m*M*—the highest final concentration of the antioxidants in reaction buffer—100 μ*M* H_2O_2 is completely reduced by GPx1 within the first minute (Fig. 7.2B, left panel). Therefore, we recommend using intermediate concentrations of ~200 μ*M* γGC or GSH for analysis in the proposed conditions.

5. ANALYSIS IN BIOLOGICAL SAMPLES

The previous steps are reliable for the use of purified enzymes and antioxidants in kinetic analysis. However, this protocol can be used for the assessment of γGC or GSH in cell or tissue homogenates. Nonetheless, a few considerations should be taken into account to avoid possible interferences, for example, the potential problem of the coexistence of both γGC and GSH in the sample, their stability, or the presence of ferric compounds. Conditions for the specific analysis of γGC or GSH are therefore detailed.

5.1. Analysis of γGC antioxidant activity

In a physiological context, γGC is almost absent in the cell due to its rapid conversion into GSH by GSS (Ristoff et al., 2002). Genetic or pharmacological approaches, such as those suggested below, may be useful to increase γGC abundance, and the subsequent changes in H_2O_2 levels are then measured. Quantification of γGC levels—that is, by HPLC (Gegg, Clark, & Heales, 2002) or UPLC–MS/MS (New & Chan, 2008; Quintana-Cabrera et al., 2012)—are complementary tools that might be used for proper interpretation of results.

1. Silencing or inhibition of enzymes from GSH metabolism. Expression of enzymes can be genetically downregulated by short hairpin or short interfering RNAs. A preliminary screening for protein levels abundance should be performed to ensure silencing. Interference RNA may result in increased cellular γGC after GSS downregulation (Ristoff et al., 2002; Quintana-Cabrera et al., 2012). L-BSO, a well-known pharmacological inhibitor of GCL activity (Griffith, 1982), can be used to deplete both γGC and GSH. Of note, L-BSO can also inhibit γ-glutamyltransferase (GGT) (Griffith, Bridges, & Meister, 1979). 0.1–1 m*M* L-BSO can be added to the media of cells in culture at least 24 h before analysis; this will ensure >80% GSH depletion (Vasquez, Almeida, & Bolanos, 2001). To avoid degradation of either γGC or GSH by GGT—placed in the extracellular side of the plasma membrane (Griffith & Meister, 1980)—50 μ*M* acivicin should be added in the incubation buffer for a minimum of 2 h

prior the assay (Dringen, Kranich, & Hamprecht, 1997). All mentioned treatments, either alone or combined, should increase the concentrations of the endogenous antioxidants. To further help depleting GSH content cells media—commonly DMEM—may not be supplemented with serum 24 h previous to analysis or while incubating with inhibitors.

2. Targeted synthesis of γGC to mitochondria. Addition of a mitochondrial import sequence—e.g., that of ornithine transcarbamylase (Gen Bank Accession number NM_000531) (Horwich, Kalousek, Mellman, & Rosenberg, 1985)—in the N-terminal domain of GCL (catalytic subunit; Gen Bank Accession number NM_012815) targets its expression to mitochondria (Quintana-Cabrera et al., 2012). As a result, γGC is synthesized without further conversion to GSH, since mitochondria lacks GSS (Mari, Morales, Colell, Garcia-Ruiz, & Fernandez-Checa, 2009). In view that GPx1 is the only known enzyme to reduce H_2O_2 by using GSH or γGC in mitochondria (Quintana-Cabrera et al., 2012; Toppo et al., 2009), any changes in the peroxide level should be representative of GPx1 activity in the organelle. Depletion of GSH by genetic downregulation of cytosolic, but not mitochondrial GCL, will reduce interferences in the analysis of γGC reducing activity. To this end, silent third codon-base point mutations can be inserted in the protein sequence recognized by siRNA (Quintana-Cabrera et al., 2012), thus expressing an ectopic siRNA-resistant GCL in mitochondria.
3. Addition of exogenous γGC or GSH. In cell or tissue lysates, it is possible to analyze the reduction of H_2O_2 by adding a known concentration of γGC or GSH. 50–220 μ*M* Acivicin should be included in lysates to avoid thiols degradation. For experiments in intact cells, ester derivatives would be required (Colell, Fernandez, & Fernandez-Checa, 2009; Drake, Kanski, Varadarajan, Tsoras, & Butterfield, 2002) since native thiols do not enter the cells due to GGT activity (Dringen, Gutterer, & Hirrlinger, 2000; Lok et al., 2011). Genetic downregulation or pharmacological inhibition of enzymes will avoid further degradation or new synthesis of γGC or GSH.

H_2O_2 levels in cells media can be determined as described in the next section. Alternatively, cell or tissue homogenates can be used as a source of γGC or GSH, in a similar fashion to that described for purified substrates. Reactions can be performed in the presence of purified GPx1, hence determining H_2O_2-dependent reduction by endogenous thiols. To obtain significant amounts of γGC, we recommend using a large cells sample, usually ranging $1–2 \times 10^9$ cells. Wash with excess cold PBS, scrape and resuspend cells in

PBS; then, freeze–thaw cells three times using liquid nitrogen. If tissues are used, cut them in small pieces and wash with cold PBS; resuspend up to 10 μl/mg tissue in PBS and homogenize it with 20 strokes at 1600 r.p.m. in an ice-chilled tight fitting glass-teflon homogenizer. To completely break down the cells, samples can be frozen–thawed three times in liquid nitrogen. Either cell or tissue lysates are finally filtered through a 10-KDa molecular weight mass cutoff device at 12,000 *g* for 15 min at 4 °C. The filtered fraction represents the γGC or GSH samples, whereas the retained fraction can be used as protein templates to measure enzymes activity.

5.2. Samples collection and H_2O_2 determination

H_2O_2 is a diffusible molecule (Rhee et al., 2010) also found in the extracellular medium. Thus, it is possible to determine H_2O_2 concentrations in cells media to evaluate its detoxification by γGC or GSH. The experimental settings will include the determination of samples at concentrations optimized to work in the millimolar range for H_2O_2. The method is set up to obtain a sensitivity in the assay enough to detect H_2O_2 in cells media when plated in 12- or 6-well dishes.

1. Prepare a H_2O_2 standard curve in Hank's buffer (134.2 m*M* NaCl, 5.26 m*M* KCl, 0.43 m*M* KH_2PO_4, 4.09 m*M* $NaHCO_3$, 0.33 m*M* Na_2HPO_4, 5.44 m*M* glucose, 20 m*M* HEPES, 4 m*M* $CaCl_2$, pH 7.4), as detailed in Table 7.1. Complete to 100 μl with Hank's buffer. Add 50 μl of concentrated (11.8 m*M*) HCl to both H_2O_2 standard curve and sample wells.
2. Wash plated cells twice with prewarmed Hank's buffer. Incubate them at 37 °C in a 5% CO_2 atmosphere in the usual volumes of media containing 100 μ*M* H_2O_2. No H_2O_2 will be added if physiological concentrations of the peroxide are to be measured.
3. At the desired time points, collect up to 100 μl of media or adjust to this final volume with Hank's buffer when using smaller amounts of samples. Immediately, add this media to wells containing HCl; otherwise, keep samples on ice until the measurement is performed.
4. Add 50 μl of $(NH_4)_2Fe(SO_4)_2$ and KSCN mix (R3) to both samples and standard curve. Incubate for 10 min at room temperature, protected from light exposure.
5. Read absorbance at 492 nm.

While Fe^{2+} oxidizing or chelating compounds can be present in cell or tissue homogenates, it is unlikely their release to extracellular medium from cultured cells (Dringen et al., 1998). However, the addition of up to 1 m*M* EDTA diK^+ to Hank's buffer can be used to counteract the interferences by iron compounds.

6. FURTHER APPLICATIONS: H_2O_2 PRODUCED BY NOS

The proposed method can be further employed in enzymatic determinations involving either H_2O_2 production or detoxification. Accordingly, Heinzel, John, Klatt, Bohme, and Mayer (1992) adapted the protocol to analyze the formation of H_2O_2 by purified (Mayer, John, & Bohme, 1990) brain NOS. NOS-derived production of H_2O_2 is measured after incubating 1 μg purified NOS in 100 μl reaction buffer (0.1 m*M* NADPH(H^+), 3 μ*M* $CaCl_2$, 2 μ*M* calmodulin, 200 m*M* HEPES, pH 7). NOS activity can be also performed with cytosolic extracts obtained from activated astrocytes. Lyse astrocytes in H_2O and centrifuge at 12,000 *g* for 15 min at 4 °C keeping supernatants. Mix equal amounts of cytosol and reaction buffer to a final volume of 100 μl per well. H_2O_2 standard curve is prepared as in Section 4.1 (see also Table 7.1). Incubations are performed at 37 °C and finished by addition of 50 μl concentrated HCl. Note that constant production of H_2O_2 by brain NOS usually lasts for 10–15 min (Heinzel et al., 1992). Absorbance is determined after 10 min incubation with 50 μl of the $(NH_4)_2Fe(SO_4)_2$ plus KSCN (R3) solution, as in step 7 from Section 4.2. As negative controls, 125×10^{-3} U/ml catalase or 100 μ*M* L-arginine can be used to reduce H_2O_2 or inhibit its production by NOS, respectively. NOS-dependent H_2O_2 production does not rely on $O_2^{\cdot-}$ formation; conversely, it relies on calmodulin, arginine, and tetrahydrobiopterin concentrations; depending on these factors, the synthesis of citrulline is not fully coupled with that of NO (Heinzel et al., 1992). Thus, 1 m*M* of the NOS inhibitor N^{ω}-monomethyl-L-arginine (NMMA) can also be employed to inhibit the production of citrulline in favor of H_2O_2 formation.

7. CONCLUSIONS

Spectrophotometric quantification of H_2O_2 described herein allows fast, simple, and reliable determination of γGC or GSH antioxidant activity in the presence of GPx1. Protocols used for *in vitro* H_2O_2 quantification usually rely on coupled enzymatic reactions to form chemiluminescent, fluorescent, or colored derivatives (Dringen et al., 1998; Rapoport et al., 1994; Rhee et al., 2010). Whilst, in these determinations, horseradish peroxidase has been widely used (Barja, 2002; Desagher et al., 1996; Pick & Mizel, 1981), and it is not exempt of artifacts (Van Scott, Miles, & Castranova, 1984). In contrast, our proposed method relies on direct Fe^{2+} oxidation by H_2O_2 to form a thiocyanate derivative. Other colorimetric protocols—that is, Fe^{3+} binding to xylenol orange—are also directly primed by H_2O_2. However, longer incubation times are required to form colored derivatives (Dringen et al., 1998). The proposed method does not allow continuous determination; instead, it is possible to stop reactions at demand and measure H_2O_2 concentration at final time points. Furthermore, it is possible to quantify H_2O_2 levels in cellular or tissue homogenates or use them as templates to perform *in vitro* reactions. These approaches account for analysis of γGC and GSH antioxidant role in several paradigms, as proposed herein.

The versatility of the assay allows *in vitro* quantification of H_2O_2 reduction by other redox couples and enzymes. For instance, it has been adapted to quantify H_2O_2 production by NOS activity in brain tissue (Heinzel et al., 1992). Xantine/xantine oxidase production of H_2O_2 and/or $O_2^{\cdot -}$ (Kelley et al., 2010), or $O_2^{\cdot -}$ reduction to H_2O_2 by superoxide dismutase (McCord & Fridovich, 1969) could be also considered as potential candidates for determination with this protocol. Furthermore, the method is also suitable to determine GPx activity, regardless the classical measurement of NADPH(H^+) oxidation from the coupled reaction with GSR (Handy et al., 2009). In conclusion, determination of ferric thiocynate formation by H_2O_2 is a recommended method for *in vitro* quantification of γGC and GSH antioxidant activity.

ACKNOWLEDGMENTS

J. P. B. is funded by the Ministerio de Economía y Competitividad (SAF2010-20008), European Regional Development Fund, Instituto de Salud Carlos III (RETICEF, RD12/0043/0021), and Junta de Castilla y Leon (SA112A12-2).

REFERENCES

Avery, S. V. (2011). Molecular targets of oxidative stress. *The Biochemical Journal*, *434*, 201–210.

Awasthi, Y. C., Dao, D. D., Lal, A. K., & Srivastava, S. K. (1979). Purification and properties of glutathione peroxidase from human placenta. *The Biochemical Journal*, *177*, 471–476.

Barja, G. (2002). The quantitative measurement of H_2O_2 generation in isolated mitochondria. *Journal of Bioenergetics and Biomembranes*, *34*, 227–233.

Belousov, V. V., Fradkov, A. F., Lukyanov, K. A., Staroverov, D. B., Shakhbazov, K. S., Terskikh, A. V., et al. (2006). Genetically encoded fluorescent indicator for intracellular hydrogen peroxide. *Nature Methods*, *3*, 281–286.

Cannon, M. B., & Remington, S. J. (2008). Redox-sensitive green fluorescent protein: Probes for dynamic intracellular redox responses. A review. *Methods in Molecular Biology, 476*, 51–65.
Carlberg, I., & Mannervik, B. (1977). Purification by affinity chromatography of yeast glutathione reductase, the enzyme responsible for the NADPH-dependent reduction of the mixed disulfide of coenzyme A and glutathione. *Biochimica et Biophysica Acta, 484*, 268–274.
Colell, A., Fernandez, A., & Fernandez-Checa, J. C. (2009). Mitochondria, cholesterol and amyloid beta peptide: A dangerous trio in Alzheimer disease. *Journal of Bioenergetics and Biomembranes, 41*, 417–423.
D'Autreaux, B., & Toledano, M. B. (2007). ROS as signalling molecules: Mechanisms that generate specificity in ROS homeostasis. *Nature Reviews. Molecular Cell Biology, 8*, 813–824.
Desagher, S., Glowinski, J., & Premont, J. (1996). Astrocytes protect neurons from hydrogen peroxide toxicity. *The Journal of Neuroscience, 16*, 2553–2562.
Drake, J., Kanski, J., Varadarajan, S., Tsoras, M., & Butterfield, D. A. (2002). Elevation of brain glutathione by gamma-glutamylcysteine ethyl ester protects against peroxynitrite-induced oxidative stress. *Journal of Neuroscience Research, 68*, 776–784.
Dringen, R. (2000). Metabolism and functions of glutathione in brain. *Progress in Neurobiology, 62*, 649–671.
Dringen, R., Gutterer, J. M., & Hirrlinger, J. (2000). Glutathione metabolism in brain metabolic interaction between astrocytes and neurons in the defense against reactive oxygen species. *European Journal of Biochemistry, 267*, 4912–4916.
Dringen, R., Kranich, O., & Hamprecht, B. (1997). The gamma-glutamyl transpeptidase inhibitor acivicin preserves glutathione released by astroglial cells in culture. *Neurochemical Research, 22*, 727–733.
Dringen, R., Kussmaul, L., & Hamprecht, B. (1998). Detoxification of exogenous hydrogen peroxide and organic hydroperoxides by cultured astroglial cells assessed by microtiter plate assay. *Brain Research. Brain Research Protocols, 2*, 223–228.
Faulkner, M. J., Veeravalli, K., Gon, S., Georgiou, G., & Beckwith, J. (2008). Functional plasticity of a peroxidase allows evolution of diverse disulfide-reducing pathways. *Proceedings of the National Academy of Sciences of the United States of America, 105*, 6735–6740.
Fernandez-Fernandez, S., Almeida, A., & Bolanos, J. P. (2012). Antioxidant and bioenergetic coupling between neurons and astrocytes. *The Biochemical Journal, 443*, 3–11.
Gegg, M. E., Clark, J. B., & Heales, S. J. (2002). Determination of glutamate-cysteine ligase (gamma-glutamylcysteine synthetase) activity by high-performance liquid chromatography and electrochemical detection. *Analytical Biochemistry, 304*, 26–32.
Grant, C. M., MacIver, F. H., & Dawes, I. W. (1997). Glutathione synthetase is dispensable for growth under both normal and oxidative stress conditions in the yeast Saccharomyces cerevisiae due to an accumulation of the dipeptide gamma-glutamylcysteine. *Molecular Biology of the Cell, 8*, 1699–1707.
Griffith, O. W. (1982). Mechanism of action, metabolism, and toxicity of buthionine sulfoximine and its higher homologs, potent inhibitors of glutathione synthesis. *The Journal of Biological Chemistry, 257*, 13704–13712.
Griffith, O. W., Bridges, R. J., & Meister, A. (1979). Transport of gamma-glutamyl amino acids: Role of glutathione and gamma-glutamyl transpeptidase. *Proceedings of the National Academy of Sciences of the United States of America, 76*, 6319–6322.
Griffith, O. W., & Meister, A. (1980). Excretion of cysteine and gamma-glutamylcysteine moieties in human and experimental animal gamma-glutamyl transpeptidase deficiency. *Proceedings of the National Academy of Sciences of the United States of America, 77*, 3384–3387.

Handy, D. E., Lubos, E., Yang, Y., Galbraith, J. D., Kelly, N., Zhang, Y. Y., et al. (2009). Glutathione peroxidase-1 regulates mitochondrial function to modulate redox-dependent cellular responses. *The Journal of Biological Chemistry, 284*, 11913–11921.

Heinzel, B., John, M., Klatt, P., Bohme, E., & Mayer, B. (1992). Ca^{2+}/calmodulin-dependent formation of hydrogen peroxide by brain nitric oxide synthase. *The Biochemical Journal, 281*(3), 627–630.

Hildebrandt, A. G., Roots, I., Tjoe, M., & Heinemeyer, G. (1978). Hydrogen peroxide in hepatic microsomes. *Methods in Enzymology, 52*, 342–350.

Horwich, A. L., Kalousek, F., Mellman, I., & Rosenberg, L. E. (1985). A leader peptide is sufficient to direct mitochondrial import of a chimeric protein. *The EMBO Journal, 4*, 1129–1135.

Kelley, E. E., Khoo, N. K., Hundley, N. J., Malik, U. Z., Freeman, B. A., & Tarpey, M. M. (2010). Hydrogen peroxide is the major oxidant product of xanthine oxidase. *Free Radical Biology & Medicine, 48*, 493–498.

Lok, J., Leung, W., Zhao, S., Pallast, S., van, L. K., Guo, S., et al. (2011). Gamma-glutamylcysteine ethyl ester protects cerebral endothelial cells during injury and decreases blood-brain barrier permeability after experimental brain trauma. *Journal of Neurochemistry, 118*, 248–255.

Maddipati, K. R., & Marnett, L. J. (1987). Characterization of the major hydroperoxide-reducing activity of human plasma. Purification and properties of a selenium-dependent glutathione peroxidase. *The Journal of Biological Chemistry, 262*, 17398–17403.

Mari, M., Morales, A., Colell, A., Garcia-Ruiz, C., & Fernandez-Checa, J. C. (2009). Mitochondrial glutathione, a key survival antioxidant. *Antioxidants & Redox Signaling, 11*, 2685–2700.

Mavis, R. D., & Stellwagen, E. (1968). Purification and subunit structure of glutathione reductase from bakers' yeast. *The Journal of Biological Chemistry, 243*, 809–814.

Mayer, B., John, M., & Bohme, E. (1990). Purification of a Ca^{2+}/calmodulin-dependent nitric oxide synthase from porcine cerebellum. Cofactor role of tetrahydrobiopterin. *FEBS Letters, 277*, 215–219.

McCord, J. M., & Fridovich, I. (1969). Superoxide dismutase. An enzymic function for erythrocuprein (hemocuprein). *The Journal of Biological Chemistry, 244*, 6049–6055.

Mills, G. C. (1959). The purification and properties of glutathione peroxidase of erythrocytes. *The Journal of Biological Chemistry, 234*, 502–506.

New, L. S., & Chan, E. C. (2008). Evaluation of BEH C18, BEH HILIC, and HSS T3 (C18) column chemistries for the UPLC-MS-MS analysis of glutathione, glutathione disulfide, and ophthalmic acid in mouse liver and human plasma. *Journal of Chromatographic Science, 46*, 209–214.

Njalsson, R. (2005). Glutathione synthetase deficiency. *Cellular and Molecular Life Sciences, 62*, 1938–1945.

Pick, E., & Mizel, D. (1981). Rapid microassays for the measurement of superoxide and hydrogen peroxide production by macrophages in culture using an automatic enzyme immunoassay reader. *Journal of Immunological Methods, 46*, 211–226.

Quintana-Cabrera, R., Fernandez-Fernandez, S., Bobo-Jimenez, V., Escobar, J., Sastre, J., Almeida, A., et al. (2012). Gamma-glutamylcysteine detoxifies reactive oxygen species by acting as glutathione peroxidase-1 cofactor. *Nature Communications, 3*, 718.

Rapoport, R., Hanukoglu, I., & Sklan, D. (1994). A fluorimetric assay for hydrogen peroxide, suitable for NAD(P)H-dependent superoxide generating redox systems. *Analytical Biochemistry, 218*, 309–313.

Rhee, S. G., Chang, T. S., Jeong, W., & Kang, D. (2010). Methods for detection and measurement of hydrogen peroxide inside and outside of cells. *Molecules and Cells, 29*, 539–549.

Ristoff, E., Hebert, C., Njalsson, R., Norgren, S., Rooyackers, O., & Larsson, A. (2002). Glutathione synthetase deficiency: Is gamma-glutamylcysteine accumulation a way to cope with oxidative stress in cells with insufficient levels of glutathione? *Journal of Inherited Metabolic Disease*, *25*, 577–584.
Roma, L. P., Duprez, J., Takahashi, H. K., Gilon, P., Wiederkehr, A., & Jonas, J. C. (2012). Dynamic measurements of mitochondrial hydrogen peroxide concentration and glutathione redox state in rat pancreatic beta-cells using ratiometric fluorescent proteins: Confounding effects of pH with HyPer but not roGFP1. *The Biochemical Journal*, *441*, 971–978.
Saraste, M. (1999). Oxidative phosphorylation at the fin de siecle. *Science*, *283*, 1488–1493.
Toppo, S., Flohe, L., Ursini, F., Vanin, S., & Maiorino, M. (2009). Catalytic mechanisms and specificities of glutathione peroxidases: Variations of a basic scheme. *Biochimica et Biophysica Acta*, *1790*, 1486–1500.
Van Scott, M. R., Miles, P. R., & Castranova, V. (1984). Direct measurement of hydrogen peroxide release from rat alveolar macrophages: Artifactual effect of horseradish peroxidase. *Experimental Lung Research*, *6*, 103–114.
Vasquez, O. L., Almeida, A., & Bolanos, J. P. (2001). Depletion of glutathione up-regulates mitochondrial complex I expression in glial cells. *Journal of Neurochemistry*, *76*, 1593–1596.
Wu, J. T., Wu, L. H., & Knight, J. A. (1986). Stability of NADPH: Effect of various factors on the kinetics of degradation. *Clinical Chemistry*, *32*, 314–319.

CHAPTER EIGHT

Peroxiredoxin-6 and NADPH Oxidase Activity

Daniel R. Ambruso[*,†,1]
[*]Department of Pediatrics, University of Colorado Denver, Anschutz Medical Campus, Aurora, Colorado, USA
[†]Center for Cancer and Blood Disorders, Children's Hospital Colorado, Aurora, Colorado, USA
[1]Corresponding author: e-mail address: daniel.ambruso@ucdenver.edu

Contents

1. Introduction 146
2. Experimental Components and Considerations 149
 2.1 Protein purification 149
 2.2 Assays for peroxiredoxin activity 150
 2.3 Cell isolation and preparation of subcellular fractions 151
 2.4 NADPH oxidase activity 152
 2.5 Suppression of Prdx6 in transgenic, oxidase-competent K562 cells, and PLB985 cells with siRNA and shRNA techniques 153
3. Peroxiredoxin Activity of Prdx6 155
 3.1 Glutamine synthetase protection 155
 3.2 Direct oxidation of H_2O_2 156
4. Effect of Prdx6 on NADPH Oxidase Activity 157
 4.1 Cell-free system of oxidase activity 157
 4.2 siRNA suppression of Prdx6 in transgenic K562 cells 159
 4.3 shRNA suppression of Prdx6 in PLB-985 cells 160
5. Summary 163
References 165

Abstract

Peroxiredoxins (Prdxs) are a family of proteins which catalyze the reduction of H_2O_2 through the interaction of active site cysteine residues. Conserved within all plant and animal kingdoms, the function of these proteins is related to protection from oxidation or participation of signaling through degradation of H_2O_2. Peroxiredoxin 6 (Prdx6), a protein belonging to the class of 1-cys Prdxs, was identified in polymorphonuclear leukocytes or neutrophils, defined by amino acid sequence and activity, and found associated with a component of the NADPH oxidase (Nox2), p67^{phox}. Prdx6 plays an important role in neutrophil function and supports the optimal activity of Nox2. In this chapter, methods are described for determining the Prdx activity of Prdx6. In addition, the approach for assessing the effect of Prdx6 on Nox2 in the SDS-activated, cell-free system of NADPH oxidase activity is presented. Finally, the techniques for

Methods in Enzymology, Volume 527
ISSN 0076-6879
http://dx.doi.org/10.1016/B978-0-12-405882-8.00008-8

suppressing Prdx6 expression in phox-competent K562 cells and cultured myeloid cells with siRNA and shRNA methods are described. With these approaches, the role of Prdx6 in Nox2 activity can be explored with intact cells. The biochemical mechanisms of the Prdx6 effect on the NADPH oxidase can be investigated with the experimental strategies described.

1. INTRODUCTION

Neutrophils and monocytes provide the first line of defense against microorganisms (Dinauer, 2003). In response to infection, they adhere to the endothelial surface at the site of inflammation, move across the vascular endothelial barrier, migrate to the site of microbial invasion, and ingest microorganisms. During phagocytosis, there is activation of the respiratory burst and fusion of specific and azurophilic granules releasing both toxic oxygen metabolites and granule contents into the developing phagolysosome. The microbe is destroyed by a variety of oxygen-dependent and -independent mechanisms.

Production of oxygen metabolites is critical to the function of the neutrophil. The respiratory burst is initiated by the NADPH oxidase, Nox2, an enzyme system with several components (Ambruso & Johnston, 2012). In the unstimulated cell, the oxidase is dormant with components occupying different compartments within the neutrophil. Membrane-bound constituents of cytochrome b_{558}, the gp91^{phox} and p22^{phox}, reside in the plasma membrane and membrane of the specific granules (Abo, Boyhan, West, Thrasher, & Segal, 1992; Ambruso, Bolscher, Stokman, Verhoeven, & Roos, 1990; Clark, Volpp, Leidal, & Nauseef, 1990; Parkos, Allen, Cochrane, & Jesaitis, 1987; Volpp, Nauseef, & Clark, 1988). The remaining components, p47^{phox}, p67^{phox}, Rac2, and p40^{phox}, are found in the cytosol. Upon stimulation of the neutrophil, cytosolic components move to the plasma membrane resulting in assembly of the oxidase complex and expression of enzyme activity (Ambruso et al., 1990; Clark et al., 1990; Parkos et al., 1987; Volpp et al., 1988). Oxidation of NADPH is coupled with reduction of oxygen to generate superoxide anion (O_2^-) (Ambruso & Johnston, 2012). Subsequently, generation of other reactive oxygen species (ROS) such as H_2O_2 and hypochlorous acid provide oxygen-dependent microbicidal activity.

Over the past 25 years, great progress has been made in defining the structural components of the oxidase as well as the role each plays in the function of this enzyme system. The requirement for Rac, p47^{phox}, p67^{phox}, gp91^{phox}, and p22^{phox} has been demonstrated by a variety of studies with intact neutrophils and cell-free systems activated by anionic detergents (Bromberg & Pick, 1985; Curnutte, 1985; McPhail, Shirley, Clayton, & Snyderman, 1985). Cytochrome b_{558}, with its heme, NADPH, and flavin-binding sites, acts as the engine for transfer of electrons from NADPH to oxygen (Quinn, Mullen, & Jesaitis, 1992; Segal et al., 1992). The p47^{phox} binds the p22^{phox} component and p67^{phox} interacts with gp91^{phox} allowing optimal electron transfer (Dang, Cross, & Babior, 2001; Huang & Kleinberg, 1999). In the cytosol, p47^{phox} interacts with p67^{phox} and is required for translocation of p67^{phox} (DeLeo & Quinn, 1996). Discovered by its binding to p67^{phox}, the p40^{phox} also translocates to the plasma membrane with the other components of the complex. The p40^{phox} can bind to phospholipid products of the PI(3) kinase and may function in associate signaling events (Matute, Arias, Dinauer, & Patino, 2005; Ueyama et al., 2011). While it is not required in the cell-free system of Nox2 activation, p40^{phox} may play a critical role in activation of Nox2 during phagocytosis (Matute et al., 2005, 2009; Ueyama et al., 2011).

Recently described, peroxiredoxins (Prdxs) are a family of antioxidant proteins of molecular size 20–30 kDa found throughout the animal and plant kingdoms (Rhee, Kang, Chang, Jeong, & Kim, 2001). Multiple isoforms of Prdx exist in all eukaryotic cells. Although the various isoforms differ in their developmental expression patterns, distribution in the cell, and their reaction intermediates, their function is related to either protection against oxidation or participation in signaling by degrading H_2O_2.

Initially discovered in yeast, Prdxs reduce H_2O_2 through hydrogens provided by a thiol (Chae, Uhm, & Rhee, 1994). These proteins are capable of reducing peroxide in the presence of dithiothreitol (DTT). For many Prdxs, interaction with H_2O_2 results in oxidation of a cysteine to form cysteine sulfenic acid. Reaction with a second cysteine residue forms an intermolecular disulfide (Chae, Chung, & Rhee, 1994). The specificity of Prdxs is attributable to reduction of this disulfide by a thiol, not ascorbic acid. All Prdxs can complete the reaction with DTT. Physiologically, some homologs use thioredoxin to complete the reaction (Chae, Chung, et al., 1994).

All Prdx proteins contain a conserved cysteine residue at the amino-terminal end of the molecule corresponding to Cys47 in the first peroxiredoxin homolog sequenced (yc-TPxI) (Rhee et al., 2001). The majority of Prdxs contain a second cysteine at the carboxy-terminal end corresponding to Cys170 in yc-TPxI (Chae, Robison, et al., 1994). The members of the family may be divided into three groups: (a) 2-Cys Prx proteins containing both C-terminal and NH_2-terminal Cys residues and requiring both for catalytic function, (b) atypical 2-Cys proteins which contain the conserved NH_2-terminal Cys but require an additional nonconserved Cys residue for catalytic activity, and (c) 1-Cys Prx which contains only the conserved NH_2-terminal Cys and may contain a second Cys but require only the conserved one for catalytic function (Rhee et al., 2001). The presence or absence of the second conserved Cys is correlated with the conservation of amino acid residues surrounding the first Cys represented by a signature motif, FTFVCPTEI or FTPVCTTEL in 2-Cys Prx and 1-Cys Prx members, respectively (Rhee et al., 2001).

1-Cys Pdrx members are found in a variety of plant and animal species including bacteria, yeast, nematodes, and mammals (Chae, Robison, et al., 1994; Kang, Baines, & Rhee, 1998). Upon exposure to H_2O_2, the residue corresponding to Cys47 is oxidized to Cys-SOH (Kang, Baines, et al., 1998). A disulfide is putatively not formed because of the absence of a second, active Cys. The Cys-SOH can be reduced by nonphysiologic thiols such as DTT but its physiologic electron donor is not known (Rhee et al., 2001). Although GSH has been proposed as the physiologic donor for 1-Cys Prdx, this remains controversial (Rhee et al., 2001).

We recently demonstrated the presence of Prdx6 in neutrophils (Leavey et al., 2002) as a 29-kDa protein (originally designated p29) with amino acid sequence identical to Prdx6 and with Prdx enzyme activity and exhibit binding to $p67^{phox}$ (Leavey et al., 2002). This 1-Cys Prx, Prdx6, isolated from neutrophils has been shown to express calcium-independent phospholipase A_2 activity with a pH optimum of 4 (Kim et al., 1998; Leavey et al., 2002). Subsequently, we showed that Prdx6 translocated to plasma membrane neutrophils after stimulation of neutrophils by western blot and confocal microscopy (Ambruso, Ellison, Thurman, & Leto, 2012). Translocation of Prdx6 in phox-competent K562 cells required both $p67^{phox}$ and $p47^{phox}$ (Ambruso et al., 2012). In intact phox-competent K562 cells and a myeloid cell line, PLB985, using siRNA and shRNA techniques to suppress Prdx6, a role for Prdx6 in optimal activity was demonstrated (Ambruso et al., 2012; Ellison, Thurman, & Ambruso, 2012). In addition, enhancement of O_2^- production by recombinant Prdx6 was shown in a cell-free system of oxidase activity (Ambruso et al., 2012). Prdx6 serves a structural or biochemical role in supporting optimal oxidase activity, enhances NADPH oxidase activity, and plays an important role in neutrophil function (Ambruso et al., 2012; Ellison et al., 2012; Leavey et al., 2002).

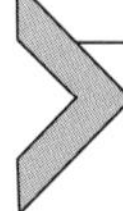

2. EXPERIMENTAL COMPONENTS AND CONSIDERATIONS

2.1. Protein purification

Sufficient quantities of protein are required to perform assays in the cell-free system of oxidase activity. There are several options but isolation of the native protein with standard gel filtration chromatography or immuno-chromatography may be inefficient because of the quantity of Prdx6 in starting cell lysates and/or characteristics of antibodies available. cDNA for Prdx6 may be subcloned into *Eco*RI site of the baculovirus expression vector pBlueBacHis for infection and protein production in sf9 cells or incorporation into pcDNA 3.0 and expression in *E. coli* as previously described (Ambruso et al., 2012; Ellison et al., 2012; Leavey et al., 2002). The poly-His tags may be expressed on the N-terminal or C-terminal domains as noted (Chen, Dodia, Feinstein, Jain, & Fisher, 2000; Leavey et al., 2002).

Cell lysates are prepared by freeze–thaw lysis (−80 to 37 °C), passage through an 18-gauge needle, and centrifugation at 10,000 × *g* for 10 min (Kim et al., 1998). The resultant supernatant is applied to a nickel affinity gel (ProBond, Invitrogen) and incubated for 2 h at 4 °C on a rotator. The beads are washed with aliquots of binding buffer (0.5 *M* NaCl, 20 m*M* sodium phosphate, pH 7.8) and then wash buffer (0.5 *M* NaCl, 20 m*M* sodium phosphate, pH 6.0). After each step, the supernatant is removed and the gel was retained for the next step. After washing, the resin can be packed into a small column. The recombinant protein can be eluted from the column with stepwise additions (1 bed volume) of wash buffer with 50 and 500 m*M* imidazole. Alternatively, elution can be performed with a continuous gradient of imidazole with the recombinant protein coming off the column at 500 m*M*. Samples may be analyzed by Bradford protein assay, BCA, or optical density (OD) 280 nm for detection of peak protein. Fractions with the protein are pooled and dialyzed against 50 m*M* Tris–HCl with 2 m*M* EDTA (pH 8.0). After dialysis, final protein concentration may be determined as noted above and further analysis performed by separation of proteins on SDS-PAGE and identification of the recombinant protein by Western blot with the various antibodies. The recombinant proteins may be stored at −70 °C until use and are stable indefinitely at −70 °C or for several weeks at 4 °C. Some preparations may require 50–100 μ*M* NaCl to assure stability at very high protein concentrations.

2.2. Assays for peroxiredoxin activity

Peroxiredoxin activity may be measured by a variety of techniques. One approach utilizes the protection of glutamine synthetase activity from inactivation by a H_2O_2-generating system containing ferric chloride and DTT (Chae, Kang, & Rhee, 1999). Varying amounts of Prdx6 are added to a reaction mixture (50 μl) containing 5 μ*M* $FeCl_3 \cdot 6H_2O$, 20 m*M* DTT, and 0.5 μg of glutamine synthetase in 0.05 *M* HEPES (pH7.0). Catalase or heat-inactivated proteins may be used as controls. After incubation for 10 min at 37 °C, 100 μl of a solution containing 0.4 m*M* ADP, 0.4 m*M* $MnCl_2$, 10 m*M* potassium arsenate, 20 m*M* hydroxylamine, and 100 m*M* glutamine is added to the initial reaction and incubated for 10 min at 37 °C. The reaction is stopped by addition of 50 μl of a solution containing 55 mg of $FeCl_3 \cdot 6H_2O$, 20 mg of trichloroacetic acid, and 21 μl of HCl per ml volume. After centrifugation for 30 s at 10,000 × *g*, 150 μl of supernatant is added to wells of a microtiter plate and OD at 550 nm in an enzyme-linked immunosorbent assay (ELISA) reader is determined. Protection-afforded Prdx6 is determined by the following equation:

$$\%\text{Protection} = 100 \times \frac{(\text{activity GS} + \text{Fe/DTT} + \text{protectant}) - (\text{activity GS} + \text{Fe/DTT})}{(\text{activity GS}) - (\text{activity GS} + \text{Fe/DTT})}$$

A second method directly measures the removal of H_2O_2 (Thurman, Ley, & Scholz, 1972). In this assay, contaminating iron is removed from water and HEPES buffer using Chelex chromatography (Thurman et al., 1972). A standard amount of H_2O_2 is generated by addition of DTT and H_2O_2. DTT (2 m*M*), H_2O_2 (12 nmol), and various concentrations of Prdx6 or other protein controls in a volume of 100 μl are incubated for 10 min at 37 °C. Then, 20 μl of 1 N HCl, 20 μl of 10 m*M* $Fe(NH_4)SO_4$, and 10 μl of 2.5 *M* potassium thiocyanate are added. After centrifuging the mixture at 10,000 × *g* for 30 s, 100 μl of the supernatant is added to wells of a microtiter plate and the OD was measured in an ELISA reader at 450 nm. The amount of H_2O_2 remaining in the reaction is determined by comparison with a standard curve. Prdx activity is calculated as % of H_2O_2 removed, based on the ratio of the amount of H_2O_2 generated in the presence and absence of added proteins.

$$\%H_2O_2 \text{ removed} = \frac{H_2O_2 \text{ presence of protein}}{H_2O_2 \text{ absence of protein}} \times 100$$

2.3. Cell isolation and preparation of subcellular fractions

Heparinized (1 U/ml) peripheral blood is obtained by phlebotomy from healthy adult volunteers. Neutrophils are isolated from peripheral blood by standard techniques of dextran sedimentation, Ficoll-hypaque density gradient centrifugation, and hypotonic lysis of red blood cells (Ambruso et al., 2000; Leavey et al., 1998; Rosenthal et al., 1996). Cultured cells are harvested from primary culture media and diluted to appropriate concentrations with appropriate buffers. For studies with intact cells, a Krebs Ringers phosphate buffer with dextrose (KRPD) (12.5 m*M* Na_2HPO_4, 3.0 m*M* NaH_2PO_4, 4.8 m*M* KCl, 120 m*M* NaCl, 1.3 m*M* $CaCl_2$, 1.2 m*M* $MgSO_4 \cdot 7\ H_2O$, and 0.2% dextrose, pH 7.34) is used for assays with intact cells.

Subcellular fractions of neutrophils are isolated by sucrose density centrifugation after disruption of the cells with nitrogen cavitation in the presence of diisopropyl fluorophosphate (1 m*M*), leupeptin (1 μg/ml), and phenylmethylsulfonyl fluoride (0.2 m*M*) as previously described (Ambruso et al., 1990, 2000; Leavey et al., 1998; Rosenthal et al., 1996). Intact cells are centrifuged at 800 × *g* and resuspended to a concentration of 5×10^7–10^8 cells/ml in cavitation buffer (100 m*M* KCl, 3 m*M* NaCl, 3.5 m*M* $MgCl_2$, 10 m*M* PIPES, pH 7.3) with proteolytic inhibitors mentioned earlier. After adding the cells to a nitrogen cavitation bomb and flushing for 5 min with nitrogen gas, the exhaust port is closed allowing the pressure to increase to 225–250 psi and the cells remain in this environment at 4 °C for 20 min. The exhaust port is then opened slowly and the cavitate is collected dropwise into an equal volume of cavitation buffer with 0.34 *M* sucrose and 2.5 m*M* EDTA. These parameters were chosen to give the most efficient disruption of cells with minimum breakage of subcellular organelles and nuclei (Ambruso, Stork, Gibson, & Thurman, 1987). The cavitate is then centrifuged at 800 × *g* for 10 min to remove unbroken cells and cellular debris. The resultant supernatant is layered over a sucrose gradient constructed as follows (from bottom to top: 3 ml of 60% sucrose (wt./vol.), 18 ml of a linear sucrose gradient 40–55%, and 3 ml of 15% sucrose) (Ambruso et al., 1990, 2000; Leavey et al., 1998; Rosenthal et al., 1996). All sucrose solutions contain 1 m*M* EGTA and the proteolytic inhibitors mentioned earlier. The gradients are centrifuged at 100,000 × *g* for 2 h. The subcellular fractions may be unloaded by hand using visual inspection to determine the subcellular fractions or with a small bore needle through the bottom of the tube using a peristaltic pump and fraction collector.

Cytosol and plasma membrane fractions obtained in this manner can be pooled, protein concentration determined, and frozen immediately at −70 °C for use in subsequent studies.

2.4. NADPH oxidase activity

Production of O_2^- in a cell-free system of NADPH oxidase activity is performed as previously described (Ambruso et al., 1990, 2000; Leavey et al., 1998; Rosenthal et al., 1996), but with further optimization to provide maximal oxidase activity as follows: 100 n*M* p67^{phox}, 100 n*M* p47^{phox}, 32 n*M* constitutively active Rac1 plus or minus varying amounts of Prdx6 or buffer were combined in buffer to 100 μl. Unless otherwise noted, phox proteins and Rac1 are recombinant proteins. After 10-min incubation at RT, 1 μg of plasma membrane from unstimulated neutrophils, 75 μ*M* cytochrome *c*, and SDS (112.5 μ*M*) are introduced. After 3 min, NADPH (varying concentrations in 15 μl) is added to give a total assay volume of 150 μl. Assays are completed in triplicate with one well containing 5 μg superoxide dismutase (SOD) from a stock solution of 1 mg/ml. The initial rate of O_2^- production is determined as SOD-inhibitable cytochrome *c* reduction in an ELISA reader at 550 nm using an extinction coefficient of 8.4 nmol/L^{-1} experimentally determined for a 150 μl reaction volume as previously reported (Ambruso et al., 1990, 2000, 2012; Leavey et al., 1998; Rosenthal et al., 1996).

O_2^- production by intact cells is determined with two techniques: Neutrophils or tissue culture cells (3.75×10^5) in KRPD and 75 μ*M* cytochrome *c* are added to microtiter wells and incubated at 37 °C for 3 min. PMA (200 ng/ml) or fMLP (1 μ*M*) for neutrophils or PMA for tissue culture cells is added to give a final reaction volume of 150 μl and the OD at 550 nm is measured every 20 s for 5 min. Assays are run in triplicate with one well containing SOD as mentioned earlier. O_2^- production is determined as SOD-inhibitable cytochrome *c* reduction as noted earlier (Ambruso et al., 1990, 2000, 2012; Leavey et al., 1998; Rosenthal et al., 1996). For tissue culture cells stimulated with fMLP, the same cell numbers in KRPD as described previously are added to 30 μl Diogenes (National Diagnostics) and 1 μ*M* fMLP included to start the reaction (total volume of 150 μl). Chemiluminescence is measured in a FB12 Luminometer (Berthold Detection Systems, Pforzheim, Germany) as relative light units (RLUs). The maximum RLUs at 30 s can be used to measure the output of the respiratory burst; >95% of this chemiluminescence is SOD-inhibitable indicating the reactivity is O_2^-.

O_2^- production may also be determined by luminescence using a LUMIstar Optima luminometer (BMG Labtech) (Ellison et al., 2012). To wells of microtiter plates are added cells in KRPD prewarmed to 37 °C as described above. fMLP or other appropriate agonists and Diogenes (prewarmed to 37 °C) are then injected in 1 s by the fluid-handling system of the instrument or added manually. After shaking, the amount of light detected over the course of 1 s may be measured every second for 4 min (fMLP) or every 20 s (after additional shaking) for at least 30 min (PMA). Peak areas were determined by summing all light measurements made during assay.

The magnitude of luminescence produced in response to an agonist by a given cell type may vary quite widely between experiments. However, the results within an experiment done on a specific day are quite uniform. A technique of normalizing results for each experiment may be helpful to compare results from multiple experiments done over time (Ellison et al., 2012). For example, on each day that assays are done comparing K562 cells (wild type, WT) with K562 cells with Prdx6 suppressed (suppressed or knockdown, KD), the average peak area for WT cells (*n* of at least 3) is set at 1 and all peak areas recorded on that day are normalized to this value. Choice of the normalized cells will vary by the types of comparisons done, but this will help organize the data in a uniform way.

2.5. Suppression of Prdx6 in transgenic, oxidase-competent K562 cells, and PLB985 cells with siRNA and shRNA techniques

NADPH oxidase-competent K562 cells are grown at 37 °C in 5% CO_2 with PMI and 10% fetal calf serum(de Mendez, Adams, Sokolic, Malech, & Leto, 1996; de Mendez & Leto, 1995; Leto et al., 2007). These cells have been stably transduced with p47^{phox}, p67^{phox}, gp91^{phox} (K+++), and subsequently transfected with the fMLP receptor (FPR); this reconstituted line has sufficient levels of endogenous p22^{phox}, p40^{phox}, and Rac proteins to support NADPH oxidase activity in response to stimulation by PMA or fMLP (Leto et al., 2007). Transient suppression of Prdx6 is completed with small interfering RNA (siRNA) molecules designed to degrade the mRNA for Prdx6. The siRNAs are uniquely targeted to hybridize 19-nucleotide sense-strand sequences of Prdx6 (Elbashir et al., 2001; Hannon, 2002; Paddison, Caudy, & Hannon, 2002; Sharp, 2001; Tuschl, Zamore, Lehmann, Bartel, & Sharp, 1999) and obtained commercially (Ambion, Austin, TX and Qiagen, Valencia, CA). The sequences for the siRNAs are: si1 (Qiagen), UUG GUG AAG ACU CCU UUC GGG; si2 (Ambion), GCA AUC AAC UUA ACA UUC Ctc; si3 (Ambion), UGG GUA GAG GAU AGA CAG Ctt; and Nsi (Qiagen), UUC UCC GAA CGU GUC ACG UdT dT. A plasmid expressing GFP assessed transfection efficiency.

Transfections are conducted with Amaxa Nucleofector technology (Lonza, Basel, Switzerland) (Ambruso et al., 2012). Transgenic K562 cells (5×10^6, passage < 12) are grown for 24 h and are resuspended in 100 μl Solution L. GFP plasmid (to assess transfection efficiency), si1, si2, si3, and Nsi are added at various concentrations (0.5, 1, 2, and 5 μ*M*) to optimize the knockdown and transfection completed with the T020 protocol. RPMI with 10% (500 μl) FCS is added, and the cells are incubated for 15 min at 37 °C. Cells are transferred to larger plates, brought up in 2 ml RPMI with 10% FCS, and cultured for various times before being harvested for assays. Cell viability is determined by trypan blue exclusion and for all experiments is $\geq 90\%$.

For the PLB-985 cell line, a stable knockdown is produced as follows: 2×10^6 cells growing in log phase cells are transfected (solution V, program C-023) with 2 μg of plasmid DNA using the Amaxa Nucleofector according to the manufacturer's instructions (Ellison et al., 2012). The plasmid DNA is one of four different pRS-based plasmids encoding shRNA molecules predicted to silence Prdx6 (Origene). Cell lines stably transfected with a pRS-based plasmid (TR30003) encoding a noneffective, scrambled shRNA (Origene) are established in parallel. Transfected cells are maintained in the absence of penicillin and streptomycin for 7 days and then moved to the standard culture media supplemented with a selective antibiotic, puromycin (1 μg/ml). The cells are transferred to 96-well tissue culture plates at approximately 3×10^4 cells per well and antibiotic-resistant colonies appear after about 5 weeks. Cells from wells in which a single colony had developed are expanded in the same media used for initial selection and screened by Prdx6 western blot for suppression of Prdx6. Cell lines arising from the transfection with plasmid TI340760 (Ellison et al., 2012) demonstrate the greatest suppression of Prdx6. A cell line obtained with this plasmid can be used along with the nonsilenced control cells transfected with TR30003 in studies below.

PLB-985 cell lines were differentiated into neutrophil-like cells with the addition of 1.3% (vol./vol.) DMSO for 4 days (Ellison et al., 2012). Cell viability decreases during the maturation but is always >75%. Unmodified, matured cells retain functions similar to neutrophils from peripheral blood.

3. PEROXIREDOXIN ACTIVITY OF Prdx6

3.1. Glutamine synthetase protection

Most Prdxs contain two cysteine residues and reduce H_2O_2 with electrons provided by thioredoxin (Chae, Chung, et al., 1994; Kang, Baines, et al., 1998; Kang, Chae, et al., 1998; Netto, Chae, Kang, Rhee, & Stadtman, 1996). However, Prdx6 contains only one conserved cysteine and a partner which provides electrons physiologically has not been completely defined. Prdx6 and other 1-cys Prdxs can reduce H_2O_2 by electrons provided by DTT (Kang, Baines, et al., 1998). The first assay to investigate the peroxidase activity by recombinant Prdx6 is the protection of glutamine synthetase

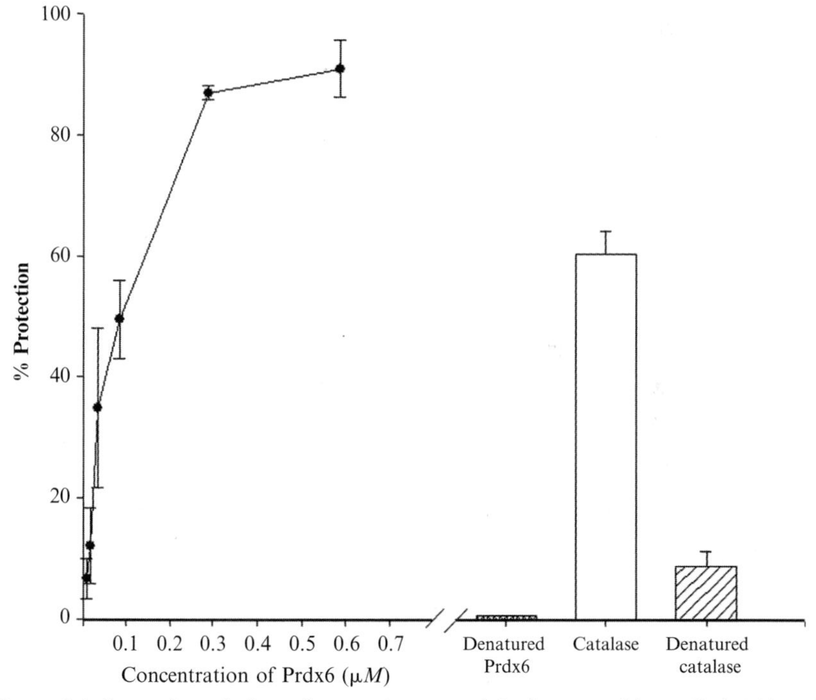

Figure 8.1 Protection of glutamine synthetase activity by recombinant Prdx6. Experiments were completed as described in Section 2.2. Circles and bars or columns and bars represent mean and SEM for 12 experiments; concentrations of Prdx6 $>0.05\,\mu M$ different from no protein added, $p<0.005$ by two tailed t-test. Assays with catalase (20 μg/ml), heat-denatured catalase, and heat-denatured Prdx6 (0.30 μM) are shown on the right. *This figure was adapted from studies originally published in Leavey et al. (2002) © The American Society for Biochemistry and Molecular Biology.*

from the oxidative effects of H_2O_2. In the presence of DTT, $FeCl_3$ catalyzes the formation of H_2O_2 that destroys the activity of glutamine synthetase (Fig. 8.1). Prdx6 protects glutamine synthetase by eliminating H_2O_2 (Chae, Chung, et al., 1994; Kang, Baines, et al., 1998; Netto et al., 1996). Addition of Prdx6 provides concentration-dependent inhibition of damage to glutamine synthetase with the optimum effect at 0.30 μ*M* (Fig. 8.1). Catalase, which also inactivates H_2O_2, similarly protects glutamine synthetase activity. Heat-denatured Prdx6 and catalase do not significantly protect glutamine synthetase activity.

3.2. Direct oxidation of H_2O_2

To determine whether the protective effect of Prdx6 as noted earlier was related to inactivation of H_2O_2, direct measurement of H_2O_2 degradation is employed. In the presence of DTT, H_2O_2 concentrations decrease with addition of recombinant Prdx6 (Fig. 8.2). The effect is concentration dependent with optimum reduction in H_2O_2 seen between 0.30 and 0.58 μ*M* Prdx6. Catalase has an effect similar to Prdx6 eliminating 95% of the H_2O_2 (data not shown). Denatured Prdx6 and catalase have no effect on inactivating H_2O_2 (data not shown). Thus, Prdx6 inactivates H_2O_2 directly.

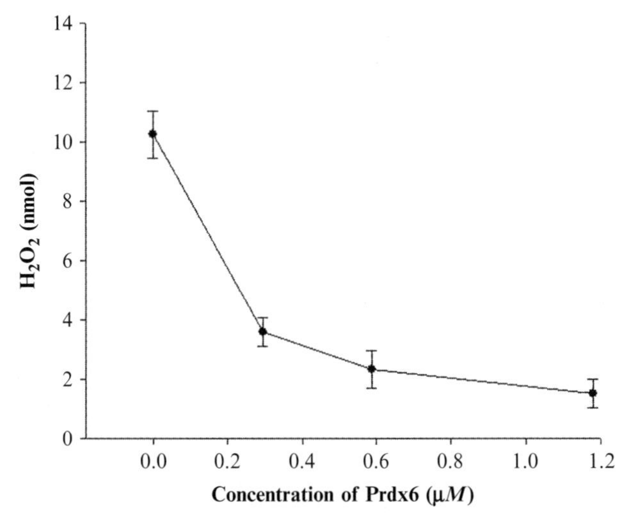

Figure 8.2 Inactivation of H_2O_2 by recombinant Prdx6. Experiments were performed as described in Section 2.2. Circles and bars represent mean and SEM results for 12 experiments; concentrations >0.2 μ*M* different from no protein added, $p < 0.005$, two-tailed *t*-test. *This figure was adapted from studies originally published in Leavey et al. (2002) © The American Society for Biochemistry and Molecular Biology.*

4. EFFECT OF Prdx6 ON NADPH OXIDASE ACTIVITY

4.1. Cell-free system of oxidase activity

The effect of Prdx6 on Nox2 activity can be measured in the SDS-activated, cell-free NADPH oxidase system. In this system, cytosolic phox proteins, recombinant Rac1, $p67^{phox}$, and $p47^{phox}$ are combined with a source of cytochrome b_{558} (neutrophil plasma membrane) in the presence of SDS and NADPH, and oxidase activity is measured as SOD-inhibitable cytochrome *c* reduction. Addition of recombinant Prdx6 directly to the assay mixture and initiation with NADPH has little effect on oxidase activity. However, preincubation of Prdx6 with the recombinant oxidase proteins, $p47^{phox}$, $p67^{phox}$, and Rac1, with subsequent addition of these to the remainder of the assay constituents and initiation of the system results in enhancement of oxidase activity (Leavey et al., 2002). The optimum effect occurs after 10 min of preincubation. Table 8.1 shows activity of this SDS-activated cell-free system. The system is optimized as described in Section 2.4 to further evaluate the effects on the kinetics of the system.

Two types of studies define the effect of Prdx6 on Nox2 activity. In one series of experiments, Prdx6 (0, 0.27, 0.55, 1.10, and 2.20 μ*M*) is added to the reaction mixture and NADPH (0–200 μ*M*) started the reaction. Initial rates of O_2^- production are determined, the data fit to the Michaelis–Menten equation and kinetic parameters, K_m (for NADPH) and V_{max}, calculated using commercial software (Prism, GraphPad Software Inc., La Jolla, CA). Results from this series of studies are shown in Fig. 8.3A. All the Prdx6 concentrations increase the rate of O_2^- generation but do not change the

Table 8.1 Production of O_2^- in the basic SDS-activated cell-free system of oxidase activity

Assay constituents	O_2^- production (% of system alone)[a]
System alone	100
System + Prdx6 (0.50 μ*M*)	198 ± 12 ($n = 15$)[b,c]
System + heat-inactivated Prdx6	132 ± 14 ($n = 4$)[b,c,d]

[a]Assay of oxidase activity as described in Section 2. Results expressed as percentage of basic system alone.
[b]Numbers are mean ± SEM with number of experiments in parentheses.
[c]Different from basic system alone, $p < 0.025$, two-tailed *t*-test.
[d]Different from Prdx6, $p < 0.025$.
Data was adapted from studies originally published in Leavey et al. (2002) © The American Society for Biochemistry and Molecular Biology.

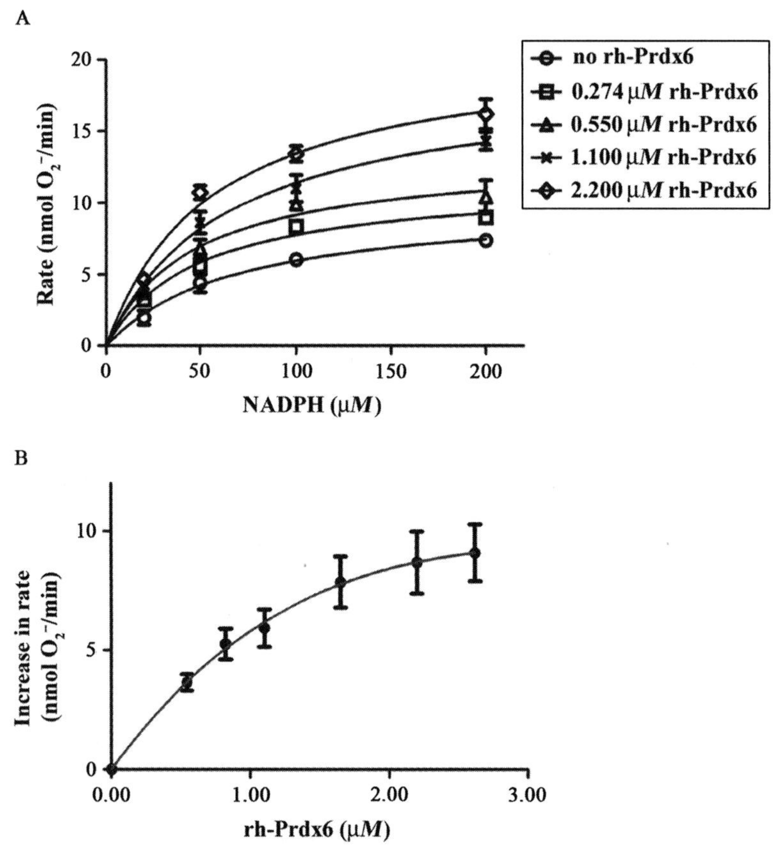

Figure 8.3 Effect of recombinant Prdx6 on oxidase activity in the SDS-activated cell-free system. Production of O_2^- was measured as described in Section 2. (A) O_2^- production (nmol/min) in the presence of varying concentrations of Prdx6 and varying concentrations of NADPH (20–200 μ*M*). Symbols and bars are mean ± SEM for four separate experiments. (B) Concentration-dependent effects of Prdx6 in the cell-free system. The increase (Δ) in rate of O_2^- production with the addition of recombinant Prdx6 compared to its absence plotted against the concentration of Prdx6 in the presence of 200 μ*M* NADPH. The plot represents four separate experiments; symbols and bars are mean ± SEM. *Reprinted from Biochimica et Biophysica Acta (2012), vol. 1823. DR Ambruso, MA Ellison, GW Thurman, and TL Leto. Peroxiredoxin 6 translocates to the plasma membrane during neutrophil activation and is required for optimal NADPH oxidase activity. pp. 306–315, with permission from Elsevier.*

concentration of NADPH at half maximal rate. Kinetic parameters of the oxidase are calculated with these data and the results are presented in Table 8.2. Addition of Prdx6 at concentrations up to 2.2 μ*M* had no effect on the K_m for NADPH. However, the V_{max} was significantly increased at concentrations above 0.55 μ*M* Prdx6.

Table 8.2 Kinetic parameters for NADPH at various concentrations of Prdx6 in the reaction mixture

Prdx6 (μM)	K_m for NADPH (μM)[a]	V_{max} (nmol O_2^-/min)[a]
0	68 ± 20	10.0 ± 1.0
0.27	49 ± 11	11.5 ± 0.9
0.55	46 ± 14	13.4 ± 1.4
1.10	65 ± 15	18.9 ± 1.7[b]
2.20	56 ± 11	21.0 ± 1.5[b]

[a]Numbers represent mean ± SEM of four separate experiments at each concentration.
[b]Significantly different from no rh-p29 Prdx6 added $p < 0.05$ by unpaired *t*-test.
Data was adapted from studies originally published and reprinted from Biochimica et Biophysica Acta (2012), vol. 1823. DR Ambruso, MA Ellison, GW Thurman, and TL Leto. Peroxiredoxin 6 translocates to the plasma membrane during neutrophil activation and is required for optimal NADPH oxidase activity. pp. 306–315, with permission from Elsevier.

In a second set of experiments, a wider range of Prdx6 concentrations are added to the cell-free system, 200 μM NADPH initiated the reaction, and O_2^- production is determined. The difference in rate of O_2^- production between the system with and without the recombinant protein is calculated for each concentration of Prdx6 added. Changes in velocity from the cell-free system alone are plotted against the concentration of Prdx6 (Fig. 8.3B). Prdx6 produces a saturable, concentration-dependent increase in oxidase activity. Thus, enhanced production of O_2^- in the cell-free system appears to be due to a specific effect on the oxidase rather than increasing O_2^- production by another means such as chemical ROS generation.

4.2. siRNA suppression of Prdx6 in transgenic K562 cells

The studies in the cell-free system of oxidase activity may be extended to intact cells. Because neutrophils are terminally differentiated, short-lived, and difficult to transfect, we employ a K562 cell model for siRNA experiments. Production of genetically modified cultured cells has provided powerful strategies to study the activity of the oxidase (de Mendez et al., 1996; de Mendez & Leto, 1995; Leto et al., 2007; Price et al., 2002). The K+++ cells used are a clonally derived K562 cell line stably transduced with retroviral p47^{phox}, p67^{phox}, gp91^{phox}, and have been shown to express endogenous p22^{phox} and Rac (de Mendez et al., 1996; de Mendez & Leto, 1995; Leto et al., 2007). They were subsequently transfected with the FPR (Leto et al., 2007). Three siRNA molecules have been developed specifically to degrade the mRNA of Prdx6 and decrease expression of this protein in these cells. A nonsilencing siRNA (Nsi) which did not interfere with any specific mRNA in the cell serves as a control. Transfection with the active siRNA (si1, si2, and si3) and Nsi is achieved as described in Section 2.5. Transfections containing 5×10^6 K+++ cells and 2 μ*M* si or Nsi, followed by 48 h in culture give maximal knockdown of Prdx6, with reversal of suppression occurring another 72 h later (data not shown) and are judged $\geq 90\%$ efficient based on separate experiments in which a GFP expressing plasmid was transfected and cells monitored for fluorescence.

Quantitation of Prdx6, phox proteins and actin is performed by western blotting after transfection of K+++ cells with si or Nsi molecules. A representative experiment is shown in Fig. 8.4A. Expression of Prdx6 is reduced by si1, si2, and si3 as compared to Nsi. The phox proteins including $p67^{phox}$, $p47^{phox}$, $p40^{phox}$, or $p22^{phox}$ are not decreased; and actin, used as a loading control, is also unaffected. $gp91^{phox}$ is also not decreased (data not shown). Figure 8.4B summarizes quantification of these proteins by western blot densitometry. Prdx6 is reduced to 26–43% depending on the specific si used and the results were significant ($p < 0.05$, unpaired *t*-test). O_2^- production is determined in the si-treated K562 cells after stimulation with PMA (200 ng/ml) as cytochrome *c* reduction or with fMLP (1 μ*M*) using chemiluminescence. Figure 8.4C shows O_2^- production in response to PMA. These data summarize results for four separate experiments. The response to PMA in Prdx6-targeted, si-treated cells is decreased to 50–65% of the Nsi control cells and this is significant ($p < 0.025$, paired *t*-test). The suppressive effects of Prdx6 knockdown on oxidase activation are greater when using fMLP as an agonist (Fig. 8.4D), where O_2^- production for each Prdx6-targeted si was 16–38% of Nsi control ($p < 0.025$, paired *t*-test). The insert demonstrates results of fMLP-stimulated oxidase kinetics from one experiment. These studies demonstrate the requirement of Prdx6 for optimal oxidase activity in intact cells using K562 transgenic cells as a model.

4.3. shRNA suppression of Prdx6 in PLB-985 cells

To evaluate Prdx6 effects on Nox2 in myeloid cells, PLB-985 cells, with knockdown of Prdx6 and a nonsilenced, wild type, control cell line, have been established, as detailed in Section 2.5 by stable transfection with a plasmid expressing a Prdx6-targeting shRNA (for knockdown, KD) or a nonsilencing shRNA (for wild type, WT). The KD and WT cells are differentiated with DMSO (1.3%, 4 days), and western blot analysis demonstrates that Prdx6 expression was suppressed in the KD cells but the phox components $gp91^{phox}$, $p67^{phox}$, $p47^{phox}$, $p40^{phox}$, $p22^{phox}$, and rac2 were unchanged (Fig. 8.5).

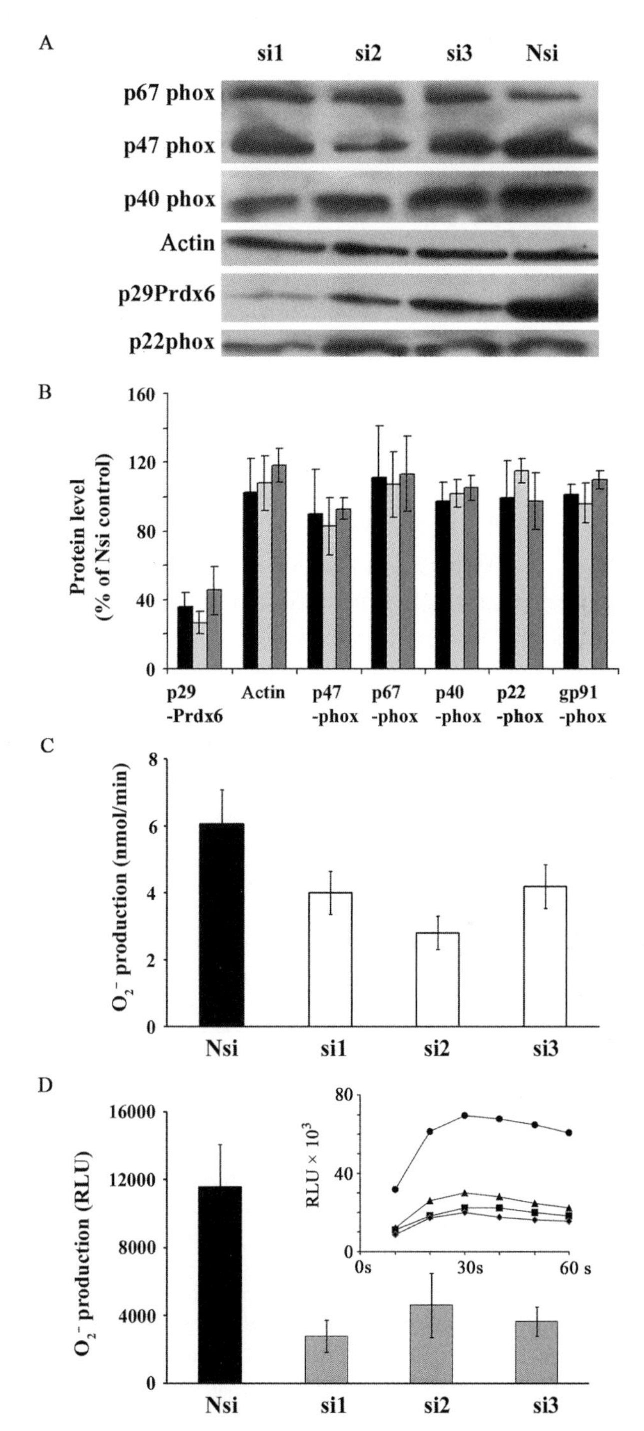

Figure 8.4 siRNA knockdown of Prdx6 suppresses whole-cell NADPH oxidase activity. (A) Representative western blot of lysates from phox-competent K562 (K+++) cells expressing the fMLP receptor transfected with Prdx6-targeted si1, si2, si3, or Nsi.

Densitometry of western blots from seven independently differentiated WT and KD cultures shows Prdx6 was suppressed by $68 \pm 6\%$ (mean $\pm$ SEM).

Knockdown of Prdx6 does not affect terminal differentiation induced by DMSO as assessed by two measures. First, the rate at which phox proteins were expressed during differentiation is not altered by suppression of Prdx6. During 4 days of DMSO exposure, $p67^{phox}$, $p47^{phox}$, and $p40^{phox}$ rose from undetectable levels ($p67^{phox}$ and $p40^{phox}$) or low levels that are equal in WT and KD cells ($p47^{phox}$, data not shown) to the higher levels that are the same in WT and KD cells (Fig. 8.5). Second, while DMSO-mediated terminal differentiation causes a progressive increase in the fraction of dead cells as measured by permeability to trypan blue, the increase in dead cells over time is not different between the WT and KD lines, even when they were exposed to DMSO for up to 7 days rather than the 4 days used to prepare cells for measurement of phox activity. Less than 1% of WT and KD cells are permeable before differentiation whereas $40 \pm 6\%$ of WT and $36 \pm 8\%$ of KD (mean $\pm$ SEM, $n=5$) are permeable after 7 days of differentiation. This is a major difficulty for evaluating neutrophils isolated from peripheral blood as well as those matured in cell culture; these cells rapidly enter an apoptotic death pathway.

Stimulation of the WT and KD cells with 1 μM fMLP results in transient bursts of O_2^- production that are detected by SOD-inhibitable luminescence of Diogenes reagent. As can be seen (Fig. 8.6A and B), these bursts are smaller in the KD cells. The difference in O_2^- production by the WT and KD cells is

Detection of $p67^{phox}$, $p47^{phox}$, $p40^{phox}$, $p22^{phox}$, and actin shows no difference between Nsi and si molecules, whereas Prdx6 is decreased. (B) Quantitation of phox proteins, actin, and Prdx6 in cells transfected with siRNA molecules was expressed as a % of Nsi control (light gray, si1; black, si2; and dark gray, si3). Prdx6 is diminished to 26–43% of control depending on the specific siRNA, whereas no significant changes in phox proteins and actin are noted (85–120% of Nsi). These are results from four separate experiments and are represented as mean $\pm$ SEM. (C) O_2^- production in response to PMA detected by cytochrome *c* reduction (nmol/min). Results presented as mean $\pm$ SEM for four separate experiments. (D) O_2^- production in response to fMLP. Insert shows a representative experiment of chemiluminescence changes following addition of stimulus for cells pretreated with siRNA molecules. Responses of cells treated with si1 (square), si2 (triangle), and si3 (diamond) are all decreased compared to Nsi control (circle). Larger graph summarizes four separate experiments measuring O_2^- production in response to fMLP by chemiluminescence (RLUs) at 30 s. Numbers represent mean $\pm$ SEM. *Reprinted from Biochimica et Biophysica Acta (2012), vol. 1823. DR Ambruso, MA Ellison, GW Thurman, and TL Leto. Peroxiredoxin 6 translocates to the plasma membrane during neutrophil activation and is required for optimal NADPH oxidase activity. pp. 306–315, with permission from Elsevier.*

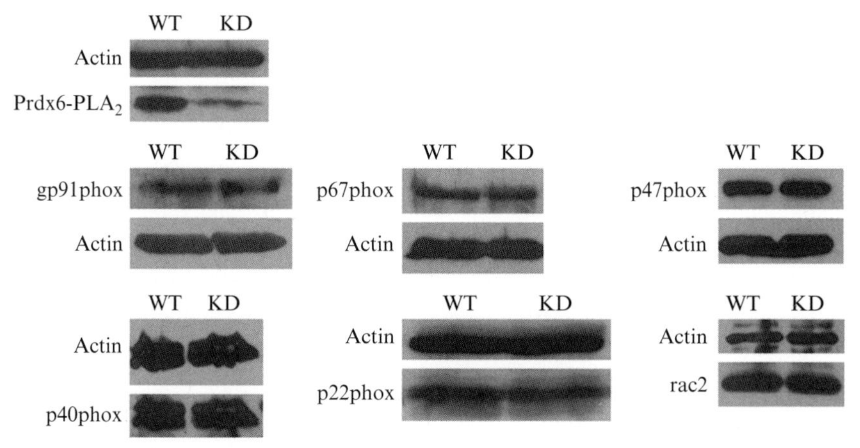

Figure 8.5 Prdx6 suppression using shRNA. PLB-985 cells with a stable knockdown of Prdx6-PLA_2 (KD) and a nonsilenced, wild type (WT), control cell line were established. Following DMSO-mediated differentiation, western blot analysis was used to confirm suppression of Prdx6 in the KD cell line (top panel) and no change in expression of other oxidase components. *Reprinted from MA Ellison, GW Thurman, and DR Ambruso. Phox activity of differentiated PLB-985 cells is enhanced, in an agonist specific manner, by the PLA_2 activity of Prdx6-PLA_2. European Journal of Immunology (2012). 42: 1609–1617. Copyright Wiley-VCH Verlag GmbH & co. KGaA. Reproduced with permission.*

significant ($p < 0.001$). Reduced O_2^- production by the KD cells is also seen when the maximal rate of SOD-inhibitable cytochrome *c* reduction was used to measure O_2^- formation (Fig. 8.6C) and the difference was again significant ($p < 0.05$). Results for PMA are not as dramatic or significant (Fig. 8.6D). PMA and fMLP are agonists that promote assembly and activation of phox; fMLP binds cell surface receptors triggering downstream signaling pathways that promote phosphorylation of phox components and activation of the complex. PMA, however, promotes assembly and activation of the Nox2 complex by direct activation of intracellular protein kinase C enzymes that carry out the necessary phosphorylation. Thus, Prdx6 may be most effective with Nox2 agonists that have receptor-linked signal pathways.

5. SUMMARY

This chapter describes general methods for evaluating Prdx activity of Prdx6. As this protein in neutrophils appears to have a specific effect on Nox2 activity, techniques are presented to further define a role for this protein. Included are studies of recombinant protein in the SDS cell-free system of oxidase activation. Also shown are experiments in intact cells with

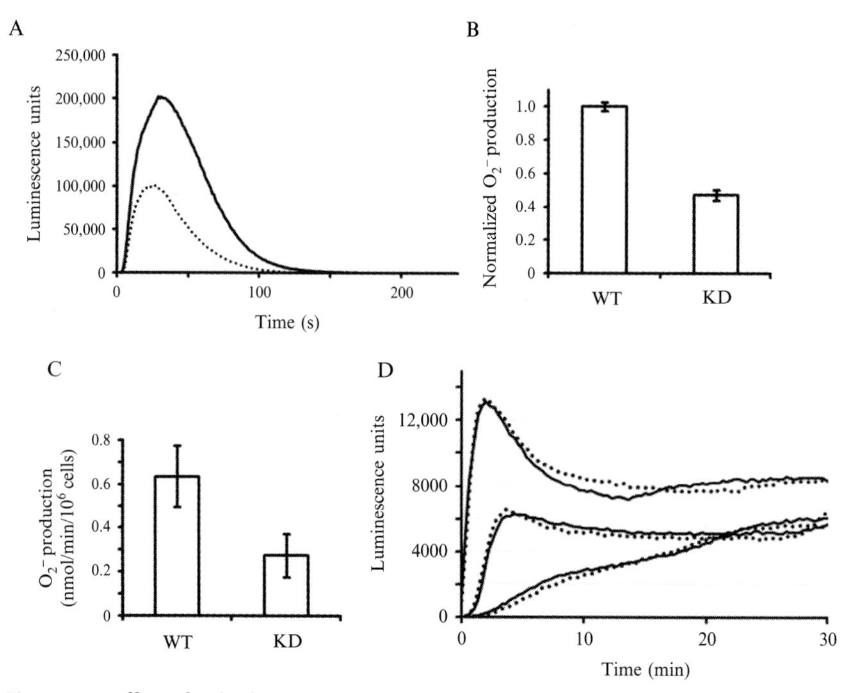

Figure 8.6 Effect of Prdx6 knockdown on O_2^- production in response to fMLP and PMA by PLB-985 cells. (A) DMSO-differentiated WT or KD cells were stimulated with 1 μ*M* fMLP and O_2^- was monitored by luminescence (WT, black line; KD, dotted line). The data are representative of multiple experiments. (B) The average peak areas for the WT and KD cells, obtained in the multiple experiments like that in (A), are shown with normalization such that the average peak area for the WT cells is 1. Data are plotted as mean ± SEM of $n = 34$ measurements pooled from seven independently differentiated cell cultures. (C) The average maximum rate of O_2^- production in response to stimulation by 1 μ*M* fMLP, as measured by SOD-inhibitable cytochrome *c* reduction, is shown as mean ± SEM of $n = 16$ measurements pooled from three independently differentiated cell cultures. (D) A representative experiment in which DMSO-differentiated WT (black line) or KD (dotted line) cells were stimulated with PMA (1 μg/ml, 200 ng/ml, or 10 ng/ml; the upper, middle, and lower pair of curves, respectively) and O_2^- was monitored by luminescence. The data are representative of six experiments from two independently differentiated cell cultures. *Reprinted from MA Ellison, GW Thurman, and DR Ambruso. Phox activity of differentiated PLB-985 cells is enhanced, in an agonist specific manner, by the PLA_2 activity of Prdx6-PLA_2. European Journal of Immunology (2012). 42: 1609-1617. Copyright Wiley-VCH Verlag GmbH & co. KGaA. Reproduced with permission.*

suppression of the expression of Prdx6 with small RNA techniques. These confirm the requirement of Prdx6 for optimal Nox2 activity. Experiments using these strategies with variations of phox-competent transgenic K562 cells document the requirement for $p47^{phox}$ and $p67^{phox}$ for incorporation of Prdx6 into the active Nox2 complex (Ambruso et al., 2012;

Kim et al., 2008). Further, separate experiments demonstrate Prdx6-associated enhancement of Nox2 activity is related to an associated increase in stability of the complex imparted by Prdx6 (Ambruso et al., 2012) extending studies in the cell-free system showing that Prdx6 increases the V_{max} of the oxidase. Documentation of the importance of the PLA_2 activity on the effect on Nox2 (Ellison et al., 2012) implies a more complicated picture, particularly in light of data demonstrating a requirement of oxidation of the cysteine active site in order to have PLA_2 activity in the physiologic pH range of cytosol (Kim et al., 2008). Further studies utilizing the methods described here will be critical to defining the specific processes and biochemical mechanisms by which Prdx6 affects the activity of Nox2 in the neutrophil.

REFERENCES

Abo, A., Boyhan, A., West, I., Thrasher, A. J., & Segal, A. W. (1992). Reconstitution of neutrophil NADPH oxidase activity in the cell-free system by four components: p67-phox, p47-phox, p21rac1, and cytochrome b-245. *The Journal of Biological Chemistry, 267*, 16767–16770.

Ambruso, D. R., Bolscher, B. G., Stokman, P. M., Verhoeven, A. J., & Roos, D. (1990). Assembly and activation of the NADPH:O2 oxidoreductase in human neutrophils after stimulation with phorbol myristate acetate. *The Journal of Biological Chemistry, 265*, 924–930.

Ambruso, D. R., Ellison, M. A., Thurman, G. W., & Leto, T. L. (2012). Peroxiredoxin 6 translocates to the plasma membrane during neutrophil activation and is required for optimal NADPH oxidase activity. *Biochimica et Biophysica Acta, 1823*, 306–315.

Ambruso, D. R., & Johnston, R. B., Jr. (2012). Primary immunodeficiency: Chronic granulomatous disease and common variable immunodeficiency disorders. In R. W. Wilmott, T. F. Boat, A. Bush, V. Chernick, R. R. Deterding, & F. Ratjen (Eds.), *Kendig and Chernick's disorders of the respiratory tract in children* (pp. 886–898). Philadelphia: Elsevier Saunders.

Ambruso, D. R., Knall, C., Abell, A. N., Panepinto, J., Kurkchubasche, A., Thurman, G., et al. (2000). Human neutrophil immunodeficiency syndrome is associated with an inhibitory Rac2 mutation. *Proceedings of the National Academy of Sciences of the United States of America, 97*, 4654–4659.

Ambruso, D. R., Stork, L. C., Gibson, B. E., & Thurman, G. W. (1987). Increased activity of the respiratory burst in cord blood neutrophils: Kinetics of the NADPH oxidase enzyme system in subcellular fractions. *Pediatric Research, 21*, 205–210.

Bromberg, Y., & Pick, E. (1985). Activation of NADPH-dependent superoxide production in a cell-free system by sodium dodecyl sulfate. *The Journal of Biological Chemistry, 260*, 13539–13545.

Chae, H. Z., Chung, S. J., & Rhee, S. G. (1994). Thioredoxin-dependent peroxide reductase from yeast. *The Journal of Biological Chemistry, 269*, 27670–27678.

Chae, H. Z., Kang, S. W., & Rhee, S. G. (1999). Isoforms of mammalian peroxiredoxin that reduce peroxides in presence of thioredoxin. *Methods in Enzymology, 300*, 219–226.

Chae, H. Z., Robison, K., Poole, L. B., Church, G., Storz, G., & Rhee, S. G. (1994). Cloning and sequencing of thiol-specific antioxidant from mammalian brain: Alkyl hydroperoxide reductase and thiol-specific antioxidant define a large family of antioxidant enzymes. *Proceedings of the National Academy of Sciences of the United States of America, 91*, 7017–7021.

Chae, H. Z., Uhm, T. B., & Rhee, S. G. (1994). Dimerization of thiol-specific antioxidant and the essential role of cysteine 47. *Proceedings of the National Academy of Sciences of the United States of America, 91*, 7022–7026.

Chen, J. W., Dodia, C., Feinstein, S. I., Jain, M. K., & Fisher, A. B. (2000). 1-Cys peroxiredoxin, a bifunctional enzyme with glutathione peroxidase and phospholipase A2 activities. *The Journal of Biological Chemistry*, *275*, 28421–28427.

Clark, R. A., Volpp, B. D., Leidal, K. G., & Nauseef, W. M. (1990). Two cytosolic components of the human neutrophil respiratory burst oxidase translocate to the plasma membrane during cell activation. *The Journal of Clinical Investigation*, *85*, 714–721.

Curnutte, J. T. (1985). Activation of human neutrophil nicotinamide adenine dinucleotide phosphate, reduced (triphosphopyridine nucleotide, reduced) oxidase by arachidonic acid in a cell-free system. *The Journal of Clinical Investigation*, *75*, 1740–1743.

Dang, P. M., Cross, A. R., & Babior, B. M. (2001). Assembly of the neutrophil respiratory burst oxidase: A direct interaction between p67PHOX and cytochrome b558. *Proceedings of the National Academy of Sciences of the United States of America*, *98*, 3001–3005.

DeLeo, F. R., & Quinn, M. T. (1996). Assembly of the phagocyte NADPH oxidase: Molecular interaction of oxidase proteins. *Journal of Leukocyte Biology*, *60*, 677–691.

de Mendez, I., Adams, A. G., Sokolic, R. A., Malech, H. L., & Leto, T. L. (1996). Multiple SH3 domain interactions regulate NADPH oxidase assembly in whole cells. *The EMBO Journal*, *15*, 1211–1220.

de Mendez, I., & Leto, T. L. (1995). Functional reconstitution of the phagocyte NADPH oxidase by transfection of its multiple components in a heterologous system. *Blood*, *85*, 1104–1110.

Dinauer, M. (2003). The phagocyte system and disorders of granulopoiesis and granulocyte function. In D. G. Nathan, S. H. Orkin, D. Ginsburg, & A. T. Look (Eds.), *Hematology of infancy and childhood* (pp. 923–1009). Philadelphia: W.B. Saunders Co.

Elbashir, S. M., Harborth, J., Lendeckel, W., Yalcin, A., Weber, K., & Tuschl, T. (2001). Duplexes of 21-nucleotide RNAs mediate RNA interference in cultured mammalian cells. *Nature*, *411*, 494–498.

Ellison, M. A., Thurman, G. W., & Ambruso, D. R. (2012). Phox activity of differentiated PLB-985 cells is enhanced, in an agonist specific manner, by the PLA2 activity of Prdx6-PLA2. *European Journal of Immunology*, *42*, 1609–1617.

Hannon, G. J. (2002). RNA interference. *Nature*, *418*, 244–251.

Huang, J., & Kleinberg, M. E. (1999). Activation of the phagocyte NADPH oxidase protein p47(phox). Phosphorylation controls SH3 domain-dependent binding to p22(phox). *The Journal of Biological Chemistry*, *274*, 19731–19737.

Kang, S. W., Baines, I. C., & Rhee, S. G. (1998). Characterization of a mammalian peroxiredoxin that contains one conserved cysteine. *The Journal of Biological Chemistry*, *273*, 6303–6311.

Kang, S. W., Chae, H. Z., Seo, M. S., Kim, K., Baines, I. C., & Rhee, S. G. (1998). Mammalian peroxiredoxin isoforms can reduce hydrogen peroxide generated in response to growth factors and tumor necrosis factor-alpha. *The Journal of Biological Chemistry*, *273*, 6297–6302.

Kim, T. S., Dodia, C., Chen, X., Hennigan, B. B., Jain, M., Feinstein, S. I., et al. (1998). Cloning and expression of rat lung acidic Ca(2+)-independent PLA2 and its organ distribution. *The American Journal of Physiology*, *274*, L750–L761.

Kim, S. Y., Jo, H. Y., Kim, M. H., Cha, Y. Y., Choi, S. W., Shim, J. H., et al. (2008). H2O2-dependent hyperoxidation of peroxiredoxin 6 (Prdx6) plays a role in cellular toxicity via up-regulation of iPLA2 activity. *The Journal of Biological Chemistry*, *283*, 33563–33568.

Leavey, P. J., Gonzalez-Aller, C., Thurman, G., Kleinberg, M., Rinckel, L., Ambruso, D. W., et al. (2002). A 29-kDa protein associated with p67phox expresses both peroxiredoxin and phospholipase A2 activity and enhances superoxide anion production by a cell-free system of NADPH oxidase activity. *The Journal of Biological Chemistry*, *277*, 45181–45187.

Leavey, P. J., Sellins, K. S., Thurman, G., Elzi, D., Hiester, A., Silliman, C. C., et al. (1998). In vivo treatment with granulocyte colony-stimulating factor results in divergent effects on neutrophil functions measured in vitro. *Blood*, *92*, 4366–4374.

Leto, T. L., Lavigne, M. C., Homoyounpour, N., Lekstrom, K., Linton, G., Malech, H. L., et al. (2007). The K-562 cell model for analysis of neutrophil NADPH oxidase function. *Methods in Molecular Biology*, *412*, 365–383.

Matute, J. D., Arias, A. A., Dinauer, M. C., & Patino, P. J. (2005). p40phox: The last NADPH oxidase subunit. *Blood Cells, Molecules & Diseases*, *35*, 291–302.

Matute, J. D., Arias, A. A., Wright, N. A., Wrobel, I., Waterhouse, C. C., Li, X. J., et al. (2009). A new genetic subgroup of chronic granulomatous disease with autosomal recessive mutations in p40 phox and selective defects in neutrophil NADPH oxidase activity. *Blood*, *114*, 3309–3315.

McPhail, L. C., Shirley, P. S., Clayton, C. C., & Snyderman, R. (1985). Activation of the respiratory burst enzyme from human neutrophils in a cell-free system. Evidence for a soluble cofactor. *The Journal of Clinical Investigation*, *75*, 1735–1739.

Netto, L. E. S., Chae, H. Z., Kang, S. W., Rhee, S. G., & Stadtman, E. R. (1996). Removal of hydrogen peroxide by thiol-specific antioxidant enzyme (TSA) is involved with its antioxidant properties. TSA possesses thiol peroxidase activity. *The Journal of Biological Chemistry*, *271*, 15315–15321.

Paddison, P. J., Caudy, A. A., & Hannon, G. J. (2002). Stable suppression of gene expression by RNAi in mammalian cells. *Proceedings of the National Academy of Sciences of the United States of America*, *99*, 1443–1448.

Parkos, C. A., Allen, R. A., Cochrane, C. G., & Jesaitis, A. J. (1987). Purified cytochrome b from human granulocyte plasma membrane is comprised of two polypeptides with relative molecular weights of 91,000 and 22,000. *The Journal of Clinical Investigation*, *80*, 732–742.

Price, M. O., McPhail, L. C., Lambeth, J. D., Han, C. H., Knaus, U. G., & Dinauer, M. C. (2002). Creation of a genetic system for analysis of the phagocyte respiratory burst: High-level reconstitution of the NADPH oxidase in a nonhematopoietic system. *Blood*, *99*, 2653–2661.

Quinn, M. T., Mullen, M. L., & Jesaitis, A. J. (1992). Human neutrophil cytochrome b contains multiple hemes. Evidence for heme associated with both subunits. *The Journal of Biological Chemistry*, *267*, 7303–7309.

Rhee, S. G., Kang, S. W., Chang, T. S., Jeong, W., & Kim, K. (2001). Peroxiredoxin, a novel family of peroxidases. *IUBMB Life*, *52*, 35–41.

Rosenthal, J., Thurman, G. W., Cusack, N., Peterson, V. M., Malech, H. L., & Ambruso, D. R. (1996). Neutrophils from patients after burn injury express a deficiency of the oxidase components p47-phox and p67-phox. *Blood*, *88*, 4321–4329.

Segal, A. W., West, I., Wientjes, F., Nugent, J. H., Chavan, A. J., Haley, B., et al. (1992). Cytochrome b-245 is a flavocytochrome containing FAD and the NADPH-binding site of the microbicidal oxidase of phagocytes. *The Biochemical Journal*, *284*(Pt 3), 781–788.

Sharp, P. A. (2001). RNA interference—2001. *Genes & Development*, *15*, 485–490.

Thurman, R. G., Ley, H. G., & Scholz, R. (1972). Hepatic microsomal ethanol oxidation. Hydrogen peroxide formation and the role of catalase. *European Journal of Biochemistry*, *25*, 420–430.

Tuschl, T., Zamore, P. D., Lehmann, R., Bartel, D. P., & Sharp, P. A. (1999). Targeted mRNA degradation by double-stranded RNA in vitro. *Genes & Development*, *13*, 3191–3197.

Ueyama, T., Nakakita, J., Nakamura, T., Kobayashi, T., Kobayashi, T., Son, J., et al. (2011). Cooperation of p40(phox) with p47(phox) for Nox2-based NADPH oxidase activation during Fcgamma receptor (FcgammaR)-mediated phagocytosis: Mechanism for acquisition of p40(phox) phosphatidylinositol 3-phosphate (PI(3)P) binding. *The Journal of Biological Chemistry*, *286*, 40693–40705.

Volpp, B. D., Nauseef, W. M., & Clark, R. A. (1988). Two cytosolic neutrophil oxidase components absent in autosomal chronic granulomatous disease. *Science*, *242*, 1295–1297.

CHAPTER NINE

Study of the Signaling Function of Sulfiredoxin and Peroxiredoxin III in Isolated Adrenal Gland: Unsuitability of Clonal and Primary Adrenocortical Cells

In Sup Kil, Soo Han Bae, Sue Goo Rhee[1]

Yonsei Biomedical Research Institute, Yonsei University College of Medicine, Seodaemun-gu, Seoul, South Korea

[1]Corresponding author: e-mail address: rheesg@yuhs.ac

Contents

1. Introduction 170
2. Hyperoxidation of PrxIII by H_2O_2 Generated During Corticosterone Synthesis 171
3. Induction of Srx by ACTH 173
4. Unsuitability of Clonal and Primary Adrenocortical Cells for Studies of the Srx–PrxIII Regulatory Pathway 174
5. Adrenal Gland Organ Culture as an *In Vitro* Model for the Srx–PrxIII Regulatory Pathway 176
6. Concluding Remarks 179

Acknowledgment 179

References 179

Abstract

Members of the peroxiredoxin (Prx) family of antioxidant enzymes are inactivated via hyperoxidation of the active site cysteine by the substrate H_2O_2 and are reactivated via an ATP-consuming process catalyzed by sulfiredoxin (Srx). PrxIII is reversibly inactivated by H_2O_2 produced by cytochrome P450 11B1 (CYP11B1) in mitochondria during corticosterone synthesis in the adrenal gland of mice injected with adrenocorticotropic hormone (ACTH). Inactivation of PrxIII triggers a sequence of events including accumulation of H_2O_2, activation of p38 mitogen-activated kinase (MAPK), inhibition of cholesterol transfer, and suppression of corticosterone synthesis. Srx expression is significantly induced by ACTH injection. The coupling of CYP11B1 activity to PrxIII inactivation and Srx induction provides a feedback regulatory mechanism for steroidogenesis that functions independently of the hypothalamic–pituitary–adrenal axis. Furthermore, the PrxIII–Srx regulatory pathway is critical for the circadian rhythm of corticosterone production.

Methods in Enzymology, Volume 527
ISSN 0076-6879
http://dx.doi.org/10.1016/B978-0-12-405882-8.00009-X

Although adrenocortical tumor cell lines such as Y-1 and H295R have been used extensively for studying the mechanism of steroidogenesis, those clonal cells were found to be unsuitable as an *in vitro* model for redox signaling because the amount of Srx in the cell lines is much higher than that in mouse adrenal gland and not affected by ACTH stimulation. Furthermore, the levels of PrxIII in the clonal cells are greatly reduced compared to that in the adrenal gland, and ACTH does not induce PrxIII hyperoxidation in the clonal cells. Primary adrenocortical cells isolated from the mouse adrenal gland were also found to be an invalid model because Srx levels are increased, along with decreased levels of hyperoxidized PrxIII, soon after isolation of these cells. Organ culture system is, however, appropriate for studying the PrxIII–Srx regulatory function as the levels of hyperoxidized PrxIII and Srx in the adrenal glands maintained overnight in culture medium are not changed.

1. INTRODUCTION

Peroxiredoxins (Prxs) are a family of peroxidases that reduce peroxides, with a conserved cysteine residue serving as the site of oxidation by peroxides. Mammalian cells express six different Prx isoforms: four 2-Cys Prx isoforms (PrxI–IV), one atypical 2-Cys Prx isoform (PrxV), and one 1-Cys Prx isoform (PrxVI). These isoforms vary in subcellular localization, with PrxI, II, and VI being localized mainly in cytosol, PrxIII being restricted in the matrix of mitochondria, PrxIV being found mainly in the lumen of endoplasmic reticulum, and PrxV being in cytosol, mitochondria, and peroxisomes. During catalysis, peroxides oxidize the catalytic cysteine (Cys–SH) to sulfenic acid (Cys–SOH), which then reacts with another cysteine residue to form a disulfide that is subsequently reduced by an appropriate electron donor like thioredoxin (Trx). Cys–SOH of PrxI to III is occasionally further oxidized to sulfinic acid (Cys–SO_2H) before disulfide formation, resulting in inactivation of peroxidase activity. This hyperoxidation is reversed by sulfiredoxin (Srx) in a slow process that requires ATP hydrolysis and reducing equivalents such as Trx and GSH (Chang et al., 2004). Given that disulfide bond is insensitive to oxidation by H_2O_2 and needs to be reduced to Cys–SH before oxidized again to Cys–SOH, hyperoxidation of Prx occurs only during catalytic cycle (Yang et al., 2002). Prokaryotic Prx enzymes are insensitive to oxidative inactivation, and prokaryotes do not express Srx (Biteau, Labarre, & Toledano, 2003; Wood, Schroder, Robin Harris, & Poole, 2003). Reversible inactivation through hyperoxidation has therefore been speculated to be a eukaryotic adaptation that allows H_2O_2 to accumulate to substantial levels under certain circumstances (Wood, Schroder, Robin Harris, & Pools, 2003).

We showed that the reversible hyperoxidation of PrxIII provides a feedback regulatory mechanism for steroidogenesis that functions independently of the hypothalamus–pituitary–adrenal axis (Kil et al., 2012). During this investigation, we found that neither clonal steroidogenic tumor cell lines nor primary adrenal cells, but isolated whole adrenal organs, are suitable for the studies on the physiological function of Srx and PrxIII.

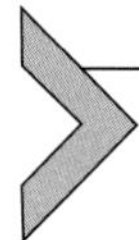

2. HYPEROXIDATION OF PrxIII BY H_2O_2 GENERATED DURING CORTICOSTERONE SYNTHESIS

When we examined various mouse tissues, sulfinic PrxIII was detected only in the adrenal cortex, whereas the sulfinic form of PrxI and PrxII was not detected in any of the tissues examined (Kil et al., 2012). In addition to PrxIII, glutathione peroxidase 1 (GPx1) and PrxV are known to be present in mitochondria. However, PrxIII is by far the most abundant antioxidant enzyme in the adrenal gland, and ~10–20% of total PrxIII was found in sulfinic form in the adrenal cortex of mice. Given that hyperoxidation of 2-Cys Prxs occurs only during the catalytic cycle (Yang et al., 2002), these results suggested that the amount of H_2O_2 produced in the adrenal gland under normal physiological conditions is sufficient to maintain the abundant PrxIII continuously engaged in H_2O_2 reduction and that the level of mitochondrial Srx is not enough to fully counteract the hyperoxidation of PrxIII.

The mammalian adrenal gland consists of two endocrine tissues of different embryological origin: the outer, mesodermally derived, steroid-producing cortex and the central medulla that harbor the neural crest-derived, catecholamin-producing chromaffin cells. Psychological or physical stresses cause the release of corticotropin-releasing hormone (CRH) and arginine-vasopressin (AVP) from the hypothalamus. CRH and AVP are transported to the pituitary gland, where they act synergistically to stimulate the secretion of adrenocorticotropic hormone (ACTH). ACTH is transported by the blood to the adrenal cortex, where it rapidly stimulates the biosynthesis and secretion of glucocorticoids (mainly corticosterone in rodents and cortisol in primates). ACTH stimulates the glucocorticoid synthesis mainly through activation of the cAMP-dependent pathway (Jefcoate, Lee, Cherradi, Takemori, & Duan, 2011; Manna, Dyson, & Stocco, 2009; Manna, & Stocco, 2011). The biosynthesis of glucocorticoid from cholesterol requires the sequential actions of cytochrome p450 (CYP) enzymes in the mitochondria and smooth endoplasmic reticulum. The first step, the oxidative cleavage of cholesterol side-chain is catalyzed by CYP11A1 in the inner mitochondrial membrane to produce pregnenolone (Jefcoate, 2002). Pregnenolone is converted to progesterone by 3-β-hydroxysteroid dehydrogenase and then to 11-deoxycorticosterone by CYP21 in the endoplasmic reticulum. The synthesis is completed in the mitochondrial inner membrane, where 11-deoxycorticosterone is hydroxylated by CYP11B1 to produce corticosterone. The oxygenation reactions catalyzed by CYP11A1 and CYP11B1 require donation of electrons from NADPH via the intermediary of mitochondrial adrenodoxin reductase (AdR) and adrenodoxin (Ad), which are located on the matrix side of the mitochondrial inner membrane (Hanukoglu, 2006; Miller, 2005). The process of electron transfer (NADPH → AdxR → Adx → CYP → substrate) during enzyme activity is not perfectly coupled and leaks electrons that react with O_2 to produce superoxide anions as is the case with electron leaks from the mitochondrial respiratory chain (Hanukoglu, 2006; Zangar, Davydov, & Verma, 2004). Superoxide anions are converted to H_2O_2 by superoxide dismutase (SOD). Though each of the three protein components (AdxR, Adx, and CYP) can be a site of electron leakage, CYP11A1 or CYP11B1 leaks most (Hanukoglu, Rapoport, Weiner, & Sklan, 1993).

We monitored the production of corticosterone and PrxIII–SO_2 in the adrenal gland of mice injected with ACTH. The plasma concentration of corticosterone reached a maximum within 1 h and decreased concurrently thereafter (Fig. 9.1A).

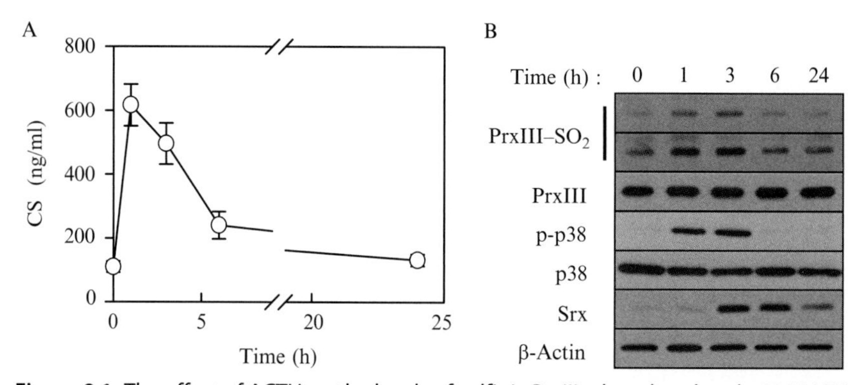

Figure 9.1 The effect of ACTH on the levels of sulfinic PrxIII, phosphorylated p38 MAPK, and Srx in the adrenal gland. Mice were injected intraperitoneally (i.p.) with saline or ACTH (10 μg/kg) and killed at the indicated times thereafter. Plasma CS levels were measured (A; means ± SD, $n = 10–12$), and adrenal gland homogenates were prepared and subjected to immunoblot analysis (B). *Data adapted from Kil et al. (2012), with permission.*

The abundance of PrxIII–SO_2 in the adrenal gland was increased approximately twofold at 1–3 h after ACTH injection and decreased thereafter, whereas the total amount of PrxIII did not change (Fig. 9.1B). The amount of phosphorylated (activated) p38 mitogen-activated protein kinase (MAPK) in the adrenal gland was markedly increased at 1–3 h after ACTH injection, the same time period at which the accumulation of PrxIII–SO_2 was maximal (Fig. 9.1B). The nearly concurrent increases in the levels of corticosterone, PrxIII–SO_2, and phosphorylated p38 in ACTH-injected mice suggest that ACTH-induced steroidogenesis is accompanied by increased H_2O_2 production in mitochondria, which results in increased hyperoxidation of PrxIII and consequent accumulation of H_2O_2 in mitochondria and that H_2O_2 molecules then overflow into the cytosol to trigger signaling pathway leading to p38 phosphorylation. The ACTH-induced increases of PrxIII–SO_2, phosphorylated p38, and corticosterone were prevented by prior treatment of mice with a CYP11B1 inhibitor metyrapone, suggesting that H_2O_2 produced by CYP11B1 is required for PrxIII hyperoxidation. The activated p38 inhibited StAR synthesis through an unknown mechanism, resulting in downregulation of corticosterone production and consequent inhibition of H_2O_2 generation (Kil et al., 2012).

3. INDUCTION OF Srx BY ACTH

ACTH also induced upregulation of Srx with its levels peaking at 3–6 h after injection of ACTH (Fig. 9.1B). ACTH did not substantially affect the expression of antioxidant proteins such as PrxI to PrxVI, SOD1, SOD2, catalase, Trx1, Trx2, and GPx1 or that of steroidogenic proteins such as StAR, CYP11A1, and CYP11B1 in the adrenal gland (Kil et al., 2012). Srx expression is known to be regulated via two distinct transcriptional factors, AP-1 and nuclear factor erythroid 2-related factor (Nrf2). AP-1 is activated downstream of a variety of signaling pathways including MAPKs, while Nrf2 responds to oxidative stress induced by reactive oxygen species and organizes the expression of various antioxidant enzymes (Soriano et al., 2009). We found that injection of ACTH increased the abundance of Srx in the adrenal glands of wild-type and Nrf2 knockout (KO) mice by similar extents, whereas induction of HO-1, whose gene is a well-characterized target of Nrf2, was apparent in wild-type mice but not in Nrf2 KO mice (Kil et al., 2012). ACTH activates ERK via a cAMP-dependent pathway in Y-1 mouse adrenocortical tumor cells (Le & Schimmer, 2001), and ERK activation leads to AP-1 activation. We found that Srx was induced by forskolin (an activator of adenylyl cyclase) in Y-1 cells and that this effect was blocked by U0126 and PD98059, two inhibitors of the ERK signaling pathway, but not by the p38 inhibitor SB202190, suggesting that ACTH increases Srx expression via a cAMP–ERK–AP-1 pathway (Kil et al., 2012).

Srx is a cytosolic protein that translocates into mitochondria in oxidatively stressed cells. Srx is found more in the mitochondria than in the cytosol of the adrenal gland of unstressed mouse, suggesting that mitochondria of such mouse adrenal gland are under constant oxidative stress; and treatment with ACTH caused further translocation of Srx to the mitochondria.

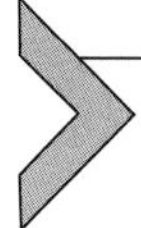

4. UNSUITABILITY OF CLONAL AND PRIMARY ADRENOCORTICAL CELLS FOR STUDIES OF THE Srx–PrxIII REGULATORY PATHWAY

Clonal cells such as the mouse Y-1 and human H295R adrenocortical tumor cell lines have been used extensively for studying the mechanism of steroidogenesis including the acute regulation of steroidogenic enzymes (Abidi et al., 2008; Clark, Pezzi, Stocco, & Rainey, 1995; Lalli, Melner, Stocco, & Sassone-Corsi, 1998). We attempted to demonstrate the Srx–PrxIII regulatory system in the *in vitro* cell system. However, these cell lines were found to be unsuitable for study of the Srx–PrxIII regulatory system. The level of Srx in Y-1 cells under basal conditions was thus much higher than that in the unstimulated mouse adrenal gland and was similar to that in the ACTH-stimulated adrenal gland (Fig. 9.2A). H295R human adrenocortical carcinoma cells, which produce cortisol on stimulation with forskolin, also expressed Srx at a high level, and they did not produce more in response to forskolin (Fig. 9.2A). Furthermore, the levels of PrxIII in both Y-1 and H295R cells were greatly reduced compared with that in the adrenal gland, and forskolin did not induce PrxIII hyperoxidation in either cell line (Fig. 9.2A). The abundance of various antioxidant enzymes in the mouse adrenal gland and Y-1 cells was also compared in more detail (Fig. 9.2B). The amounts of the cytosolic enzymes PrxI, PrxII, and Trx reductase 1 (TrxR1) in Y-1 cells were similar to those in the adrenal gland, whereas the amount of the mitochondrial TrxR (TrxR2) was much smaller in Y-1 cells, as is the case for PrxIII. TrxR2 is required to support the peroxidase function of PrxIII. These results demonstrate fundamental limitations in using clonal steroidogenic cell lines to study H_2O_2 signaling involving Srx and PrxIII, although Y-1 cells were useful to study the mechanism of Srx induction downstream of cAMP elevation.

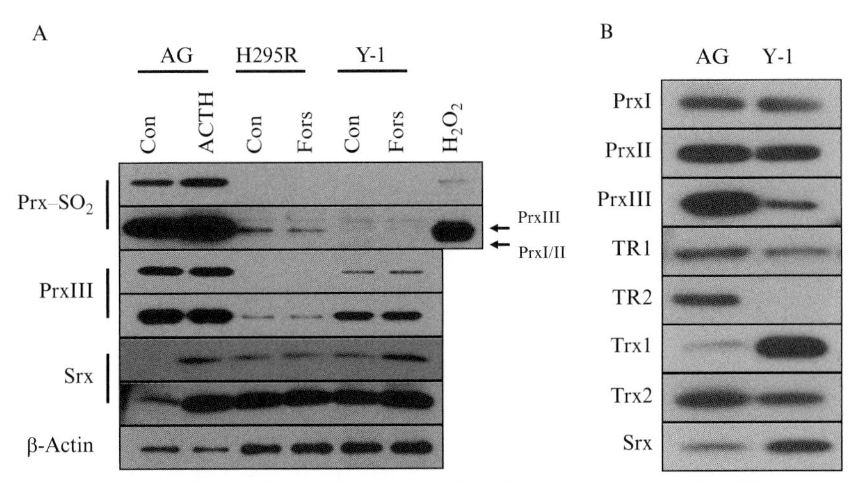

Figure 9.2 Unsuitability of clonal cells for studies of PrxIII and Srx signaling. (A) Homogenates of the adrenal glands of mice that had been injected with ACTH (10 μg/kg, i.p.) or saline (Con) and killed 3 h later were subjected to immunoblot analysis with antibodies to the indicated proteins. H295R and Y-1 cells were incubated in the absence (Con) or presence of 5 μ*M* forskolin (Fors) for 6 h, after which cell lysates were similarly analyzed. Lysates of Y-1 cells that had been exposed to 500 μ*M* H_2O_2 for 10 min were analyzed as a positive control (H_2O_2) for sulfinic PrxI/II and sulfinic PrxIII (rightmost lane). (B) Mouse adrenal gland homogenates from (A) and Y-1 cell lysates were subjected to immunoblot analysis with antibodies to the indicated proteins.

Primary adrenocortical cells isolated from the mouse adrenal gland (Tsai, Lu, Lin, & Wang, 2003) were also found to be unsuitable for study of the Srx–PrxIII regulatory system. Because of the marked oxidative stress that accompanies steroidogenesis, these cells express high levels of Srx at baseline. However, Srx production increased further, along with decreased levels of PrxIII–SO_2 and StAR, as early as 12 h after isolation of these cells; PrxIII levels remained unchanged (Fig. 9.3). Similar Srx induction and PrxIII insensitivity to hyperoxidation were also observed in several other primary cell types isolated from mice. For example, Srx was undetectable in freshly isolated mouse hepatocytes, but its abundance was greatly increased and PrxIII consequently became insensitive to hyperoxidation after culture of the cells overnight in ambient atmosphere (Bae et al., 2011). Even adrenocortical cells that already express high levels of Srx produce more of the protein in order to survive under the ambient atmosphere, likely to counteract the hyperoxidation of PrxI and PrxII in the cytosol.

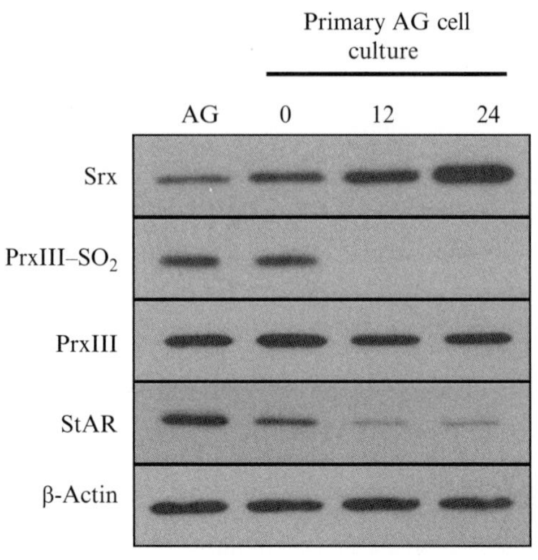

Figure 9.3 Unsuitability of primary adrenocortical cells for studies of PrxIII and Srx signaling. Homogenates of whole adrenal glands or of primary adrenal cells cultured for the indicated times were subjected to immunoblot analysis with antibodies to the indicated proteins.

5. ADRENAL GLAND ORGAN CULTURE AS AN *IN VITRO* MODEL FOR THE Srx–PrxIII REGULATORY PATHWAY

Adrenal tissue maintained in organ culture was shown to secrete corticosterone into the culture medium according to a daily cycle, and such a secretion was increased by exogenous stimulants such as ACTH. We tested if the organ culture system is a valid model for *in vivo* redox signaling. Mouse adrenal glands were cultured using a modified version of a previously described method (Gwosdow, Kumar, & Bode, 1990; Mohn et al., 2005).

Adrenal glands of mice were excised, defatted, and cultured at 37 °C in 20% O_2 and 5% CO_2 incubator. The adrenal glands were cultured for 1 h in Ham's F-12K medium before incubation for the indicated times with test agents in 1 ml of fresh medium per gland. The medium was then collected and immediately frozen at −80 °C until assay for measurement of corticosterone concentration. The concentrations of corticosterone (5 μl) in culture medium were determined with the use of enzyme-linked immunosorbent assay kits (DRG Diagnostics). The adrenal glands were homogenized in a solution containing 50 m*M* Tris–HCl (pH 7.5), 150 m*M* NaCl, 1 m*M* EDTA, 1 m*M* 4-(2-aminoethyl)-benzenesulfonyl fluoride, aprotinin (10 μg/ml), leupeptin (10 μg/ml), and 1% Nonidet P-40, and the homogenates were centrifuged to remove cell debris. The protein extracts were then subjected to SDS–polyacrylamide gel electrophoresis, the separated proteins were transferred to a polyvinylidene difluoride membrane, and the membrane was incubated first with primary antibodies and then with horseradish peroxidase-conjugated secondary antibodies and enhanced chemiluminescence reagents. Rabbit antibodies to StAR phosphorylated at Ser^{194} were prepared as described (Jo, King, Khan, & Stocco, 2005) and subjected to affinity purification with the use of Sepharose 4B resin (GE Healthcare) conjugated with the corresponding nonphosphorylated form of the antigenic phosphopeptide.

Unlike in isolated cells, the levels of PrxIII–SO_2, StAR, and Srx were not changed in adrenal glands maintained overnight in organ culture, likely because the oxygen tension in the zona fasciculata was reduced substantially by the outer zone (zona glomerulosa) of the cortex to a level similar to that encountered in the *in vivo* condition. To examine the effect of p38 activation on corticosterone synthesis, we incubated mouse adrenal glands with the p38 inhibitor SB202190 before stimulation with ACTH. The p38 inhibitor increased ACTH-induced corticosterone production by ~50% as well as enhanced the basal level of corticosterone synthesis (Fig. 9.4A). In addition, SB202190 potentiated the ACTH-induced hyperoxidation of PrxIII (Fig. 9.4B), as would be expected from the increased H_2O_2 production accompanying corticosterone synthesis. Regulation of StAR function is a complex process that involves multiple signaling pathways that coordinate the transcriptional machinery as well as posttranscriptional mechanisms. It is generally accepted, however, that cholesterol delivery to the inner mitochondrial membrane depends largely on StAR molecules that are newly synthesized and phosphorylated at Ser^{194} (Artemenko, Zhao, Hales, Hales, & Jefcoate, 2001; Manna et al., 2009). Consistent with this notion, the amount of phosphorylated StAR was increased in adrenal glands treated with ACTH and SB202190 (Fig. 9.4B).

When Srx was ablated specifically in steroidogenic tissues to generate a mouse model ($Srx^{\Delta SC}$), plasma corticosterone levels were reduced by ~50% in $Srx^{\Delta SC}$ mice compared with those in wild-type (Srx^{WT}) mice (Kil et al., 2012). The increase in the plasma corticosterone concentration in response to ACTH injection was smaller in $Srx^{\Delta SC}$ mice than in Srx^{WT} mice at all time points examined. ACTH injection further increased an already elevated abundance of PrxIII–SO_2 and phosphorylated p38 in the adrenal gland of $Srx^{\Delta SC}$ mice, suggesting that the regulatory circuit consisting of PrxIII hyperoxidation, p38 activation, and suppression of corticosterone synthesis

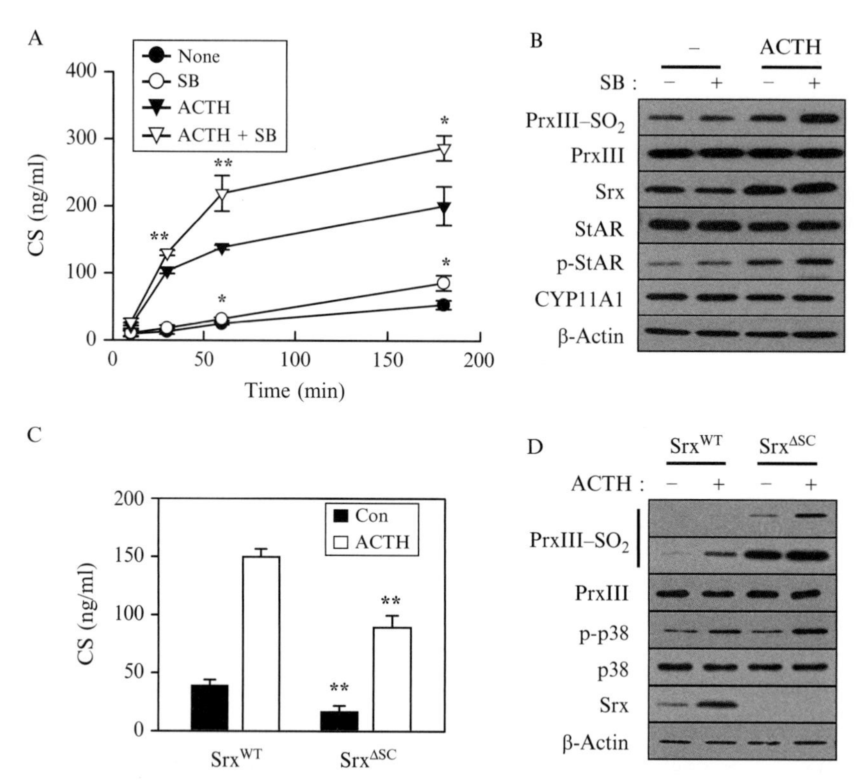

Figure 9.4 ACTH-induced corticosterone production in adrenal glandculture and suppression of corticosterone production by ablation of Srx. (A and B) Mouse adrenal glands were incubated in the absence or presence of 10 μ*M* SB202190 (SB) for 1 h and then in the additional absence or presence of 500 n*M* ACTH for the indicated times, after which the concentration of corticosterone in the medium was measured (A). Immediately after the last time point for corticosterone measurement in A, adrenal glandhomogenates were prepared and subjected to immunoblot analysis with antibodies to the indicated proteins (B). Corticosterone data are means ± SD ($n=6$). $^{*}p<0.05$, $^{**}p<0.01$ versus control (none). (C) Adrenal glands from Srx^{WT} and $Srx^{\Delta SC}$ mice were incubated in the absence (Con) or presence of 500 n*M* ACTH for 1 h, after which corticosterone levels in the culture medium were measured. Data are means ± SD ($n=6$). $^{**}p<0.01$ versus the corresponding value for Srx^{WT} mice. (D) Homogenates prepared from the adrenal glands in (A) were subjected to immunoblot analysis with antibodies to the indicated proteins. *Data adapted from Kil et al. (2012), with permission.*

operates with an increased sensitivity in $Srx^{\Delta SC}$ mice (Kil et al., 2012). When the adrenal glands were isolated and incubated in the absence or presence of ACTH, the amounts of corticosterone produced by the adrenal gland of $Srx^{\Delta SC}$ mice were also smaller than those produced by the adrenal gland of Srx^{WT} mice (Fig. 9.4C). In addition, the levels of sulfinic PrxIII and phosphorylated p38 in adrenal glands stimulated with ACTH *in vitro* were greater for $Srx^{\Delta SC}$ mice than for Srx^{WT} mice (Fig. 9.4D). These results further support the validity of organ culture model for the Srx–PrxIII regulatory system.

6. CONCLUDING REMARKS

It appears that animal cells, which normally do not see 21% O_2, essentially do not express Srx under normal conditions. Adrenal cortex, which has to confront strong oxidative stress arising from steroidogenesis, expresses relatively high levels of Srx compared to other tissues. Upon isolation from mouse tissues, however, adrenocortical cells rapidly produce more Srx under the ambient atmosphere, and become insensitive to ACTH-dependent redox regulation. Equally not useful for the study are adrenocortical tumor cells (Y-1 and H295R cells) as they express high levels of Srx and do not produce more in response to cAMP elevation. However, when adrenal glands are maintained in culture medium, Srx levels remain low and PrxIII is readily hyperoxidized in response to ACTH stimulation. Experiments with adrenal glands derived from Srx-deficient mice suggest the normal operation of the Srx–PrxIII regulatory system in the tissue. Thus, the organ culture model can replace the expensive and time-consuming animal experiments on the PrxIII–Srx regulatory pathway.

ACKNOWLEDGMENT

This study was supported by grants from the Korean Science and Engineering Foundation (National Honor Scientist program grant 2006-05106 and Bio R&D program grant M10642040001-07N4204-00110 to S. G. R.).

REFERENCES

Abidi, P., Zhang, H., Zaidi, S. M., Shen, W. J., Leers-Sucheta, S., Cortez, Y., et al. (2008). Oxidative stress-induced inhibition of adrenal steroidogenesis requires participation of p38 mitogen-activated protein kinase signaling pathway. *The Journal of Endocrinology, 198*, 193–207.

Artemenko, I. P., Zhao, D., Hales, D. B., Hales, K. H., & Jefcoate, C. R. (2001). Mitochondrial processing of newly synthesized steroidogenic acute regulatory protein (StAR), but not total StAR, mediates cholesterol transfer to cytochrome P450 side chain cleavage enzyme in adrenal cells. *The Journal of Biological Chemistry, 276*, 46583–46596.

Bae, S. H., Sung, S. H., Cho, E. J., Lee, S. K., Lee, H. E., Woo, H. A., et al. (2011). Concerted action of sulfiredoxin and peroxiredoxin I protects against alcohol-induced oxidative injury in mouse liver. *Hepatology, 53*, 945–953.

Biteau, B., Labarre, J., & Toledano, M. B. (2003). ATP-dependent reduction of cysteine-sulphinic acid by *S. cerevisiae* sulphiredoxin. *Nature, 425*, 980–984.

Chang, T. S., Jeong, W., Woo, H. A., Lee, S. M., Park, S., & Rhee, S. G. (2004). Characterization of mammalian sulfiredoxin and its reactivation of hyperoxidized peroxiredoxin through reduction of cysteine sulfinic acid in the active site to cysteine. *The Journal of Biological Chemistry, 279*, 50994–51001.

Clark, B. J., Pezzi, V., Stocco, D. M., & Rainey, W. E. (1995). The steroidogenic acute regulatory protein is induced by angiotensin II and K+ in H295R adrenocortical cells. *Molecular and Cellular Endocrinology, 115*, 215–219.

Gwosdow, A. R., Kumar, M. S., & Bode, H. H. (1990). Interleukin 1 stimulation of the hypothalamic-pituitary-adrenal axis. *The American Journal of Physiology, 258*, E65–E70.

Hanukoglu, I. (2006). Antioxidant protective mechanisms against reactive oxygen species (ROS) generated by mitochondrial P450 systems in steroidogenic cells. *Drug Metabolism Reviews, 38*, 171–196.

Hanukoglu, I., Rapoport, R., Weiner, L., & Sklan, D. (1993). Electron leakage from the mitochondrial NADPH-adrenodoxin reductase-adrenodoxin-P450scc (cholesterol side chain cleavage) system. *Archives of Biochemistry and Biophysics, 305*, 489–498.

Jefcoate, C. (2002). High-flux mitochondrial cholesterol trafficking, a specialized function of the adrenal cortex. *The Journal of Clinical Investigation, 110*, 881–890.

Jefcoate, C. R., Lee, J., Cherradi, N., Takemori, H., & Duan, H. (2011). cAMP stimulation of StAR expression and cholesterol metabolism is modulated by co-expression of labile suppressors of transcription and mRNA turnover. *Molecular and Cellular Endocrinology, 336*, 53–62.

Jo, Y., King, S. R., Khan, S. A., & Stocco, D. M. (2005). Involvement of protein kinase C and cyclic adenosine 3,5′-monophosphate-dependent kinase in steroidogenic acute regulatory protein expression and steroid biosynthesis in Leydig cells. *Biology of Reproduction, 73*, 244–255.

Kil, I. S., Lee, S. K., Ryu, K. W., Woo, H. A., Hu, M. C., Bae, S. H., et al. (2012). Feedback control of adrenal steroidogenesis via H2O2-dependent, reversible inactivation of peroxiredoxin III in mitochondria. *Molecular Cell, 46*, 584–594.

Lalli, E., Melner, M. H., Stocco, D. M., & Sassone-Corsi, P. (1998). DAX-1 blocks steroid production at multiple levels. *Endocrinology, 139*, 4237–4243.

Le, T., & Schimmer, B. P. (2001). The regulation of MAPKs in Y1 mouse adrenocortical tumor cells. *Endocrinology, 142*, 4282–4287.

Manna, P. R., Dyson, M. T., & Stocco, D. M. (2009). Regulation of the steroidogenic acute regulatory protein gene expression: Present and future perspectives. *Molecular Human Reproduction, 15*, 321–333.

Manna, P. R., & Stocco, D. M. (2011). The role of specific mitogen-activated protein kinase signaling cascades in the regulation of steroidogenesis. *Journal of Signal Transduction, 2011*, 821615.

Miller, W. L. (2005). Minireview: Regulation of steroidogenesis by electron transfer. *Endocrinology, 146*, 2544–2550.

Mohn, C. E., Fernandez-Solari, J., De Laurentiis, A., Prestifilippo, J. P., de la Cal, C., Funk, R., et al. (2005). The rapid release of corticosterone from the adrenal induced by ACTH is mediated by nitric oxide acting by prostaglandin E2. *Proceedings of the National Academy of Sciences of the United States of America, 102*, 6213–6218.

Soriano, F. X., Baxter, P., Murray, L. M., Sporn, M. B., Gillingwater, T. H., & Hardingham, G. E. (2009). Transcriptional regulation of the AP-1 and Nrf2 target gene sulfiredoxin. *Molecules and Cells, 27*, 279–282.

Tsai, S. C., Lu, C. C., Lin, C. S., & Wang, P. S. (2003). Antisteroidogenic actions of hydrogen peroxide on rat Leydig cells. *Journal of Cellular Biochemistry, 90*, 1276–1286.

Wood, Z. A., Schroder, E., Robin Harris, J., & Poole, L. B. (2003). Structure, mechanism and regulation of peroxiredoxins. *Trends in Biochemical Sciences, 28*, 32–40.

Yang, K. S., Kang, S. W., Woo, H. A., Hwang, S. C., Chae, H. Z., Kim, K., et al. (2002). Inactivation of human peroxiredoxin I during catalysis as the result of the oxidation of the catalytic site cysteine to cysteine-sulfinic acid. *The Journal of Biological Chemistry*, *277*, 38029–38036.

Zangar, R. C., Davydov, D. R., & Verma, S. (2004). Mechanisms that regulate production of reactive oxygen species by cytochrome P450. *Toxicology and Applied Pharmacology*, *199*, 316–331.

SECTION II

H_2O_2 in the Regulation of Cellular Processes in Plants

CHAPTER TEN

The Use of HyPer to Examine Spatial and Temporal Changes in H_2O_2 in High Light-Exposed Plants

Marino Exposito-Rodriguez[*,1], Pierre Philippe Laissue[*,1], George R. Littlejohn[†], Nicholas Smirnoff[†], Philip M. Mullineaux[*,2]

[*]School of Biological Sciences, University of Essex, Colchester, United Kingdom
[†]Biosciences, College of Life and Environmental Sciences, University of Exeter, Exeter, United Kingdom
[1]These authors contributed equally
[2]Corresponding author: e-mail address: mullin@essex.ac.uk

Contents

1. Introduction 186
   1.1 Hydrogen peroxide and signaling in response to changes in light intensity 186
   1.2 A spatial component to H_2O_2 signaling 187
   1.3 Methods currently used for H_2O_2 localization 188
   1.4 HyPer, a GFP-based genetically encoded biosensor for H_2O_2 189
   1.5 Silencing of HyPer expression in plant tissues 190
2. Experimental Procedures 191
   2.1 In planta expression pattern of cytosolic HyPer (cHyPer) 191
   2.2 Mounting live seedlings for experimental manipulation 192
   2.3 HyPer dynamics in epidermal cells 194
   2.4 Confocal microscopy 194
   2.5 Image processing and analysis 195
3. Pilot Experiments Using HL Stress 196
   3.1 Experimental setup 196
   3.2 Chlorophyll fluorescence imaging 196
   3.3 Pilot experiment 196
4. Conclusions 198
Acknowledgments 198
References 198

Abstract

Exposure of photosynthetic cells of leaf tissues of *Arabidopsis thaliana* (Arabidopsis) to high light intensities (HL) may provoke a rapid rise in hydrogen peroxide (H_2O_2) levels in chloroplasts and subcellular compartments, such as peroxisomes, associated with photosynthetic metabolism. It has been hypothesized that when H_2O_2 is contained at or near its site of production then it plays an important role in signaling to induce acclimation to HL. However, should this discrete containment fail and H_2O_2 levels exceed the

Methods in Enzymology, Volume 527
ISSN 0076-6879
http://dx.doi.org/10.1016/B978-0-12-405882-8.00010-6

capacity of antioxidant systems to scavenge them, then oxidative stress ensues which triggers cell death. To test this hypothesis, the spatiotemporal accumulation of H_2O_2 needs to be quantified in different subcellular compartments. In this chapter, preliminary experiments are presented on the use of Arabidopsis seedlings transformed with a nuclear-encoded cytosol-located yellow fluorescent protein-based sensor for H_2O_2, called HyPer. HyPer allows ratiometric determination of its fluorescence at two excitation wavelengths, which frees quantification of H_2O_2 from the variable levels of HyPer *in vivo*. HyPer fluorescence was shown to have the potential to provide the necessary spatial, temporal, and quantitative resolution to study HL responses of seedlings using confocal microscopy. Chlorophyll fluorescence imaging was used to quantify photoinhibition of photosynthesis induced by HL treatment of seedlings on the microscope staging. However, several technical issues remain, the most challenging of which is the silencing of HyPer expression beyond the seedling stage. This limited our pilot studies to cotyledon epidermal cells, which while not photosynthetic, nevertheless responded to HL with 45% increase in cytosolic H_2O_2.

1. INTRODUCTION

1.1. Hydrogen peroxide and signaling in response to changes in light intensity

In common with all nonphotosynthetic aerobic eukaryotic cells, plants produce and accumulate hydrogen peroxide (H_2O_2) and other reactive oxygen species (ROS) at different times and in different subcellular locations. In addition, plant and algal cells produce ROS as a byproduct of photosynthesis in chloroplasts (Galvez-Valdivieso & Mullineaux, 2010; Suzuki, Koussevitzky, Mittler, & Miller, 2012; Waring, Klenell, Bechtold, Underwood, & Baker, 2010) and processes allied to this, such as the photorespiratory cycle, which results in production of millimolar quantities of H_2O_2 in peroxisomes (Costa et al., 2010; Queval et al., 2007). Many perturbations of photosynthesis promote H_2O_2 production of which one of the most studied is exposure to fluctuating light intensity (Galvez-Valdivieso & Mullineaux, 2010; Pogson, Woo, Förster, & Small, 2008; Suzuki et al., 2012). Sudden exposure to very high light intensities (often called excess light, EL) can provoke irreversible photoinhibition caused by the rapid accumulation of ROS, including H_2O_2, in chloroplasts. This leads to oxidative damage to the photosynthetic apparatus, which can trigger cell death. Visibly, this is often manifested as the bleaching of leaf tissues and production of discrete microscopic lesions (Karpinski et al., 1999; Mühlenbock et al., 2008; Mullineaux & Baker, 2010). Lesion formation is partly due to direct tissue damage but mostly by the initiation of programmed cell death initiated by the accumulation of ROS in the chloroplast (Kim et al., 2012; Mullineaux & Baker, 2010).

Under moderate increases in light intensity, typically <10-fold over growth light intensity (referred to as high light; HL), which elicit largely reversible photoinhibition and initiation of acclimation to HL (Mullineaux & Baker, 2010). H_2O_2 accumulation in leaf tissues also occurs, but in a more specific manner. In *Arabidopsis thaliana* (Arabidopsis) under HL, H_2O_2 accumulation was detected primarily in the chloroplasts of bundle sheath cells (BSCs), which are located adjacent to the vascular parenchyma (VP; Fryer et al., 2003; Galvez-Valdivieso et al., 2009). In addition, H_2O_2 was detected in the apoplast of BSCs, possibly dismutated from superoxide anion produced in a reaction catalyzed by plasma membrane-associated NADPH oxidases. NADPH oxidase activity was stimulated by the hormone abscisic acid (ABA) whose synthesis was activated in the VP in a HL- and humidity-dependent manner. It was proposed that H_2O_2 produced in BSC chloroplasts augmented this ABA signal allowing HL responses from multiple chloroplasts present in this cell type to be integrated into a single signal transduction pathway (Galvez-Valdivieso et al., 2009; Galvez-Valdivieso & Mullineaux, 2010; Mullineaux & Baker, 2010).

1.2. A spatial component to H_2O_2 signaling

From the above studies, it was hypothesized that the H_2O_2-initiated signaling from BSC chloroplasts may be converted to a non-ROS signal so that signal transduction could traverse the reducing environment of the cytosol (Galvez-Valdivieso & Mullineaux, 2010; Mullineaux, Karpinski, & Baker, 2006). The alternative view is that H_2O_2 may diffuse out of chloroplasts and accumulate in a localized manner to trigger a cytosol-based signaling system (Mubarakshina et al., 2010; Mullineaux et al., 2006). This might occur in a manner similar to the localized accumulation of H_2O_2 for signal transduction in microdomains associated with microbody membranes and plasma membranes of mammalian cells challenged with growth factors or by wounding (Ushio-Fukai, 2006; Woo et al., 2010) and perhaps also the containment and Ca^{2+}-associated signaling associated with H_2O_2 scavenging in plant peroxisomes (Costa et al., 2010). In general terms, when such containment of H_2O_2 occurs in plant cells, it was suggested that signaling would lead to an acclimation in response to changes in the environment. In contrast, a general diffusion of H_2O_2 away from its site of production to sufficient levels to promote oxidative damage could trigger cell death-associated signaling (Mullineaux & Baker, 2010; Mullineaux et al., 2006). However, observations with isolated chloroplasts and protoplasts have suggested that H_2O_2 can potentially leak out of chloroplasts in a light intensity-dependent manner (Mubarakshina et al., 2010). These observations suggest that H_2O_2 could act directly as a retrograde signal from the chloroplast to possibly activate regulatory proteins such as the protein phosphatase 2C isoform ABI2 (Miao et al., 2006) and/or heat shock transcription factors implicated in HL responses (Miller & Mittler, 2006; Nishizawa-Yokoi et al., 2011).

To answer these important questions it has become imperative to determine at the subcellular level in plant cells and tissues, the spatial and temporal patterns of the accumulation of H_2O_2 and other ROS in response to changes in light intensity and other environmental perturbations. In this chapter, we describe some of the problems to be overcome and technical developments needed to carry out further research.

1.3. Methods currently used for H_2O_2 localization

Of all the ROS, H_2O_2 is the most stable (Halliwell & Gutteridge, 1999) and can be estimated by a variety of methods based on acid extraction from tissues (Queval, Hager, Gakière, & Noctor, 2008). However, the major drawback of these methods is that they are limited for temporal resolution and completely lack any spatial resolution. Fluorescent and colorimetric chemical dyes used for detection of ROS allow high spatial resolution using microscopic imaging techniques. However, a critical review concluded that none can be used without very careful assessment of artifacts and that most measure what they term "reactive species," that is, various forms of ROS and reactive nitrogen species (e.g., nitric oxide, peroxynitrite) (Halliwell & Whiteman, 2004). For H_2O_2 dye probes, it is evident that the currently used methods suffer from a number of disadvantages (Snyrychova, Ferhan, & Eva, 2009). These problems include the degree of chemical specificity (fluorescein- and rhodamine-based compounds), the influence of peroxidase activity (Amplex Red, 2′,3′-diaminobenzidine [DAB]), degree of toxicity (DAB, Amplex Red), and the sacrifice of tissue (DAB, cerium trichloride). However, there is a more universal problem; all the probes react irreversibly with H_2O_2 so that they provide poor temporal resolution. This is because oxidized probe accumulates and takes time to dissipate. Related to this problem, none of these dye-based methods indicate the concentration of peroxide. Indeed, accumulation of highly fluorescent oxidized probe could even be giving a false impression that high H_2O_2 concentrations occur in plant cells.

1.4. HyPer, a GFP-based genetically encoded biosensor for H_2O_2

The recent development of genetically encoded green fluorescent protein (GFP)-based sensors has made it possible to develop powerful methods for monitoring noninvasively the dynamics of small molecules and atoms *in vivo*. In principle, GFP sensors allow excellent subcellular and spatiotemporal resolution, with sensitivities in the nanomolar to millimolar range and are reversible in their interactions with their targets (Markvicheva et al., 2011; Okumoto, Alexander, & Frommer, 2012). GFP-based probes have been successfully used *in planta* and have simplified measurements of Ca^{2+} (YC3.6; Krebs et al., 2011; Monshausen, Messerli, & Gilroy, 2008), cellular redox state (RoGFP; Meyer & Brach, 2009), pH (pHusion; Gjetting, Ytting, Schulz, & Fuglsang, 2012), and the hormone auxin (DII-VENUS; Brunoud et al., 2012). The development of a sensor for H_2O_2 named HyPer, has changed the situation completely for measuring the *in vivo* concentrations of this molecule (Belousov et al., 2006; Markvicheva et al., 2011). HyPer consists of the regulatory domain of an *Escherichia coli* (*E. coli*) transcription factor OxyR, which is used by this bacterium to monitor H_2O_2 levels (Choi et al., 2001), inserted into a circularly permuted yellow fluorescent protein (cpYFP; Nagai, Sawano, Park, & Miyawaki, 2001). A pair of redox-active cysteine residues in the OxyR domain of HyPer is located in a hydrophobic pocket, accessible only to amphiphilic molecules such as H_2O_2 and inaccessible to charged oxidants such as superoxide anion or reactive nitrogen species such as nitric oxide and peroxynitrite (Belousov et al., 2006). Importantly, the excitation spectrum of HyPer has two maxima (420 and 500 nm) which show differential changes in fluorescence emission (516 nm) after oxidation. Both forms can be visualized by laser excitation in a confocal system or with widefield fluorescence microscopy. The measurement of emission at two excitation maxima means that the redox state of HyPer can be determined ratiometrically. This allows the calculation of a dimensionless value, which avoids artifacts associated with differences in the levels of HyPer expression. This is important as HyPer levels will vary from cell to cell, tissue to tissue, and transgenic line to transgenic line and avoids issues where cells move. In summary, the properties of the H_2O_2-sensing domain of HyPer, dictate a high selectivity of the probe, high sensitivity and, importantly, good reversibility in the intracellular environment (Belousov et al., 2006; Malinouski, Zhou, Belousov, Hatfield, & Gladyshev, 2011).

The ability of HyPer to report increases in H_2O_2 *in vivo* has been demonstrated in *E. coli*, mammalian cells (Belousov et al., 2006) and in the wound response of zebrafish larvae (Niethammer, Grabher, Look, & Mitchison, 2009). In plant cells, Costa et al. (2010) presented the first and successful expression of HyPer in the cytoplasm and peroxisomes of Arabidopsis and tobacco cells. The authors demonstrated that the fluorescent ratio of cytosolic HyPer (cHyPer) changes upon the addition of exogenous H_2O_2 and was proportional to the amount of H_2O_2 applied. There was clear dose dependence in guard cells of epidermal peels indicating that HyPer appears to be a useful tool to measure the real-time *in vivo* spatiotemporal dynamics of H_2O_2 accumulation in plant cells.

One limitation of HyPer, that is not uncommon for GFP-based sensors (e.g., Pericam and RoGFP; Jiang et al., 2006; Nagai et al., 2001; Schwarzländer et al., 2012), is for potential artifacts to arise due to the pH sensitivity of the fluorescence which can mimic a significant part of its H_2O_2-sensing range (Belousov et al., 2006; Choi, Swanson, & Gilroy, 2012). Therefore, the local pH must be measured in each particular experiment. This is especially the case when the biosensor is targeted to a subcellular compartment where a change in pH can be anticipated, such as the chloroplast stroma in response to light. pH measurements could be made separately using a pH-sensitive dye (Niethammer et al., 2009) or dual imaging using a genetically encoded pH sensor to control this potential artifact and allow the calculation of H_2O_2 levels corrected for alterations in pH. However, in our initial experiments we have confined ourselves to measuring changes in cHyPer fluorescence in HL-exposed cells. This is because the cytosol of plant cells has been shown to maintain pH homeostasis by possession of a high passive buffering capacity. This has been demonstrated by the absence of cytosolic pH fluctuations in cells challenged with external pH changes and where apoplastic pH changed drastically (Gjetting et al., 2012). Consequently, for the purposes of our initial experiments, the effects of pH on HyPer fluorescence were assumed to be minimal for this subcellular compartment.

1.5. Silencing of HyPer expression in plant tissues

A problem that has been observed in some genetically encoded biosensors in plant tissues is a weak or absent fluorescence. For example, the glucose sensors based on the CFP–YFP pair, which has been used to measure steady state glucose levels in mammalian cells, showed transgene silencing in Arabidopsis transformants (Deuschle et al., 2006). Furthermore, an early version of the calcium sensor Yellow Cameleon used to determine [Ca^{2+}] dynamics expressed only in guard cells (Allen et al., 1999). That guard cells did not display a suppression of fluorescence suggests that posttranscriptional gene silencing (PTGS) might be playing a role. This is because guard cells are symplastically isolated from surrounding tissues and do not receive gene silencing signals from their neighbors through plasmodesmata (Himber, Dunoyer, Moissiard, Ritzenthaler, & Voinnet, 2003). This hypothesis was supported by using *sgs3* and *rdr6* transgene-silencing mutants, defective in PTGS (Mourrain et al., 2000; Peragine, Yoshikawa, Wu, Albrecht, & Poethig, 2004) to eliminate silencing of the glucose sensor genes, resulting in high fluorescence levels. The use of transgene-silencing mutants, however, is not ideal because they show phenotypic differences from the wild type and could complicate the interpretation of data.

2. EXPERIMENTAL PROCEDURES

2.1. In planta expression pattern of cytosolic HyPer (cHyPer)

A binary Ti plasmid carrying 35S:cHyPer was introduced into *Agrobacterium tumefaciens* strain GV3101:pMP90. In anticipation of potential gene silencing problems (see Section 1.5), *Arabidopsis sgs3–11* plants (Mourrain et al., 2000) were transformed with the *Agrobacterium* strain using standard procedures (Clough & Bent, 1998). In addition, a cHyPer expressed in *Arabidopsis* wild type (ecotype Col-0) was used and was kindly provided by Dr. Alex Costa (Costa et al., 2010). Seedlings were grown in peat-based compost in a controlled environment under an 8/16-h light/dark cycle at a photosynthetically active photon flux densities (PPFD) of 150 μmol m^{-2} s^{-1}, 22 ± 1 °C temperature, and relative humidity of 50%.

The seedlings expressing cHyPer were imaged with confocal laser scanning microscopy to analyze cHyPer fluorescence (Fig. 10.1). A strong fluorescence signal of cHyPer in epidermal cells, with a clear cytosolic and nuclear signal (Fig. 10.2C), was easily detected in both wild type and *sgs3–11* transgenic lines at 7 days from the date of sowing and including a 3-day stratification treatment of seeds at 4 °C to synchronize germination. Fluorescence was limited to cotyledons (stage 1.0; cotyledons fully opened; Boyes et al., 2001). No cHyPer fluorescence signal was detected in

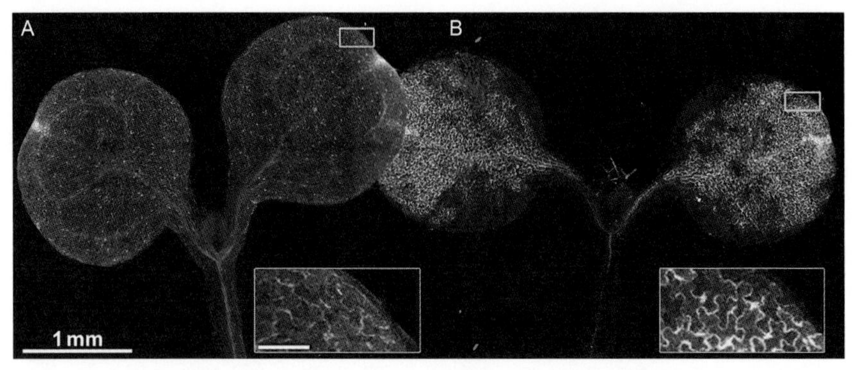

Figure 10.1 Expression of cytosolic HyPer (cHyPer) in 8-day-old *Arabidopsis* seedlings. (A) cHyPer in wild type. (B) cHyPer in *sgs3–11* background. The seedlings were placed next to each other on the same slide, ensuring image acquisition conditions were identical. Inset cyan rectangles show epidermal cells at higher magnification (scale bar: 100 μm). (See Color Insert.)

epidermal cells from roots or hypocotyls. However, from day 8, HyPer fluorescence in plants with wild-type background diminished, becoming dim and patchy (Fig. 10.1A). This problem was attenuated by using *Arabidopsis* mutant *sgs3–11* (Fig. 10.1B). The onset of silencing in seedlings of these mutants occurs much later, between days 12 and 15 (stage 1.02; rosette leaves >1 mm in length). After this time, no signal was detected in epidermal cells of cotyledons or rosette leaves in the mutant.

2.2. Mounting live seedlings for experimental manipulation

To ensure that cHyPer can dynamically report H_2O_2 within cotyledon epidermal cells, we performed pilot experiments using live cell imaging. Live microscopy at high spatial and temporal resolution depends on appropriate immobilization of the specimen. To evaluate the amplitude of the HyPer response to H_2O_2 added to intact epidermal cells, we designed two procedures to add solutions and measure H_2O_2 changes without affecting cell viability. For the first assay, a seedling was placed in a 35 mm petri dish with a #1.5 coverslip glass-bottom (ibidi GmbH, D-82152 Martinsried), covered with a mesh (Fig. 10.2A) and glued down on the sides using rubber cement ("Fixogum," Marabuwerke GmbH, D-71732 Tamm), gently pressing the seedling against the coverslip. The dish was filled with 200 μL of water and H_2O_2 solution dropped directly onto the water to obtain a final concentration of 1 m*M*. In the second procedure, the seedlings were mounted in a custom-built perfusion chamber to achieve exchange of H_2O_2 solution. Microscope slides were covered with 24 × 40 mm #1.5 coverslips (Agar Scientific, Essex, UK), placed so as to form a cross, and attached with

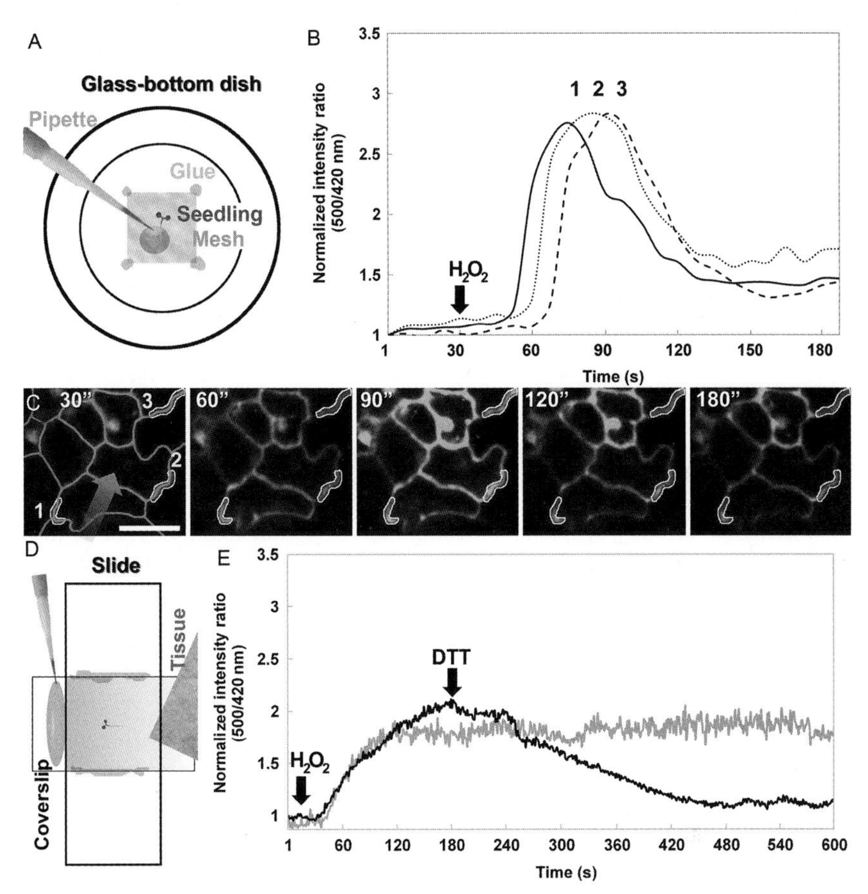

Figure 10.2 Cytosolic HyPer (cHyPer) response to added H_2O_2 in cotyledon epidermal cells. (A) Experimental setup using 35 mm glass-bottom petri dishes. (B) Time course of HyPer response to H_2O_2 addition (arrow). 1, 2, and 3 indicate the maxima in regions of interest, indicated in (C). (C) Image sequence of cotyledon epidermal cells expressing cHyPer. Only the green channel (excitation at 500 nm) is shown. Cells were exposed to 1 m*M* H_2O_2. A blue arrow shows the general direction of H_2O_2 flux. Time is indicated in the bottom right. Numbers 1, 2, 3 indicate the positions of the analyzed regions of interest. Cell borders are overlaid in red. Scale bar: 10 μm. (D) Slide-based experimental setup using custom-built perfusion chamber. (E) Time course of redox changes in cHyPer after the addition of H_2O_2 (1 m*M*) and the reducing agent dithiothreitol (DTT; 1 m*M*) using the experimental setup indicated in (D). All ratio values were normalized to a minimum value of 1.0. (See Color Insert.)

double-sided adhesive tape (Fig. 10.2D). As the coverslip sticks out at one end, solution can be positioned on it. Using a tissue at the other end of the coverslip, the liquid is sucked into the space between slide and coverslip and moves across by capillary force.

2.3. HyPer dynamics in epidermal cells

The first assay using glass-bottom petri dishes allows the analysis of HyPer at high temporal resolution in single cells. Cellular response after the addition of H_2O_2 was rapid (~30 s, Fig. 10.2B). Ratios decreased in the same amount of time, but did not fall to the same levels as before H_2O_2 addition. This may be because exogenous H_2O_2 remained present in the media surrounding the seedling after the addition. Around threefold increase of the cHyPer ratio (500/420 nm) in epidermal cells was observed using saturating amounts of H_2O_2 (1 m*M*) (Fig. 10.2B). This agrees with the maximum value of the HyPer dynamic range determined earlier (Belousov et al., 2006; Markvicheva et al., 2011). This assay enabled the recording of the cHyPer response in three different cells (Fig. 10.2C) at high temporal resolution, allowing the determination of the direction and rate of H_2O_2 diffusion. The second assay showed different H_2O_2 dynamics (Fig. 10.2D). Response was slow and sustained, and took almost 3 min to reach a maximum. These values were lower compared to the first assay, only reaching a ratio increase of around 1.8-fold. After H_2O_2 treatment, addition of the reducing agent dithiothreitol (1 m*M*) confirmed the reversible proprieties of HyPer in this system (Fig. 10.2E). Interestingly, although the initial kinetics in both systems were different, the steady state signal values attained by both mounting methods were very similar after 2 min.

2.4. Confocal microscopy

A flow chart detailing the steps of image acquisition, processing and analysis is shown in Fig. 10.3. It is important that optimized samples are acquired respecting several criteria. First, saturated pixels should be avoided, as they may represent lost information and cannot be used for quantification. Look-up tables color-coding the maximum greyscale value, commonly implemented in microscope image acquisition software, allowed the detection of saturated pixels by adjusting laser power and/or detector gain to avoid them. Second, if temporal resolution was essential (as used in the

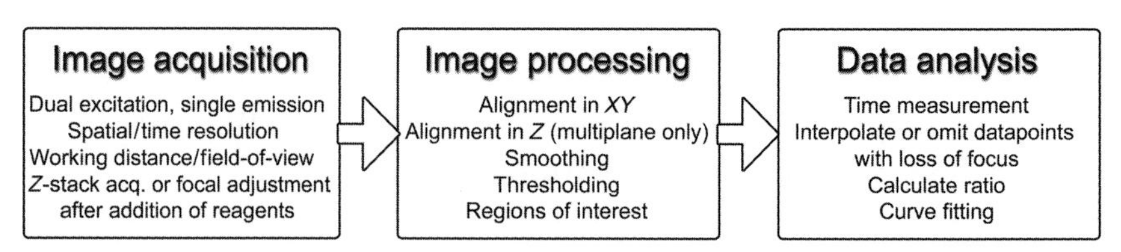

Figure 10.3 Flow chart summarizing the different steps and approaches to be considered.

glass-bottom dish assay), an image size of 256 × 256 pixels offered sufficient spatial information while greatly reducing acquisition time, allowing for ~1 frame/s using unidirectional scanning. Spectral information was separated using line scanning rather than full-frame mode. For higher spatial resolution (for slide-based assays), an image size of 512 × 512 pixels was used.

For image acquisition, a Nikon A1si confocal microscope was used with the following objectives: a CFI 10× Plan Fluor with numerical aperture (NA) 0.3, a CFI 20× Plan Apochromat violet-corrected (VC) with NA 0.75, and a CFI 60× Plan Apochromat VC oil-immersion objective with NA 1.4. Images were acquired using one-way sequential line scans of two excitation lines, with laser power at 405 nm between 15 and 33 arbitrary units (AU), and at 488 nm between 7 and 17 AU, and emission collected with one detector at 540/30 nm, with a photomultiplier tube gain of 90–120 AU. Differential interference contrast images were acquired using the transmitted light detector at a gain of 80–120 AU. No offset was used, and pinhole size was set between 1.2 and 2 times the Airy disk size of the used objective, depending on signal strength. Axial step size was 0.8–1.6 μm, with three image planes per *z*-stack.

2.5. Image processing and analysis

Invariably, samples move in the three spatial dimensions (*X*, *Y*, and *Z*) upon the addition of solutions to the media. Lateral movement in *X* and *Y* was corrected in one channel using the translational mode of the StackReg algorithm (Thévenaz, Ruttimann, & Unser, 1998) in ImageJ (Rasband, 1997), and the stored shift matrix applied to the other channel. Color separation using line scans ensured nearly zero temporal deviation between the channels. Alignment was validated by 3D visual inspection of fiduciary marks in overlaid channels. *Z*-movement was corrected by acquiring three image planes along the *Z*-axis, and the appropriate planes were selected and combined postacquisition. Alternatively, rapid manual adjustment of the focal plane after the addition of a solution was carried out and yielded good results.

For measurement of fluorescent intensity, a previous approach for ratiometry was expanded (Schwarzländer et al., 2008). Images were first smoothed using a Gaussian filter with 3 × 3 kernel size. Bright structures were then segmented in NIS-Elements (version 3.21.03, build 705 LO). The aim was to restrict measurement to brightly fluorescing structures only. Segmentation was done using the "define binary" command, smoothing and cleaning the binary area once. Thresholding was assessed by visual inspection. Within the segmented datasets, regions of interest were used, providing a second layer for restricting measurements. This allowed measuring and comparing fluorescent signals in different parts of the imaged area Fig. 10.2.

3. PILOT EXPERIMENTS USING HL STRESS

3.1. Experimental setup

HL was applied to seedlings mounted on the microscope stage. The tungsten lamp, usually used for bright field illumination, was adapted for light treatments by adjusting the field diaphragm, in order to deliver a small, brightly illuminating circular beam. This enabled application of white light to single cotyledons only (Fig. 10.4A). Intensity was measured with a light meter (SKP200, Skye, Powys, UK). The cotyledon was exposed to HL with PPFDs of 1200, 1600, and 2000 $\mu mol\ m^{-2}\ s^{-1}$. The spectrum of the light emitted from the tungsten lamp at different intensities was further assessed using a spectroradiometer (SR9910, Macam, Livingstone, UK) (Fig. 10.4B).

3.2. Chlorophyll fluorescence imaging

To ensure that illumination was locally contained, and to determine the effects that light of different PPFDs had on the photoinhibition of cotyledons, chlorophyll fluorescence (Cf) parameters were measured using a Fluorimager chlorophyll fluorescent imaging system (Fluorimager; Technologica Ltd., Colchester, Essex, UK). For the theory and use of Cf to measure photosynthetic efficiency, the reader is referred to Baker (2008). A decline in the Cf parameter F_v/F_m, which describes the maximum quantum efficiency of photosystem II photochemistry (Baker, 2008), was used to indicate the degree of photoinhibition suffered by cotyledons exposed to HL (Fig. 10.4C). The first cotyledon of each seedling was not treated with HL, serving as control with imaged F_v/F_m values around 0.76. Increasing light intensities on the second cotyledon caused a progressive decrease in F_v/F_m after 20 min in a light intensity-dependent manner (Fig. 10.4C).

3.3. Pilot experiment

The light intensity that caused a decrease in F_v/F_m of ca. 20% (PPFD 1200 $\mu mol\ m^{-2}\ s^{-1}$) was used to image the fluorescence of cHyPer in *sgs3–11* cotyledons before, during, and after exposure to HL. The two

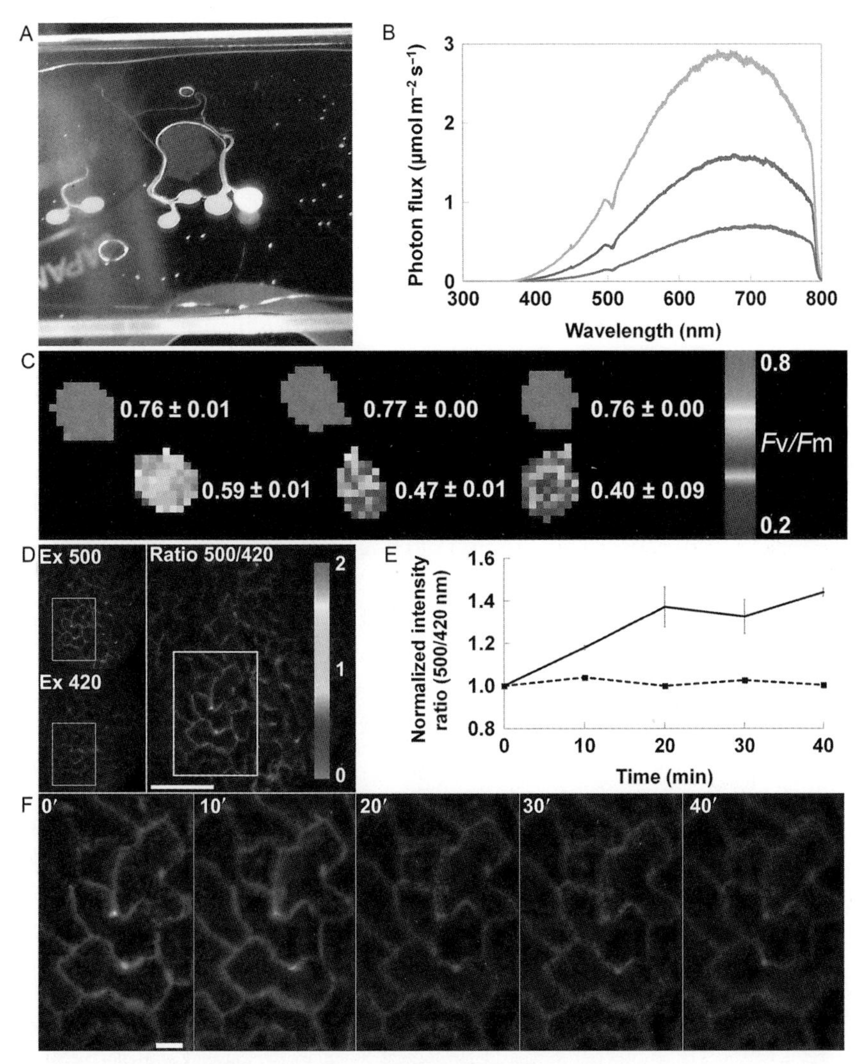

Figure 10.4 Pilot experiment inducing high light stress on the confocal microscope. (A) A precise, small area of light was applied to only one cotyledon. (B) Spectral measurement of the light used in this study. (C) Fluorescence images of cotyledons showing F_v/F_m values after 20 min exposures at 1200, 1600, and 2000 μmol m^{-2} s^{-1} (left to right, respectively). (D) Combining the two channels to produce a ratiometric image. The inset cyan rectangles demarcate the enlarged area shown in (F). (E) cHyPer response of cotyledon after high light treatments in 10 min intervals. Time point zero is a low light (LL) value taken immediately prior to exposure of the cotyledon to HL. Solid line and dashed line are values from cotyledons exposed to HL or LL parallel control, respectively. Values are the means (±SD; $n=3$). All ratios were normalized to a minimum value of 1.0. (F) False-color ratio images corresponding to (E). (See Color Insert.)

channels (excitation at 500 nm, green channel, and excitation at 420 nm, red channel) were combined and visualized as ratio images using a false-color scale (Fig. 10.4D). Using the slide-based setup described in Section 2.2, a fluorescence image of a seedling's cotyledon, grown in conditions described in Section 2.1, was taken (defined as time point zero). The cotyledon was then exposed to HL for 20 min. Light was then switched off, and a fluorescence image taken every 10 min over 40 min. After exposure to HL, the fluorescence intensity ratio of cHyper in epidermal cells increased by a maximum of 45% compared to time point zero (Fig. 10.4E).

4. CONCLUSIONS

From these preliminary experiments, it is clear that HyPer can provide the necessary spatial, temporal, and quantitative resolution to study HL responses and the compartmentation of H_2O_2 at the subcellular level. However, before experiments to examine the accumulation and diffusion of H_2O_2 in HL-exposed cells' photosynthetic tissues of true leaves, the silencing of HyPer expression has to be overcome and robust methods for codetermination of pH changes also have to be instigated. If these technical issues can be overcome, then HyPer holds out the prospect of giving us many more insights into the behavior of H_2O_2 in HL-exposed cells. This is clear from the preliminary data present in Fig. 10.4 in that HL triggers an accumulation of H_2O_2 in cotyledon epidermal cells and even more once the light has been switched off (Fig. 10.4E and F). This observation is remarkable because these cells do not contain chloroplasts.

ACKNOWLEDGMENTS

We thank Dr. Alex Costa for the cHyPer/Col-0 line used in this study. The authors are grateful to the UK Biotechnology and Biological Sciences Research Council (BBSRC) for the support of this research (Grant Number BB/I020071/1).

REFERENCES

Allen, G. J., Kwak, J. M., Chu, S. P., Llopis, J., Tsien, R. Y., Harper, J. F., et al. (1999). Cameleon calcium indicator reports cytoplasmic calcium dynamics in Arabidopsis guard cells. *The Plant Journal*, *19*, 735–747.

Baker, N. R. (2008). Chlorophyll fluorescence: A probe of photosynthesis in vivo. *Annual Review of Plant Biology*, *59*, 89–113.

Belousov, V. V., Fradkov, A. F., Lukyanov, K. A., Staroverov, D. B., Shakhbazov, K. S., Terskikh, A. V., et al. (2006). Genetically encoded fluorescent indicator for intracellular hydrogen peroxide. *Nature Methods*, *3*, 281–286.

Boyes, D. C., Zayed, A. M., Ascenzi, R., McCaskill, A. J., Hoffman, N. E., Davis, K. R., et al. (2001). Growth stage-based phenotypic analysis of Arabidopsis: A model for high throughput functional genomics in plants. *The Plant Cell, 13,* 1499–1510.

Brunoud, G., Wells, D. M., Oliva, M., Larrieu, A., Mirabet, V., Burrow, A. H., et al. (2012). A novel sensor to map auxin response and distribution at high spatio-temporal resolution. *Nature, 482,* 103–106.

Choi, H., Kim, S., Mukhopadhyay, P., Cho, S., Woo, J., Storz, G., et al. (2001). Structural basis of the redox switch in the OxyR transcription factor. *Cell, 105,* 103–113.

Choi, W.-G., Swanson, S. J., & Gilroy, S. (2012). High-resolution imaging of Ca^{2+}, redox status, ROS and pH using GFP biosensors. *The Plant Journal, 70,* 118–128.

Clough, S. J., & Bent, A. F. (1998). Floral dip: A simplified method for Agrobacterium-mediated transformation of *Arabidopsis thaliana*. *The Plant Journal, 16,* 735–743.

Costa, A., Drago, I., Behera, S., Zottini, M., Pizzo, P., Schroeder, J. I., et al. (2010). H_2O_2 in plant peroxisomes: An in vivo analysis uncovers a $Ca(^{2+})$-dependent scavenging system. *The Plant Journal, 62,* 760–772.

Deuschle, K., Chaudhuri, B., Okumoto, S., Lager, I., Lalonde, S., & Frommer, W. B. (2006). Rapid metabolism of glucose detected with FRET glucose nanosensors in epidermal cells and intact roots of Arabidopsis RNA-silencing mutants. *The Plant Cell, 18,* 2314–2325.

Fryer, M. J., Ball, L., Oxborough, K., Karpinski, S., Mullineaux, P. M., & Baker, N. R. (2003). Control of *Ascorbate Peroxidase 2* expression by hydrogen peroxide and leaf water status during excess light stress reveals a functional organisation of *Arabidopsis* leaves. *The Plant Journal, 33,* 691–705.

Galvez-Valdivieso, G., Fryer, M. J., Lawson, T., Slattery, K., Truman, W., Smirnoff, N., et al. (2009). The high light response in *Arabidopsis* involves ABA signaling between vascular and bundle sheath cells. *The Plant Cell, 21,* 2143–2162.

Galvez-Valdivieso, G., & Mullineaux, P. M. (2010). The role of reactive oxygen species in signaling from chloroplasts to the nucleus. *Physiologia Plantarum, 138,* 430–439.

Gjetting, K. S., Ytting, C. K., Schulz, A., & Fuglsang, A. T. (2012). Live imaging of intra- and extracellular pH in plants using pHusion, a novel genetically encoded biosensor. *Journal of Experimental Botany, 63,* 3207–3218.

Halliwell, B., & Gutteridge, J. M. C. (1999). *Free radicals in biology and medicine* (3rd ed.). Oxford, UK: Oxford University Press.

Halliwell, B., & Whiteman, M. (2004). Measuring reactive species and oxidative damage in vivo and in cell culture: How should you do it and what do the results mean? *British Journal of Pharmacology, 142,* 231–255.

Himber, C., Dunoyer, P., Moissiard, G., Ritzenthaler, C., & Voinnet, O. (2003). Transitivity-dependent and -independent cell-to-cell movement of RNA silencing. *The EMBO Journal, 22,* 4523–4533.

Jiang, K., Schwarzer, C., Lally, E., Zhang, S. B., Ruzin, S., Machen, T., et al. (2006). Expression and characterization of a redox-sensing green fluorescent protein (reduction–oxidation-sensitive green fluorescent protein) in Arabidopsis. *Plant Physiology, 141,* 397–403.

Karpinski, S., Reynolds, H., Karpinska, B., Wingsle, G., Creissen, G., & Mullineaux, P. (1999). Systemic signaling and acclimation in response to excess excitation energy in Arabidopsis. *Science, 284,* 654–657.

Kim, C., Meskauskiene, R., Zhang, S., Lee, K.-P., Asok, M. L., Blajecka, K., et al. (2012). Chloroplasts of *Arabidopsis* are the source and a primary target of a plant-specific programmed cell death signaling pathway. *The Plant Cell, 24,* 3026–3039.

Krebs, M., Held, K., Binder, A., Hashimoto, K., DenHerder, G., Parniske, M., et al. (2011). FRET-based genetically encoded sensors allow high-resolution live cell imaging of Ca^{2+} dynamics. *The Plant Journal, 69,* 181.

Malinouski, M., Zhou, Y., Belousov, V. V., Hatfield, D. L., & Gladyshev, V. N. (2011). Hydrogen peroxide probes directed to different cellular compartments. *PLoS One*, *6*(1), e14564. http://dx.doi.org/10.1371/journal.pone.0014564.

Markvicheva, K. N., Bilan, D. S., Mishina, N. M., Gorokhovatsky, A. Y., Vinokurov, L. M., Lukyanov, S., et al. (2011). A genetically encoded sensor for H_2O_2 with expanded dynamic range. *Bioorganic & Medicinal Chemistry*, *19*, 1079–1084.

Meyer, A. J., & Brach, T. (2009). Dynamic redox measurements with redox-sensitive GFP in plants by confocal laser scanning microscopy. *Methods in Molecular Biology*, *479*, 93–107.

Miao, Y. C., Lv, D., Wang, P. C., Wang, X. C., Chen, J., Miao, C., et al. (2006). An *Arabidopsis* glutathione peroxidase functions as both a redox transducer and a scavenger in abscisic acid and drought stress responses. *The Plant Cell*, *18*, 2749–2766.

Miller, G., & Mittler, R. (2006). Could heat shock transcription factors function as hydrogen peroxide sensors in plants? *Annals of Botany*, *98*, 279–288.

Monshausen, G. B., Messerli, M. A., & Gilroy, S. (2008). Imaging of the Yellow Cameleon 3.6 indicator reveals that elevation in cytosolic Ca^{2+} following oscillating increases in growth in root hairs of Arabidopsis. *Plant Physiology*, *147*, 1690–1698.

Mourrain, P., Béclin, C., Elmayan, T., Feuerbach, F., Godon, C., Morel, J. B., et al. (2000). Arabidopsis *SGS2* and *SGS3* genes are required for posttranscriptional gene silencing and natural virus resistance. *Cell*, *101*, 533–542.

Mubarakshina, M. M., Ivanov, B. N., Naydov, I. A., Hillier, W., Badger, M. R., & Krieger-Liszkay, A. (2010). Production and diffusion of chloroplastic H_2O_2 and its implication to signaling. *Journal of Experimental Botany*, *61*, 3577–3587.

Mühlenbock, P., Szechynska-Hebda, M., Plaszczyca, M., Baudo, M., Mateo, A., Mullineaux, P. M., et al. (2008). Chloroplast signaling and *LESION SIMULATING DISEASE1* regulate crosstalk between light acclimation and immunity in Arabidopsis. *The Plant Cell*, *20*, 2339–2356.

Mullineaux, P. M., & Baker, N. R. (2010). Oxidative stress: Antagonistic signaling for acclimation or cell death? *Plant Physiology*, *154*, 521–525.

Mullineaux, P. M., Karpinski, S., & Baker, N. R. (2006). Spatial dependence for hydrogen peroxide-directed signaling in light-stressed plants. *Plant Physiology*, *141*, 346–350.

Nagai, T., Sawano, A., Park, E. S., & Miyawaki, A. (2001). Circularly permuted green fluorescent proteins engineered to sense Ca^{2+}. *Proceedings of the National Academy of Sciences of the United States of America*, *98*, 3197–3202.

Niethammer, P., Grabher, C., Look, A. T., & Mitchison, T. J. (2009). A tissue-scale gradient of hydrogen peroxide mediates rapid wound detection in zebrafish. *Nature*, *459*, 996–999.

Nishizawa-Yokoi, A., Nosaka, R., Hayashi, H., Tainaka, H., Maruta, T., Tamoi, M., et al. (2011). HsfA1d and HsfA1e involved in the transcriptional regulation of *HsfA2* function as key regulators for the Hsf signaling network in response to environmental stress. *Plant & Cell Physiology*, *52*, 933–945.

Okumoto, S., Alexander, J., & Frommer, W. B. (2012). Quantitative imaging with fluorescent biosensors. *Annual Review of Plant Biology*, *63*, 663–706.

Peragine, A., Yoshikawa, M., Wu, G., Albrecht, H. L., & Poethig, R. S. (2004). *SGS3* and *SGS2/SDE1/RDR6* are required for juvenile development and the production of trans-acting siRNAs in Arabidopsis. *Genes & Development*, *18*, 2368–2379.

Pogson, B. J., Woo, N. S., Förster, B., & Small, I. D. (2008). Plastid signaling to the nucleus and beyond. *Trends in Plant Science*, *13*, 602–609.

Queval, G., Hager, J., Gakière, B., & Noctor, G. (2008). Why are literature data for H_2O_2 contents so variable? A discussion of potential difficulties in the quantitative assay of leaf extracts. *Journal of Experimental Botany*, *59*, 135–146.

Queval, G., Issakidis-Bourguet, E., Hoeberichts, F. A., Vandorpe, M., Gakière, B., Vanacker, H., et al. (2007). Conditional oxidative stress responses in the Arabidopsis photorespiratory mutant *cat2* demonstrate that redox state is a key modulator of daylenghth-dependent gene expression, and defined photoperiod as a crucial factor in the regulation of H_2O_2-induced cell death. *The Plant Journal*, *52*, 640–657.

Rasband, W. S. (1997) ImageJ. U. S. National Institutes of Health, Bethesda, Maryland.

Schwarzländer, M., Fricker, M. D., Müller, C., Marty, L., Brach, T., Novak, J., et al. (2008). Confocal imaging of glutathione redox potential in living plant cells. *J Microsc*, *231*(2), 299–316.

Schwarzländer, M., Murphy, M. P., Duchen, M. R., Logan, D. C., Fricker, M. D., Halestrap, A. P., et al. (2012). Mitochondrial 'flashes': A radical concept repHined. *Trends in Cell Biology*, *22*, 503–508.

Snyrychova, I., Ferhan, A., & Eva, H. (2009). Detecting hydrogen peroxide in leaves in vivo—A comparison of methods. *Physiologia Plantarum*, *135*, 1–18.

Suzuki, N., Koussevitzky, S., Mittler, R., & Miller, G. (2012). ROS and redox signaling in the response of plants to abiotic stress. *Plant, Cell & Environment*, *35*, 259–270.

Thévenaz, P., Ruttimann, U. E., & Unser, M. (1998). A pyramid approach to subpixel registration based on intensity. *IEEE Trans Image Process*, *7*, 27–41.

Ushio-Fukai, M. (2006). Localizing NADPH oxidase-derived ROS. *Science Signaling*, *2*, re8.

Waring, J., Klenell, M., Bechtold, U., Underwood, G. J. C., & Baker, N. R. (2010). Light-induced responses of oxygen photoreduction, reactive oxygen species production and scavenging in two diatom species. *Journal of Phycology*, *46*, 1206–1217.

Woo, H. A., Yim, S. H., Shin, D. H., Kang, D., Yu, D.-Y., & Rhee, S. G. (2010). Inactivation of peroxiredoxin I by phosphorylation allows localized H_2O_2 accumulation for cell signaling. *Cell*, *140*, 517–528.

CHAPTER ELEVEN

A Simple and Powerful Approach for Isolation of *Arabidopsis* Mutants with Increased Tolerance to H_2O_2-Induced Cell Death

Tsanko Gechev[*,†,1,2], **Nikolay Mehterov**[*,†,1], **Iliya Denev**[*], **Jacques Hille**[‡]

[*]Department of Plant Physiology and Plant Molecular Biology, University of Plovdiv, Plovdiv, Bulgaria
[†]Institute of Molecular Biology and Biotechnologies, Plovdiv, Bulgaria
[‡]Department Molecular Biology of Plants, University of Groningen, Groningen, The Netherlands
[1]These authors contributed equally to this work
[2]Corresponding author: e-mail address: tsangech@uni-plovdiv.bg

Contents

1. Introduction 204
2. Generation and Isolation of Mutants More Tolerant to H_2O_2-Induced Oxidative Stress 206
   2.1 Protocol for EMS mutagenesis and AT screening 207
3. Identification of Mutations in the Genome 209
   3.1 Protocol for TAIL-PCR 210
4. Analysis of the Mutants with Enhanced Tolerance to H_2O_2-Induced Oxidative Stress 211
   4.1 Evaluation of the mutants tolerance to oxidative stress 211
   4.2 Protocol for the determination of chlorophyll content in plant tissues 212
   4.3 Protocol for MDA determination in plant tissues 213
   4.4 CAT activity measurements 214
   4.5 Protocol for measuring CAT activity by native PAGE 214
   4.6 Protocol for photometric determination of CAT activity 215
   4.7 Determination of anthocyanins content 216
   4.8 Protocol for determination of anthocyanins content in plant tissues 216
5. Conclusion 217
Acknowledgments 218
References 218

Abstract

A genetic approach is described to isolate mutants more tolerant to oxidative stress. A collection of T-DNA activation tag *Arabidopsis thaliana* mutant lines was screened

Methods in Enzymology, Volume 527
ISSN 0076-6879
http://dx.doi.org/10.1016/B978-0-12-405882-8.00011-8

for survivors under conditions that trigger H_2O_2-induced cell death. Oxidative stress was induced by applying the catalase (CAT) inhibitor aminotriazole (AT) in the growth media, which results in decrease in CAT enzyme activity, H_2O_2 accumulation, and subsequent plant death. One mutant was recovered from the screening and named *oxr1* (*ox*idative stress *r*esistant *1*). The location of the T-DNA insertion was identified by TAIL-PCR. *Oxr1* exhibited lack of cell death symptoms and more fresh weight and chlorophyll content compared to wild type. The lack of cell death correlated with more prominent induction of anthocyanins synthesis in *oxr1*. These results demonstrate the feasibility of AT as a screening agent for the isolation of oxidative stress-tolerant mutants and indicate a possible protective role for anthocyanins against AT-induced cell death. The chapter includes protocols for ethyl methanesulfonate mutagenesis, mutant screening using AT, T-DNA identification by TAIL-PCR, CAT activity measurements, and determination of malondialdehyde, chlorophyll, and anthocyanins.

ABBREVIATIONS

AT aminotriazole
CAT catalase
MDA malondialdehyde
MS Murashige and Skoog
PCD programmed cell death
ROS reactive oxygen species

1. INTRODUCTION

Reactive oxygen species (ROS) are toxic by-products of metabolism in all aerobic organisms. Accumulation of ROS, referred to as oxidative stress, is observed under many unfavorable environmental factors, including drought, salinity, extreme temperatures, and pollutants, such as herbicides or heavy metals (Apel & Hirt, 2004). In most cases, oxidative stress occurs as a result of both increased production and hampered detoxification of ROS. ROS, including hydrogen peroxide (H_2O_2), superoxide radicals (O_2^-), and singlet oxygen ($^1O_2$), can also perform signaling functions, modulating a number of plant developmental processes, stress responses, and programmed cell death (PCD) (Gadjev, Stone, & Gechev, 2008; Gechev & Hille, 2005; Gechev, Van Breusegem, Stone, Denev, & Laloi, 2006). Examples of ROS-modulated developmental processes include embryo development, root hair growth, nucellar degeneration, maturation of tracheal elements, and epidermal trichomes, formation of lace leaf shape, and leaf senescence (Gechev et al., 2006). Many of these processes are also associated with ROS-dependent PCD. ROS-induced PCD is also an important component of the hypersensitive response, a defense reaction in which plant cells in and around the site of pathogen infection die in order to physically restrict the spread of the pathogen (Gechev et al., 2006). While in the above examples cell death is beneficial and/or essential for plant development and survival, some necrotrophic pathogens can secrete toxins that cause ROS-dependent cell death in healthy tissues so that the pathogens can feed on the dead tissues (Dangl & Jones, 2001). Thus, maintaining proper ROS levels at particular times is essential for normal plant growth and development as well as for adaptation to the environment.

Factors that influence the biological effects of ROS signaling include the chemical identity of ROS, sites of ROS production, amounts and duration of the elevated ROS levels, and interaction with other signaling molecules like plant hormones, nitric oxide, and lipid messengers (Gechev et al., 2006). Signaling properties have been reported for H_2O_2, O_2^-, $^1O_2$, and even for the most destructive and short-lived hydroxyl radicals. In general, low doses of ROS may induce protective mechanisms resulting in stress acclimation, while higher doses of ROS can initiate PCD (Gadjev et al., 2008; Gechev & Hille, 2005).

A vast and intricate network of ROS-producing and -metabolizing enzymes regulates ROS levels in a complex manner. ROS are detoxified by the antioxidant system of the cell, composed of antioxidant molecules and enzymes (Gechev et al., 2006). Catalase (CAT) is the main H_2O_2-detoxifying enzyme, serving as a cellular sink for H_2O_2, while superoxide dismutase is the only plant enzyme-metabolizing superoxide radicals. Reducing CAT activity by gene silencing or by CAT inhibitor aminotriazole (AT) leads to increased endogenous H_2O_2 levels, oxidative stress, and eventual cell death (Dat et al., 2003; Gechev, Minkov, & Hille, 2005; Vanderauwera et al., 2005). H_2O_2-dependent cell death is a programmed process, associated with specific alterations in gene expression, and can be compromised by an increased CO_2 concentration (Vandenabeele et al., 2003, 2004; Vanderauwera et al., 2005).

Previous studies demonstrated that H_2O_2 is involved in the regulation of many aspects of plant development as well as stress acclimation and cell death (Gechev et al., 2002, 2005, 2006). Identifying genes from the intricate H_2O_2 network would therefore be of fundamental importance for understanding the genetic mechanisms by which these processes are regulated. In this chapter, we report a method for screening and isolating mutants with enhanced tolerance to H_2O_2-dependent cell death triggered by the CAT inhibitor

AT. In the highlighted case study, double screening of 8600 mutant T-DNA lines has resulted in recovery of a mutant which survives under the lethal screening conditions. The position of the mutation is identified by TAIL-PCR, and the mutant is characterized with regard to its stress tolerance.

2. GENERATION AND ISOLATION OF MUTANTS MORE TOLERANT TO H_2O_2-INDUCED OXIDATIVE STRESS

The two most commonly used methods for generating and isolating *Arabidopsis* mutants are chemical mutagenesis by ethyl methanesulfonate (EMS) and T-DNA mutagenesis. The T-DNA insertion itself can be a simple one, allowing identification of genes that are physically disrupted, or it can be activation tagging, where an enhancer in the T-DNA construct can activate nearby gene(s). The latter allows identification of both positive and negative regulators of the studied process. There are now many mutant collections deposited in the *Arabidopsis* stock centers and most of them can be screened for particular mutants of interest. In addition, researchers can always make their own mutant collections, which in terms of EMS is relatively easy. The advantage of this approach is that one can get a much broader variety of mutations, including single amino acid substitutions that alter particular amino acids and the function of the corresponding protein, without abolishing completely its activity. We used both types of mutagenesis in order to obtain mutants with increased tolerance to oxidative stress. In addition, there are other ways to chemically mutagenize plants, as well as other mutagens such as γ-rays or transposons (Britt & Jiang, 1999; Frye & Innes, 1998).

The next critical step is the screening procedure, which allows identifying the mutant of interest, in our case - the mutants that can survive oxidative stress-induced cell death. The two most widely used ROS-inducing screening agents are paraquat (PQ), known also as methyl viologen, and AT. The first one generates mainly superoxide radicals in the chloroplasts (which are rapidly converted to H_2O_2 by the enzyme superoxide dismutase), while the second one is a CAT inhibitor, that reduces CAT activity resulting in endogenous H_2O_2 accumulation (Gechev et al., 2006). Both agents can be applied in the plant growth media, particularly useful for screening very large collections of mutants at the early plant growth stage right after germination, or by spraying mature plants.

We have previously demonstrated *in planta* that the CAT inhibitor AT leads to dramatic decrease in CAT activity, followed by H_2O_2 accumulation, transcriptional reprogramming, and eventually cell death (Gechev et al., 2002, 2005). These effects were observed when AT was applied by spraying plants at rosette leaf stage or when added in the plant growth media. We argued that AT in plant growth media can be used as a screening agent to search in large collections of chemically or T-DNA mutagenized lines for plants with increased tolerance to oxidative stress. This notion was supported by the discovery that the *atr1* and *atr7* mutants, obtained through chemical mutagenesis and isolated on media with AAL-toxin, were also more tolerant to AT when the CAT inhibitor was added in the growth media (Gechev et al., 2008; Mehterov et al., 2012).

In our case study presented here, T-DNA lines were used to evaluate the feasibility of AT as screening agent. The lines were constructed by D. Weigel with the pSKI15 vector, which encodes a phosphinothricin-resistance gene and contains the tetramerized CaMV 35S enhancer sequences, T-DNA right border, *ori*V, and T-DNA left border (Weigel et al., 2000). Screening for tolerance to AT was done by placing 8600 M3 T-DNA activation tagged mutant lines obtained from the *Arabidopsis* stock center (NASC N21995) on Murashige and Skoog (MS) plant growth media supplemented with 9 μM AT. Plants were grown in a climate room under the following conditions: 60 $\mu mol\ m^{-2}\ s^{-1}$, 22 °C. Results were scored 10 days after germination. Only one survivor was recovered 10 days after seed germination (Fig. 11.1). The screening was then repeated and the same T-DNA mutant line showed viability under the experimental screening conditions, indicating stringent screening conditions. The mutant was named *oxr1* (*ox*idative stress *r*esistant *1*). The lone survivor was then recovered on soil under standard greenhouse conditions (14-h light/10-h dark period, photosynthetic photon flux density 400 $\mu mol\ m^{-2}\ s^{-1}$, 22 °C and relative humidity 70%) to collect progeny for further analysis.

2.1. Protocol for EMS mutagenesis and AT screening

1. Immerse the seeds in 0.1%, 0.2%, or 0.3% EMS (Sigma M0880 [62-50-0]) dissolved in water. Use 50-ml Falcon tubes or other vials tightly capped. Incubate for 8 or 15 h with constant rotation in order EMS to reach all seeds. In our case, 0.1% EMS for 8 h was sufficient to induce mutations. Higher concentrations and longer exposure times can induce multiple mutations, which may be problematic later on.
2. Rinse extensively the newly obtained M1 seeds with H_2O (at least 10 times!) and put in the cold for 2–7 days for vernalization.
3. Sow the seeds in pots (as many pots as possible; if the seeds are too many, pooling may be considered in order to reduce the pots), grow,

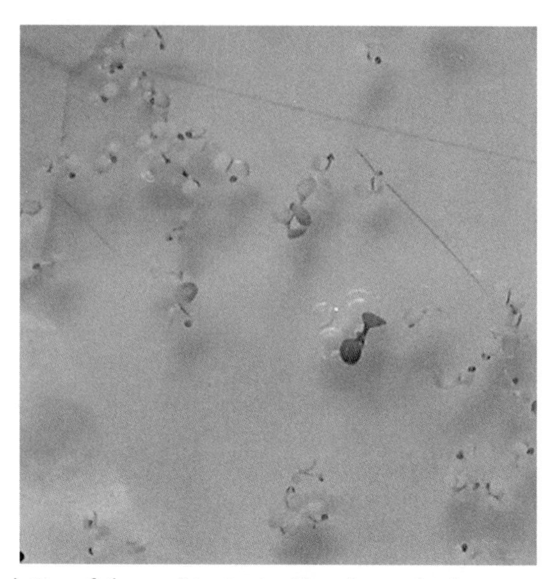

Figure 11.1 Isolation of the *oxr1* mutant with enhanced tolerance toward oxidative stress. Seeds from 8600 M3 T-DNA activation tagged mutant lines were plated on MS plant growth media supplemented with 9 μM AT. Plants were grown in a climate room under 16-h light/8-h dark photoperiod and light intensity 60 $\mu mol\ m^{-2}\ s^{-1}$, 22 °C. Results were scored 10 days after germination. (See Color Insert.)

self-pollinate, and collect the progeny (M2). This is the most labor- and time-consuming process and can last for 2–3 months. M2 progeny should contain homozygous mutants suitable for screening.

4. Sterilize the seeds (e.g., dry sterilization: placing the seeds for 2 h in a closed glass container, in eppendorf tubes with open lids, and a small beaker with 50–80 ml bleach + 2 ml HCl acid in it).
5. Perform the screening by germinating the sterile seeds on MS media supplemented with AT. A concentration of 5–9 μM AT usually works fine. We used 7 and 9 μM AT and light intensity of 60 $\mu mol\ m^{-2}\ s^{-1}$, 21 °C. Higher light intensities may result in rapid cell death, so they may work better with lower doses of AT. Alternatively, AT may also be used with the glutathione biosynthesis inhibitor buthionine sulfoximine (BSO), a specific inhibitor of γ-glutamylcysteinyl synthetase (Sukrong et al., 2012; Zhang et al., 2008). Observe the germinating seeds and the young seedlings regularly. AT can also be applied at later developmental stages on mature plants by spraying.

Caution! EMS is highly carcinogenic and volatile. EMS is also colorless and so extremely dangerous. Always work in the fume hood. Direct contact with

EMS must be completely avoided. Read a current MSDS sheet from the supplier before carrying out the experiment. Always wear lab coat, safety glasses, and a double layer of chemical-resistant gloves. All solutions containing EMS should be collected and stored in the specific containers in the fume hood until decontaminated and disposed of. Neutralize and decontaminate the EMS spills with 1 *M* Na-thiosulphate or 3 *M* NaOH. EMS-treated seeds can retain mutagen even after extensive rinsing, so the treated seeds should be handled with gloves.

3. IDENTIFICATION OF MUTATIONS IN THE GENOME

Depending on the type of mutation, alternative approaches are utilized to identify the mutated genes. In case of chemical (EMS) or γ-ray mutagenesis, map-based (positional) cloning is the method of choice. Description of this technology is outside the scope of the current chapter, but there are several useful reviews on map-based cloning (Jander et al., 2002; Peters, Cnudde, & Gerats, 2003). With the advance of the Next Generation Sequencing Technologies, the time from mutagenesis and screening to identification of the position of the mutation has decreased significantly. In case of insertion mutagenesis (T-DNA or transposons), TAIL-PCR, inverse-PCR, or other strategies that utilize the sequence information from the insert can be used to analyze the neighboring genome sequences (Liu, Mitsukawa, Oosumi, & Whittier, 1995; Liu & Whittier, 1995). This approach is still much faster compared with positional cloning. In our case study, we used TAIL-PCR to identify the position of the T-DNA insertion. Genomic DNA from *oxr1* was isolated with DNA plant minikit (Qiagen) according to the instructions of the manufacturer. TAIL-PCR was performed following the protocol described below (adopted from Liu et al., 1995; Liu & Whittier, 1995) by using three specific nested primers (SP1, SP2, and SP3: SP1 5′-TCCTGCTGAGCCTCGACATGTTGTC, SP2 5′-TCGACGTGTCTACATTCACGTCCA, SP3 5′-CCGTCGTA TTTATAGGCGAAAGC, and SP4 5′-GGAGGAAAAGAAGAGTAA TTA-3′ used for sequencing) and three arbitrary degenerated primers (AD1, AD2, and AD3: AD1 5′-NTCGASTWTSGWGTT, AD2 5′-NGTCGASWGANAWGAA, and AD3 5′-WGTGNAGWANCANAGA). The flanking sequence recovered indicated that the T-DNA insertion was in chromosome two, between gene loci At2g27270 At2g27280. There is also a third gene locus (At2g27285) in proximity close enough to be activated by the enhancer. Further molecular analysis by PCR with primer combinations from the flanking regions and inside the T-DNA revealed the presence of a single insertion.

3.1. Protocol for TAIL-PCR

1. Genomic DNA was isolated with DNA plant minikit (Qiagen, # 69104) according to the instructions of the manufacturer (other DNA isolation kits that yield high-quality DNA may also be used).
2. The TAIL-PCR reactions were carried out in a TC-512 THERMAL CYCLER (Techne) PCR apparatus. For the first TAIL-PCR reaction (TAIL1), approximately 120 ng (2 μl) of genomic DNA was taken and mixed in a 250-μl PCR tube with 0.4 μl SP1, 6 μl AD2, 12.5 μl master mix (Fermentas, # K0171), and ddH_2O up to 25 μl.
3. The following PCR program was used for TAIL1: initial denaturation at 95 °C for 3 min, five cycles of 95 °C for 1 min; 63 °C for 1 min and 72 °C for 2.5 min. Next two cycles of 94 °C for 1 min, 25 °C for 3 min followed by slow heating (0.8°C per s) up to 72 °C, 72 °C for 5.5 min. After completion of these steps, a macro cycle was used 20 times. It contained the following steps: 94 °C for 30 s, 63 °C for 1 min, 72 °C for 2.5 min, 94 °C for 45 s, 63 °C for 1 min, 72 °C for 2.5 min, 94 °C for 30 s, 40 °C for 1 min, 72 °C for 2.5 min. The final extension was for 5 min at 72 °C.
4. For the second TAIL-PCR reaction (TAIL2), the products of the first were diluted 250 times and 1.5 μl of them were used as template. The other ingredients were SP2—0.4 μl, AD2—4 μl, 12.5 μl master mix, and ddH_2O up to 25 μl.
5. The TAIL2 program was: 15 cycles of 94 °C for 30 s, 60 °C for 1 min, 72 °C for 2.5 min, 94 °C for 30 s, 60 °C for 1 min, 72 °C for 2.5 min, 94 °C for 30 s, 40 °C for 1 min, 72 °C for 2.5 min. The final extension was 72 °C for 5 min.
6. The products were diluted 500 times and 1.5 μl of them were mixed with SP3—0.4 μl, AD2—4 μl, 12.5 μl master mix, and ddH_2O up to 25 μl.
7. The TAIL 3 program was: initial denaturation at 94 °C for 1.5 min and 35 cycles of 94 °C for 40 s, 40 °C for 1 min, 72 °C for 2.5 min. Final extension was 5 min at 72 °C.

The PCR products were mixed with 4 μl of loading dye (Fermentas # R0611), loaded onto 1% agarose gel containing 0.5 mg l^{-1} ethidium bromide (final concentration) covered with 1X TBE buffer and separated by applying 7 V cm^{-1} gel length electrical currency. The size of the products was determined by comparison with a DNA ladder (Fermentas GeneRuler # SM0311). The PCR products were visualized by UV light. The PCR products were isolated from the agarose by cutting out with pure surgical blade and extracted with QIAquick Gel extraction kit (Qiagen, # 28704) following the original protocol. The purified products were sequenced in MWG Biotech AG, Frankfurt, Germany, using the SP4 primer.

4. ANALYSIS OF THE MUTANTS WITH ENHANCED TOLERANCE TO H_2O_2-INDUCED OXIDATIVE STRESS

4.1. Evaluation of the mutants tolerance to oxidative stress

The performance of the mutants under oxidative stress-generating conditions can be assessed by several parameters, such as the amount of photosynthetic pigments (chlorophyll, e.g., significantly decreases during stress), fresh weight measurements, accumulation of malondialdehyde (MDA; its increase indicates ROS-induced degradation of lipids), carbonylation of proteins, and others (Shulaev & Oliver, 2006).

Assessment for tolerance to ROS-induced PCD was done in our case by plating seeds from the T-DNA mutant and wild-type controls on media containing either 7 or 9 μ*M* AT. Plants were inspected visually after 7 days, and samples were taken for fresh weight and chlorophyll analysis.

Pictures and samples for analysis were taken 7 days after germination from 9 μ*M* AT-treated samples. On that day, *oxr1* looked green without any visible necrotic lesions, whereas WT plants were yellow and apparently dying. Measurements of trypan blue confirmed the cell death in WT plants (data not shown). Measurements of fresh weight and chlorophyll content confirmed that *oxr1* had a better performance than WT on both 7 and 9 μ*M* AT. Chlorophyll content was measured photometrically as previously described (Gechev et al., 2003). *Oxr1* retained most of its fresh weight and its chlorophyll, in contrast with the wild type (Fig. 11.2).

In addition to AT, *oxr1* was evaluated for tolerance to PQ, a herbicide that leads primarily to the generation of superoxide radicals which are then converted to H_2O_2 (Gechev et al., 2006). Unlike AT, no tolerance against PQ was observed under the conditions tested (concentration range 0.5–1.5 μ*M* PQ, data not shown).

A

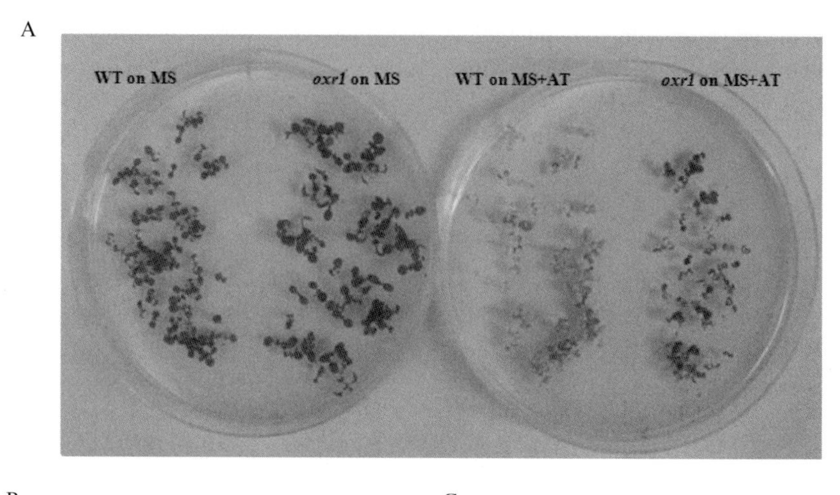

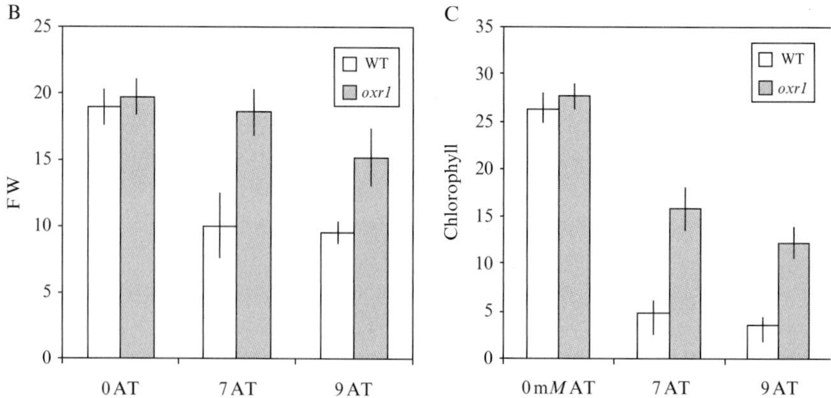

Figure 11.2 *Oxr1* exhibits enhanced tolerance to AT-induced oxidative stress. Seeds from wild-type (WT) and *oxr1* mutant plants were germinated on MS media with 0, 7, or 9 μ*M* AT. Pictures and samples for analysis were taken 7 days after germination on 9 μ*M* AT. (A) WT (*Arabidopsis thaliana* ecotype Col-0) and *oxr1* on MS media and MS media supplemented with 9 μ*M* AT, (B) fresh weight (FW) is expressed in mg per 10 seedlings, and (C) chlorophyll content expressed in μg chlorophyll per mg FW. Data are means of three-independent biological repetitions ± SD. (For color version of this figure, the reader is referred to the online version of this chapter.)

4.2. Protocol for the determination of chlorophyll content in plant tissues

Total chlorophyll, as well as chlorophyll *A*, chlorophyll *B*, and carotenoids, can be determined by a fast and simple spectrophotometric assay (Porra, Thompson, & Kriedemann, 1989). According to the Beer–Lambert–Bouguer law, the absorption of light with particular wave length that passes through a compound is proportional to the concentration of that compound (*c*), its specific extinction coefficient (ε), and the length of the light path (*l*, which is 1 cm for most photometers):

$$A = \varepsilon l c$$

1. Extract the chlorophyll pigments from approximately 50 mg plant tissue with 1 ml 80% acetone for overnight at 4 °C.
2. Transfer the supernatant in a 1-ml glass cuvette for spectrophotometric measurement.
3. Read the absorbance at both 663.6 and 646.6 nm, corresponding to Chl *a* and Chl *b*, respectively. Use 80% acetone as blank.
4. Calculate the pigment content using the following molar extinction coefficients: Chl *a*, $\varepsilon_{663.6} = 76.79$ and $\varepsilon_{646.6} = 18.58$; Chl *b*, $\varepsilon_{663.6} = 9.79$ and $\varepsilon_{646.6} = 47.04$. The total Chl content can be calculated using the following formula and normalize per fresh weight:

$$\text{Chl}\, a + b = 19.54 A_{646.6} + 8.29 A_{663.6}$$

4.3. Protocol for MDA determination in plant tissues

1. Extract the MDA from 50 mg plant material (fresh or frozen and ground in liquid nitrogen) using 1 ml 0.25% thiobarbituric acid (TBA) dissolved in 10% trichloroacetic acid (TCA).
2. Collect the extract in a 1.5-ml eppendorf tube, heat the mixture at 85 °C for 30 min, and then quickly chill it on ice. Proceed by the same way for the blank sample but without plant material.
3. Centrifuge the mixture at maximum speed for 10 min to pellet the particles.
4. Transfer the supernatant in a 1-ml plastic cuvette for spectrophotometric measurement.
5. Read the absorbance first at 532 nm (the peak of MDA–TBA complex) and second time at 600 nm (nonspecific absorption). As blank, use 1 ml 0.25% TBA in 10% TCA. Calculate $A_{(532-600)}$ (Taulavuori, Hellström, & Taulavuori, 2001).
6. Estimate the MDA concentration using the Beer–Lambert–Bouguer law, MDA extinction coefficient ε_{532} 155 mM^{-1} cm^{-1}, calculate the amount of MDA, and normalize the values to the fresh weight of each sample.

4.4. CAT activity measurements

As AT inhibits CAT, measuring CAT activity is essential to confirm that the AT treatment worked. Furthermore, CAT is one of the most important antioxidant enzymes, subjected to regulations by environmental factors and the circadian clock (Gechev et al., 2013; Lai et al., 2012). CAT activity can be studied by either in-gel assays or photometrically by following the decrease in absorbance at 240 nm (decomposition of H_2O_2) or increase in O_2 (with an oxygen electrode) (Gechev et al., 2003).

In our case, the AT-dependent inhibition of CAT activity was analyzed by in-gel assay. Protein extracts from 4-day-old wild-type and *oxr1* seedlings grown on standard MS media in the presence or absence of 9 μ*M* AT were isolated, and CATs were separated by native PAGE and then stained to visualize their activity using the protocol described below. AT inhibited CATs in both wild type and *oxr1* (Fig. 11.3).

4.5. Protocol for measuring CAT activity by native PAGE

1. Extract total protein from plant material (fresh leaves or frozen plant material ground to fine powder in liquid nitrogen works equally well)

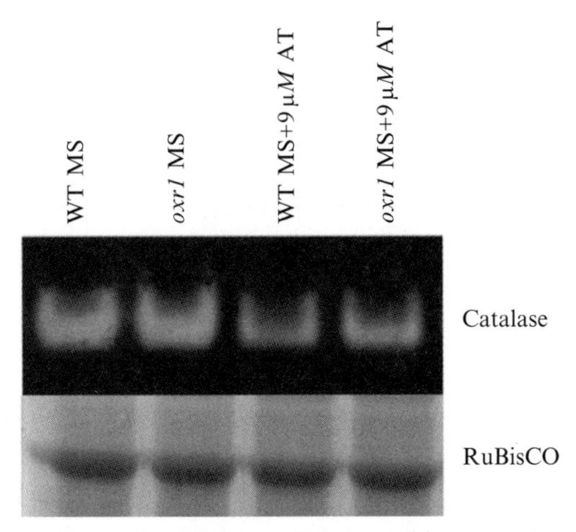

Figure 11.3 Effect of AT on the catalase activity in wild-type and *oxr1* mutant plants. Wild type and *oxr1* were grown on standard MS media in the presence or absence of 9 μ*M* AT for 4 days and total protein was isolated for enzyme activity measurements. Catalase activity determined by native PAGE (15 μg total protein loaded in each line). As a control, RuBisCO large subunit is shown below. The experiment was repeated three times showing always the same activity pattern; one representative example is shown. (For color version of this figure, the reader is referred to the online version of this chapter.)

with the following CAT extraction buffer: 50 m*M* potassium phosphate, 1 m*M* EDTA, 0.1% Triton X-100, 2% (w/v) polyvinylpolypyrrolidone, and 1 m*M* phenylmethanesulfonyl fluoride.

2. Transfer the extract in 1.5-ml eppendorf tubes, centrifuge to pellet the debris at 4 °C, and transfer the supernatant in new tubes, keeping them on ice.
3. Measure the protein concentration using Bradford reagent in order to calculate later on equal protein loading.
4. Prepare a standard native 7.5% polyacrylamide running gel (pH 8.8, Tris–HCl) with 3.5% stacking gel (pH 6.8, Tris–HCl). The running buffer is 250 m*M* glycine and 25 m*M* Tris–HCl, pH 8.3.
5. Separate the total protein by running the electrophoresis for 5 h at 80 V, at 4 °C to preserve the CAT activity. 15 μg protein extract is normally sufficient to produce good results.
6. After electrophoretic separation, immerse the gel in 0.01% of H_2O_2 solution for 10 min.
7. Wash the gel with water twice and incubate for 10 min in 2% $FeCl_3$ and 2% $K_3[Fe(CN)_6]$, mixed in 1:1 ratio immediately before use.
8. Wash the gel in water and observe the staining. Yellow to light green bands corresponding to the active CAT should emerge on a dark-green background.

4.6. Protocol for photometric determination of CAT activity

1. Extract total protein from plant material as described for the native PAGE above, transfer the extract in 1.5-ml eppendorf tubes to centrifuge, pellet the debris at 4 °C, and transfer the supernatant in new tubes, keeping them on ice. Measure the protein concentration using Bradford reagent.
2. Prepare reaction mixtures containing 20 μg of protein in 50 m*M* KPi buffer, pH 7.0, final volume 1 ml in 1.5-ml eppendorf tubes.
3. Add 10 μl of 1.76 *M* H_2O_2. This starts the reaction. Mix vigorously and transfer immediately in a quartz cuvette.
4. Read the absorbance at 240 nm for 1 min and calculate ΔA_{240}. Use protein sample in 50 m*M* KPi buffer as blank. The decrease in absorbance should be linear, which means sufficient substrate is still present in the reaction mixture (when the specific CAT activity is very high, H_2O_2 can decompose very rapidly and the reaction mixture runs out of substrate; in this case, less protein can be used).
5. Calculate the CAT activity as μmol of metabolized H_2O_2 per minute per μg protein, using an H_2O_2 extinction coefficient ε_{240} of 0.0436 mM^{-1} cm^{-1}.

4.7. Determination of anthocyanins content

The CAT deficiency resulted in visible accumulation of anthocyanins, especially in the *oxr1* mutant. To quantify it, the anthocyanins content was determined photometrically as described below. Seven-day-old *oxr1* seedlings accumulated much more anthocyanins than wild type (Fig. 11.4). In contrast to the *oxr1*, the wild-type seedlings became yellow and eventually died at this time point, suggesting that the significantly higher levels of anthocyanins may have a protective role against AT-induced oxidative stress (Fig. 11.4).

4.8. Protocol for determination of anthocyanins content in plant tissues

1. Extract the anthocyanins from 100 mg fresh or frozen in liquid nitrogen plant material using 1 ml 1% HCl in methanol.
2. Collect the extract in a 1.5-ml eppendorf tube and centrifuge to pellet the particles.
3. Transfer the supernatant in 1-ml plastic cuvettes for spectrophotometric measurements.
4. Read the absorbance first at 530 nm (anthocyanins peak) and second time at 657 nm (nonspecific absorption of photosynthetic pigments). Use 1% HCl in methanol as blank. Calculate $A_{530} - 0.25A_{657}$.

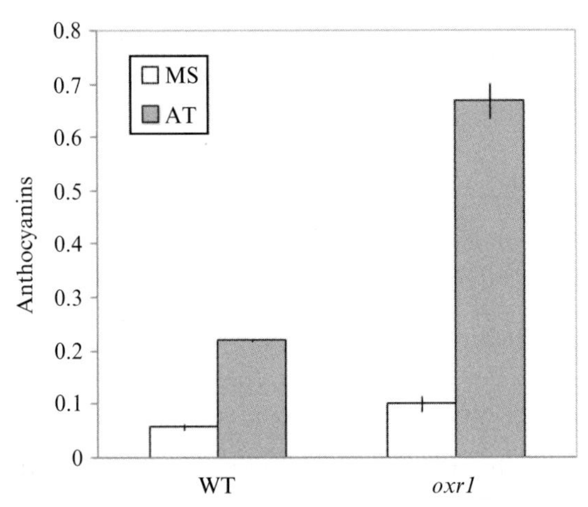

Figure 11.4 AT induces anthocyanins accumulation in *Arabidopsis*. Wild-type and *oxr1* plants were grown on MS media with or without 9 μ*M* AT and anthocyanin levels were determined in 7-day-old seedlings. Values show $A_{530} - 0.25A_{657}$ from 100 mg plant material. Values are means of three measurements ± SD.

5. Estimate the anthocyanins concentration using the Beer–Lambert–Bouguer law, calculate the amount of anthocyanins, and normalize the values to the fresh weight of each sample. Different anthocyanins have different absorption spectra with specific absorption maximums, and extinction coefficients (Fuleki & Francis, 1968). Most authors assume an extinction coefficient ε_{529} of 30,000 l mol^{-1} cm^{-1} for total anthocyanins. Alternatively, the total anthocyanins can be presented as $A_{530} - 0.25A_{657}$ per fresh weight (Fig. 11.4; Vanderauwera et al., 2005).

5. CONCLUSION

The CAT inhibitor AT is a simple and powerful tool for isolating mutants with enhanced tolerance to oxidative stress. Previously, it was shown that it can be used to characterize mutants obtained from a population of chemically mutagenized plants (Gechev et al., 2008; Mehterov et al., 2012). Furthermore, lower doses of AT in combination with the glutathione biosynthesis inhibitor BSO, a specific inhibitor of γ-glutamylcysteinyl synthetase, were successfully employed to isolate *oxt1* and *oxt6*, two mutants more tolerant to oxidative stress (Sukrong et al., 2012; Zhang et al., 2008). The *oxt1* mutation affects a gene encoding adenine phosphoribosyltransferase, while the *oxt6* mutation is in a gene encoding a subunit of the cleavage and polyadenylation specificity factor (Sukrong et al., 2012; Zhang et al., 2008). Both *oxt1* and *oxt6* were more tolerant to both AT and PQ, similar to the *atr1* and *atr7* mutants which are also tolerant to the two ROS-generating agents (Gechev et al., 2008; Mehterov et al., 2012; Zhang et al., 2008). Unlike these mutants, however, the *oxr1* mutant reported here is not tolerant to PQ, indicating that either the affected gene(s) in *oxr1* may be more specific to the H_2O_2 signaling pathway or the PQ testing conditions in the present study were too stringent to discriminate between the mutant and the wild type. An obvious advantage of screening T-DNA insertion mutants is that the mutation/gene affected can be immediately associated with the phenotype. If a gene has a function as a positive regulator of oxidative stress-induced PCD, its disruption by T-DNA insertion will result in enhanced tolerance to cell death. When the T-DNA insertion contains an activation cassette, then both knockouts of positive cell death regulators and activated negative regulators of cell death can be isolated.

The significantly reduced CAT activity in the AT-treated wild-type and *oxr1* plants confirmed the efficiency of AT as a CAT inhibitor. The *oxr1* mode of action and the reasons for its survival under these oxidative stress-generating conditions still need to be investigated in details. However, higher levels of anthocyanins in AT-treated *oxr1* may contribute to the oxidative stress tolerance. Anthocyanins were implicated in protection against oxidative stress in a number of plant species (Dinakar & Bartels, 2012; Nagata et al., 2003; Zeng, Chow, Su, Peng, & Peng, 2010). Although the anthocyanins biosynthesis was induced by AT in the wild-type plants as well, the anthocyanins levels in AT-treated *oxr1* were nearly threefold higher than in the wild type, and this quantitative difference could account for the better performance of *oxr1* under oxidative stress.

ACKNOWLEDGMENTS

Authors are grateful to B. Venema for technical assistance. This work was financially supported by the Swiss Enlargement Contribution in the framework of the Bulgarian–Swiss Research Programme, project No. IZEBZ0_143003/1, and Grant DO2-1068 from the Ministry of Education, Youth, and Science of Bulgaria.

REFERENCES

Apel, K., & Hirt, H. (2004). Reactive oxygen species: Metabolism, oxidative stress, and signal transduction. *Annual Review of Plant Biology*, *55*, 373–399.

Britt, A., & Jiang, C. Z. (1999). Generation, identification, and characterization of repair-defective mutants of Arabidopsis. *Methods in Molecular Biology*, *113*, 31–40.

Dangl, J. L., & Jones, J. D. G. (2001). Plant pathogens and integrated defence responses to infection. *Nature*, *411*, 826–833.

Dat, J. F., Pellinen, R., Beeckman, T., Van De Cotte, B., Langebartels, C., Kangasjärvi, J., et al. (2003). Changes in hydrogen peroxide homeostasis trigger an active cell death process in tobacco. *The Plant Journal*, *33*, 621–632.

Dinakar, C., & Bartels, D. (2012). Light response, oxidative stress management and nucleic acid stability in closely related Linderniaceae species differing in desiccation tolerance. *Planta*, *236*, 541–555.

Frye, C. A., & Innes, R. W. (1998). An *Arabidopsis* mutant with enhanced resistance to powdery mildew. *The Plant Cell*, *10*, 947–956.

Fuleki, T., & Francis, F. J. (1968). Quantitative methods for anthocyanins. 1. Extraction and determination of total anthocyanin in cranberries. *Journal of Food Science*, *33*, 72–77.

Gadjev, I., Stone, J. M., & Gechev, T. S. (2008). Programmed cell death in plants: New insights into redox regulation and the role of hydrogen peroxide. *International Review of Cell and Molecular Biology*, *270*, 87–144.

Gechev, T., Benina, M., Obata, T., Tohge, T., Sujeeth, N., Minkov, I., et al. (2013). Molecular mechanisms of desiccation tolerance in the resurrection glacial relic *Haberlea rhodopensis*. *Cellular and Molecular Life Sciences*, *70*, 689–709.

Gechev, T., Ferwerda, M., Mehterov, N., Laloi, C., Qureshi, K., & Hille, J. (2008). *Arabidopsis* AAL-toxin-resistant mutant *atr1* shows enhanced tolerance to programmed cell death induced by reactive oxygen species. *Biochemical and Biophysical Research Communications*, *375*, 639–644.

Gechev, T., Gadjev, I., Van Breusegem, F., Inzé, D., Dukiandjiev, S., Toneva, V., et al. (2002). Hydrogen peroxide protects tobacco from oxidative stress by inducing a set of antioxidant enzymes. *Cellular and Molecular Life Sciences*, *59*, 708–714.

Gechev, T. S., & Hille, J. (2005). Hydrogen peroxide as a signal controlling plant programmed cell death. *The Journal of Cell Biology, 168*, 17–20.

Gechev, T. S., Minkov, I. N., & Hille, J. (2005). Hydrogen peroxide-induced cell death in *Arabidopsis*: Transcriptional and mutant analysis reveals a role of an oxoglutarate-dependent dioxygenase gene in the cell death process. *IUBMB Life, 57*, 181–188.

Gechev, T. S., Van Breusegem, F., Stone, J. M., Denev, I., & Laloi, C. (2006). Reactive oxygen species as signals that modulate plant stress responses and programmed cell death. *BioEssays, 28*, 1091–1101.

Gechev, T., Willekens, H., Van Montagu, M., Inzé, D., Van Camp, W., Toneva, V., et al. (2003). Different responses of tobacco antioxidant enzymes to light and chilling stress. *Journal of Plant Physiology, 160*, 509–515.

Jander, G., Norris, S. R., Rounsley, S. D., Bush, D. F., Levin, I. M., & Last, R. L. (2002). *Arabidopsis* map-based cloning in the post-genome era. *Plant Physiology, 129*, 440–450.

Lai, A. G., Doherty, C. J., Mueller-Roeber, B., Kay, S. A., Schippers, J. H., & Dijkwel, P. P. (2012). CIRCADIAN CLOCK-ASSOCIATED 1 regulates ROS homeostasis and oxidative stress responses. *Proceedings of the National Academy of Sciences of the United States of America, 109*, 17129–17134.

Liu, Y. G., Mitsukawa, N., Oosumi, T., & Whittier, R. F. (1995). Efficient isolation and mapping of *Arabidopsis thaliana* T-DNA insert junctions by thermal asymmetric interlaced PCR. *The Plant Journal, 8*, 457–463.

Liu, Y. G., & Whittier, R. F. (1995). Thermal asymmetric interlaced PCR—Automatable amplification and sequencing of insert end fragments from P1 and Yac clones for chromosome walking. *Genomics, 25*, 674–681.

Mehterov, N., Balazadeh, S., Hille, J., Toneva, V., Mueller-Roeber, B., & Gechev, T. (2012). Oxidative stress provokes distinct transcriptional responses in the stress-tolerant *atr7* and stress-sensitive *loh2 Arabidopsis thaliana* mutants as revealed by multi-parallel quantitative real-time PCR analysis of ROS marker and antioxidant genes. *Plant Physiology and Biochemistry, 59*, 20–29.

Nagata, T., Todoriki, S., Masumizu, T., Suda, I., Furuta, S., Du, Z., et al. (2003). Levels of active oxygen species are controlled by ascorbic acid and anthocyanin in *Arabidopsis*. *Journal of Agricultural and Food Chemistry, 51*, 2992–2999.

Peters, J. L., Cnudde, F., & Gerats, T. (2003). Forward genetics and map-based cloning approaches. *Trends in Plant Science, 8*, 484–491.

Porra, R. J., Thompson, W. A., & Kriedemann, P. E. (1989). Determination of accurate extinction coefficients and simultaneous equations for assaying chlorophylls a and b extracted with four different solvents: Verification of the concentration of chlorophyll standards by atomic absorption spectroscopy. *Biochimica et Biophysica Acta, 975*, 384–394.

Shulaev, V., & Oliver, D. J. (2006). Metabolic and proteomic markers for oxidative stress. New tools for reactive oxygen species research. *Plant Physiology, 141*, 367–372.

Sukrong, S., Yun, K. Y., Stadler, P., Kumar, C., Facciuolo, T., Moffatt, B. A., et al. (2012). Improved growth and stress tolerance in the *Arabidopsis oxt1* mutant triggered by altered adenine metabolism. *Molecular Plant, 5*, 1310–1332.

Taulavuori, E., Hellström, E. K., & Taulavuori, K. (2001). Comparison of two methods used to analyse lipid peroxidation from *Vaccinium myrtillus* (L.) during snow removal, reacclimation and cold acclimation. *Journal of Experimental Botany, 52*, 2375–2380.

Vandenabeele, S., Van Der Kelen, K., Dat, J., Gadjev, I., Boonefaes, T., Morsa, S., et al. (2003). A comprehensive analysis of hydrogen peroxide-induced gene expression in tobacco. *Proceedings of the National Academy of Sciences of the United States of America, 100*, 16113–16118.

Vandenabeele, S., Vanderauwera, S., Vuylsteke, M., Rombauts, S., Langebartels, C., Seidlitz, H. K., et al. (2004). Catalase deficiency drastically affects gene expression induced by high light in *Arabidopsis thaliana*. *The Plant Journal, 39*, 45–58.

Vanderauwera, S., Zimmermann, P., Rombauts, S., Vandenabeele, S., Langebartels, C., Gruissem, W., et al. (2005). Genome-wide analysis of hydrogen peroxide-regulated gene expression in *Arabidopsis* reveals a high light-induced transcriptional cluster involved in anthocyanin biosynthesis. *Plant Physiology*, *139*, 806–821.

Weigel, D., Ahn, J. H., Blázquez, M. A., Borevitz, J. O., Christensen, S. K., Fankhauser, C., et al. (2000). Activation tagging in *Arabidopsis*. *Plant Physiology*, *122*, 1003–1013.

Zeng, X. Q., Chow, W. S., Su, L. J., Peng, X. X., & Peng, C. L. (2010). Protective effect of supplemental anthocyanins on *Arabidopsis* leaves under high light. *Physiologia Plantarum*, *138*, 215–225.

Zhang, J., Addepalli, B., Yun, K. Y., Hunt, A. G., Xu, R., Rao, S., et al. (2008). A polyadenylation factor subunit implicated in regulating oxidative signaling in *Arabidopsis thaliana*. *PloS One*, *3*(6), e2410.

CHAPTER TWELVE

Analysis of Environmental Stress in Plants with the Aid of Marker Genes for H_2O_2 Responses

Ayaka Hieno[*], Hushna Ara Naznin[*], Katsunobu Sawaki[*], Hiroyuki Koyama[*,†], Yusaku Sakai[†], Haruka Ishino[†], Mitsuro Hyakumachi[*,†], Yoshiharu Y. Yamamoto[*,†,1]

[*]The United Graduate School of Agricultural Sciences, Gifu University, Gifu, Japan
[†]Faculty of Applied Biological Sciences, Gifu University, Gifu, Japan
[1]Corresponding author: e-mail address: yyy@gifu-u.ac.jp

Contents

1. Introduction 222
2. Experimental Materials and Procedures 224
   2.1 Preparation of *Arabidopsis* plants for hydrogen peroxide treatment 224
   2.2 Preparation of *Arabidopsis* plants for abiotic stress treatment 226
   2.3 Extraction of total RNA 228
   2.4 Quantitative real-time RT-PCR 230
3. Example of Analysis 231
   3.1 Selection of primers of test marker genes for real-time PCR 231
   3.2 Expression of test marker genes to H_2O_2 treatment 231
   3.3 Analysis of abiotic stress response 232
Appendix: Recipes of Stock Solutions, Buffers, and Media 235
References 236

Abstract

Hydrogen peroxide acts as a signaling molecule mediating the acquisition of tolerance to both biotic and abiotic stresses. Identification of marker genes for H_2O_2 response could help to intercept the signaling network of stress response of plants. Here, we describe application of marker genes for H_2O_2 responses to monitoring several abiotic stress responses. *Arabidopsis* plants were treated with UV-B, high light, and cold stresses, where involvement of H_2O_2-mediated signaling is known or suggested. Monitoring of these stress responses with molecular markers using quantitative real-time RT-PCR can detect landmark events in the sequential stress responses. These methods can be used for analysis of mutants and transgenic plants to examine natural H_2O_2 responses that are involved in environmental adaptation.

Methods in Enzymology, Volume 527
ISSN 0076-6879
http://dx.doi.org/10.1016/B978-0-12-405882-8.00012-X

1. INTRODUCTION

Plants have evolved the complex and specialized transcriptional network to control their response to environmental changes. Understanding the mechanisms by which plants perceive environmental signals and transmit such signals to the cellular machinery to activate adaptive responses is a fundamental issue in plant biology, and is also vital for the continued development of breeding to improve stress tolerance in crops. Identification of marker genes expressed in different environmental stress and detailed study of the expression pattern could help us to intercept the signaling network of stress response mechanism of higher plants.

H_2O_2 plays as a signal molecule for various stress responses, including pathogen infection (Suzuki, Yano, & Shinshi, 1999), ultraviolet (UV) (Mackerness, John, Jordan, & Thomas, 2001), drought (Quan, Zhang, Shi, & Li, 2008), high light (HL) (Karpinski et al., 1999), wounding (Orozco-Cardenas & Ryan, 1999), and high or low temperature (Quan et al., 2008; Rizhsky, Davletova, Liang, & Mittler, 2004). In addition, H_2O_2 is known to mediate not only intracellular but also intercellular signaling (HL: Karpinski et al., 1999; SAR: Ryals et al., 1996). Therefore, there would be variations in the temporal and spatial profile of H_2O_2 accumulation after stress generation according to the type of stress.

Because each of the stress responses is thought to contain multiple signal transduction pathways, each of which is composed of multiple steps, setting and monitoring landmark events would be useful in understanding the complex responses. In this chapter, we introduce a method of application of molecular markers for H_2O_2 responses to dissect several environmental stress responses. Monitoring of molecular markers can be easily and flexibly achieved by quantitative real-time reverse transcriptase-polymerase chain reaction (RT-PCR). Identification of marker genes can also be easily done using public microarray data, which are now a large quantity in the case of *Arabidopsis thaliana*. A list of potentially useful markers is shown in Table 12.1. The table includes information of response to several biotic and abiotic stresses for each gene in order to facilitate selection of the markers of interest.

Because H_2O_2 is involved in almost all of the stress responses in plants, analysis of the role of H_2O_2 in stress responses requires various physiological experimental systems, including application of both biotic and abiotic stresses. Here, we also include methods for application of cold, HL, and UV-B stresses to *Arabidopsis*.

Table 12.1 List of the stress responsive genes with fold change analyzed from microarray data

	Locus	Description	3% H_2O_2 (3 h)[1]	UV-B (3 h)[2]	High light (3 h)[3]	Cold (3 h)[4]	Cold (24 h)[5]	Drought (12 h)[6]	Salt (24 h)[7]	Wounding (3 h)[8]	Pst DC3000 (24 h)[9]	SA (3 h)[10]
	AT5G12020	HSP17.6II	814.6	3.8	117.3	nd	nd	8.0	36.1	14.8	7.2	1.3
	AT1G52560	HSP26.5-P	214.9	1.2	33.3	nd	nd	nd	35.3	nd	57.2	nd
*1	AT1G53540	HSP17.6C-CI	202.5	0.9	5.8	0.9	nd	6.9	5.2	7.6	4.2	nd
	AT5G12030	AT-HSP17.6A	173.9	7.9	15.9	nd	4.1	11.7	116.9	8.9	8.5	1.1
	AT3G46230	ATHSP17.4	112.1	3.6	13.4	0.8	nd	5.8	85.0	8.5	6.9	2.1
	AT2G29500	HSP17.6B-CI	79.0	7.7	9.3	0.4	nd	4.4	3.7	7.5	5.3	1.6
	AT4G25200	ATHSP23.6-MITO	79.0	73.6	6.6	nd	nd	nd	nd	nd	nd	nd
*2	AT1G74310	ATHSP101	46.5	2.8	13.9	1.4	1.7	1.5	5.4	1.5	15.0	nd
	AT5G48570	peptidyl-prolyl cis-trans isomerase, putative	38.5	8.8	18.2	1.7	1.0	1.3	5.2	1.7	1.6	1.0
	AT5G15120	-	37.7	0.9	1.1	nd	nd	1.1	nd	nd	1.7	0.7
	AT5G59720	HSP18.2	34.8	0.9	0.8	nd	nd	nd	6.8	nd	3.8	nd
	AT4G12400	stress-inducible protein, putative	32.9	6.4	15.0	2.4	0.5	1.2	3.2	2.0	6.2	1.3
	AT5G03780	TRFL10 (TRF-LIKE 10); DNA binding	32.7	1.3	5.0	1.0	nd	1.1	nd	1.1	nd	nd
	AT5G14470	GHMP kinase-related	31.0	0.2	1.1	nd	nd	nd	nd	nd	nd	nd
	AT5G11410	protein kinase family protein	27.9	0.9	9.5	0.6	1.1	2.2	0.9	2.8	0.9	nd
	AT3G02240	-	26.8	0.8	3.1	nd	nd	nd	1.4	2.1	nd	2.5
	AT1G72660	GTP-binding protein, putative	24.5	1.1	3.6	nd	nd	nd	3.0	nd	4.6	nd
	AT1G57630	disease resistance protein (TIR class), putative	22.3	150.6	0.3	nd	nd	nd	nd	10.9	21.5	nd
	AT5G51440	HSP23.5-M	22.2	42.5	5.2	1.4	1.1	1.4	2.3	1.7	2.1	2.5
	AT2G46240	BAG6 (BCL-2-ASSOCIATED ATHANOGENE 6)	21.8	1.5	1.4	nd	1.4	1.2	4.2	1.4	2.2	nd
	AT3G60420	-	20.4	68.2	1.0	2.3	3.2	1.3	11.0	22.2	7.0	6.2
	AT1G54050	HSP17.4-CIII	20.1	0.7	1.4	0.9	nd	1.2	1.4	1.4	3.2	1.1
	AT1G16030	HSP70B	19.5	1.0	1.3	nd	0.9	1.3	1.6	2.3	2.6	0.8
	AT2G25230	MYB100	17.6	3.1	15.6	nd	nd	1.8	nd	nd	nd	nd
	AT3G28210	PMZ; zinc ion binding	17.5	20.0	1.3	2.4	24.8	4.9	nd	8.1	10.2	7.8
	AT2G38250	DNA-binding protein-related	17.1	11.0	1.5	nd	nd	nd	nd	nd	34.3	nd
	AT5G52640	HSP81-1	16.1	26.5	5.6	3.3	3.0	1.4	4.8	4.8	3.5	2.0
	AT3G25250	AGC2-1 (OXIDATIVE SIGNAL-INDUCIBLE1)	15.7	15.1	1.3	nd	nd	nd	nd	nd	8.7	2.8
	AT2G41730	-	15.2	77.4	1.4	nd	nd	nd	nd	nd	4.4	5.3
	AT2G32120	HSP70T-2	15.0	1.0	2.3	1.2	0.7	1.1	1.8	1.3	1.7	0.6
	AT3G63380	Ca(2+)-ATPase, putative (ACA12)	14.2	29.5	0.6	nd	nd	1.4	nd	3.6	7.8	1.4
	AT5G37670	HSP15.7-CI	14.0	1.6	1.8	0.9	0.7	1.0	4.8	0.8	1.3	1.1
	AT5G27610	DNA binding / transcription factor	12.9	3.5	1.1	nd	nd	nd	nd	nd	nd	nd
	AT5G10695	-	12.8	4.2	1.4	16.9	4.8	1.3	2.7	1.9	3.1	1.3
	AT1G17180	ATGSTU25; glutathione transferase	12.2	2.8	0.7	nd	nd	2.0	nd	1.6	2.6	17.5
	AT1G09080	BIP3; ATP binding	11.6	105.3	0.5	nd	nd	8.0	6.8	7.9	1.3	8.3
	AT4G24110	-	11.3	7.2	1.6	nd	nd	nd	nd	nd	29.5	1.5
	AT5G42380	CML37/CML39; calcium ion binding	11.3	36.2	0.8	6.3	nd	nd	nd	4.6	44.1	nd
	AT3G44190	pyridine nucleotide-disulphide oxidoreductase family protein	11.1	1.3	2.2	1.1	1.4	0.9	1.4	1.2	2.3	3.1
	AT3G20340	-	11.0	0.2	0.9	3.4	1.3	nd	nd	nd	2.5	0.8
	AT5G24660	-	10.9	2.2	2.6	1.0	0.7	1.5	3.4	1.2	0.5	1.4
	AT2G41380	embryo-abundant protein-related	10.7	14.5	0.9	1.6	1.5	12.9	4.1	3.4	35.8	3.7
	AT4G17240	-	10.4	1.4	4.4	nd	nd	nd	2.1	nd	nd	0.8
	AT2G15490	UGT73B4; UDP-glycosyltransferase	10.2	62.7	0.8	nd	nd	22.3	nd	nd	10.4	24.4
	AT5G25450	ubiquinol-cytochrome C reductase complex 14 kDa protein, putative	10.1	1.8	3.1	1.1	0.6	1.2	3.5	2.0	2.4	1.2
	AT1G22400	ATUGT85A1/UGT85A1 (UDP-GLUCOSYL TRANSFERASE 85A1)	10.1	9.4	1.5	1.1	0.7	1.8	1.2	1.0	11.0	1.5
*2	AT3G12580	HSP70 (heat shock protein 70); ATP binding	8.0	28.1	8.4	9.2	8.7	1.6	12.2	6.1	3.2	2.3
*3	AT4G02380	SAG21 (SENESCENCE-ASSOCIATED GENE 21)	1.5	13.8	0.8	3.7	57.9	4.5	69.9	1.9	21.4	1.1

Data were collected from different microarray data sources: [1,3]Yamamoto et al. (2004), [2,4~10]The Arabidopsis Information Resource (TAIR, http://www.arabidopsis.org/portals/expression/microarray/ATGenExpress.jsp), [2]TAIR-ME00329, [4,5]TAIR-ME00325, [6]TAIR-ME00338, [7]TAIR-ME00328, [8]TAIR-ME00330, [9]TAIR-ME00331, [10]TAIR-ME00364. Genes were arranged on the basis of H_2O_2 response and data of ≥threefold changed genes were marked with shade. ***1–3**: Marker gene using abiotic stresses assay selected by following information, ***1** Bechtold et al. (2008), ***2** Nishizawa et al. (2006), ***3** Desikan, Neill, and Hancock (2000). Microarray data of other environmental stresses were included for interested researchers for possible use following the protocol described in this chapter.

2. EXPERIMENTAL MATERIALS AND PROCEDURES

2.1. Preparation of *Arabidopsis* plants for hydrogen peroxide treatment

2.1.1 Reagents

A. thaliana (Col-0)
70% Ethanol
Seed surface sterilization solution: 10% Sodium hypochlorite (NaOCl) (kept at 4 °C)
0.2 × MGRL nutrient solution (see Appendix)
50 m*M* H_2O_2
Liquid N_2 for sampling

2.1.2 Equipments

Plastic photo slide mount (50 × 50 mm, FUJIFILM, Japan)
Nylon mesh (hole size-308 μm) (Filter-net, Sansyo, Japan)
Thin styrofoam blocks (as floats, Fig. 12.1)
Plastic tray (360 × 270 × 130 mm, Fig. 12.1A)
Growth chamber
Light-meter (LI-250A, LI-COR, Bioscience, USA) with light sensor (Pyranometer PY56584, LI-COR, Bioscience, USA) (or equivalent instrument for measuring intensity of visible light)

2.1.3 Methods

For analysis of H_2O_2, plant material should be ideally grown without any stresses. Care should be needed to avoid infection of pathogen and insects, wounding, and drought. The method of hydroponic culture of *Arabidopsis* shown here has been developed by Toda, Koyama, and Hara (1999) in order to prepare intact roots. This method is also good for stress treatment to roots by adding stressors to the media or by changing media.

1. Assemble slide mount and nylon mesh. Put the mesh like a film in a slide mount (Fig. 12.1B).
2. Take *Arabidopsis* seeds in a 50-mL Falcon tube (~0.1 g seeds/tube), rinse with approximately 50 mL of 70% Et-OH for 1 min followed by 50 mL of 10% NaOCl for 5 min, and then wash frequently with SDW more than five times.

A

B

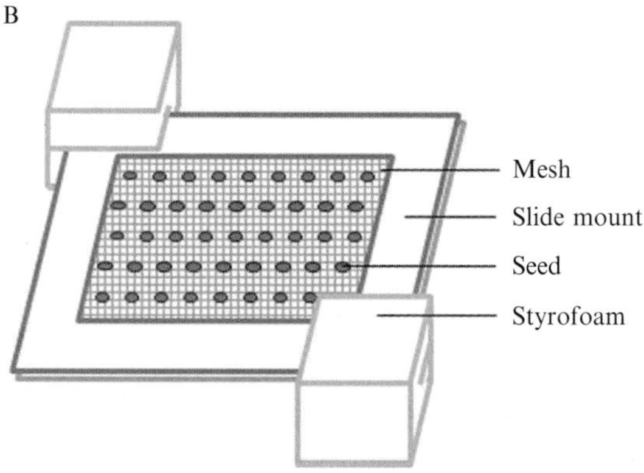

Figure 12.1 Hydroponic culture of *Arabidopsis*. (A) The mounts float on 0.2 × MGRL nutrient solution. (B) Hydroponic culture apparatus consists of a plastic photo slide mount and a nylon mesh inside. Styrofoam blocks are attached at both sides as floats.

3. Put seeds one by one on the assembled slide mount using Pipetman with an yellow tip (200 μL). Typically, 120 seeds (15 in each line) are placed on a slide (Fig. 12.1B).
4. Grow plants in hydroponic culture. Put styrofoam blocks to the slide mount (Fig. 12.1B). Float the slide mounts on a tray containing 5 L of 0.2 × MGRL nutrient solution (40 mounts/tray) (Fig. 12.1A) and keep in a growth chamber and allow to grow for 10 days (see Table 12.2).
5. The nutrient solution is renewed after 7 days of growing.
6. After 10 days, spray shoots 20 times with 1 mL of 50 m*M* H_2O_2.
7. Harvest shoots for RT-PCR at different time points and freeze them in liquid N_2. Store samples at −80 °C until extraction.

Table 12.2 Experimental conditions for stress treatments

Stress treatment	Growth condition		Plant age	Treatment
	Media	Environment		
H_2O_2	MGRL hydroponic culture	7 W/m^2 continuous light, 22 °C	10 days	50 m*M*
High light	GM	7 W/m^2 continuous light, 22 °C	8 days	150 W/m^2
UV-B	GM	7 W/m^2 continuous light, 22 °C	8 days	6.5 W/m^2 UV-B ray for 10 min
Cold	GM	20 W/m^2 continuous light, 22 °C	2 weeks	4 °C

2.2. Preparation of *Arabidopsis* plants for abiotic stress treatment

2.2.1 Reagents

A. thaliana (Col-0)

70% Ethanol

Seed surface sterilization solution: 10% Sodium hypochlorite solution (NaOCl) (kept at 4 °C)

Germination medium (GM) (see Appendix)

Liquid N_2 for sampling

2.2.2 Equipments

50-mL Falcon tubes

Round deep plastic plates (20 × 90 mm, FG-2090, NIPPON Genetics, Japan) for UV-B and cold treatment

Rectangular Plastic plates (97 × 140 mm, Eiken Chemical, Japan) for HL treatment

Micropore™ Surgical Tape (Sumitomo 3M, Tokyo, Japan)

Light-meter (LI-250A) with light sensor (Pyranometer PY56584) to measure intensity of visible light

Light sensor for UV-B ray intensity measurement (SD204B2-Cos., IRRADIAN Ltd., Scotland)

HL source: 1000 W xenon lamp (USHIO, Tokyo), set with a cold glass filter, a UV cut-off filter, and a condenser lens, and also a zoom lens for manipulation of light intensity and irradiation area. A mirror can be used to control light direction so as to irradiate seedlings in plates from the top. The light spectrum from our apparatus is described elsewhere (Kimura, Yoshizumi, Manabe, Yamamoto, & Matsui, 2001)

UV-B light source: Phillips TL-20W/12RS, Holland (fluorescent tubes)

Growth chamber for cold treatment: Temperature should be kept as low as 4 °C. Light source is also required, for example, M-130 (TAITEC, Japan) or MIR154 with the light unit (Panasonic Biomedical, Japan)

Growth chamber

Laminar airflow cabinet

2.2.3 Methods for preculture

1. Take wild-type *A. thaliana* (Col-0) seeds in a 50-mL Falcon tube (~0.1 g seeds/tube), rinse with approx. 50 mL of 70% Et-OH for 1 min followed by 50 mL of 10% NaOCl for 5 min, and then wash frequently with SDW more than five times.
2. Sow seeds on GM-containing plates using sterile pipet inside laminar flow cabinet and seal with Parafilm.
3. Keep the plates in dark at 4 °C for 48 h for vernalization.
4. Uncover the plates in a laminar flow cabinet and leave them for 1 h with illumination by fluorescent tubes. Removal of too much moist is necessary in order to avoid vitrification of seedlings.
5. Put the lid again and seal plates with Micropore™ Surgical Tape.
6. Transfer to a growth room or growth chamber and grow until stress treatments (see Table 12.2).

2.2.4 Methods for HL treatment

1. Adjust light intensity (e.g., 150 W/m^2), irradiation area, and direction of illumination using a light-meter. The xenon lamp requires prerun for ~30 min to stabilize its output.
2. Start HL treatment to 8-day-old seedlings. Before treatment, spray the plants with SDW (5 mL/plate) to keep moisture in plants during the HL treatment. Cover the plates with a cellophane sheet to avoid desiccation of the seedlings. A cellophane sheet does not fog up during the treatment.
3. Harvest aerial parts of seedlings for RT-PCR at appropriate time points. Immediately freeze harvested tissue in liquid N_2 and store them at −80 °C until use. Harvesting can be done after freezing seedlings on a plate by directly pouring liquid N_2 onto the plate.

2.2.5 Methods for UV-B treatment

1. Adjust the light intensity using a light-meter (7 W/m^2).
2. Remove the lids from plates and treat the seedlings with UV-B ray for 10 min.
3. After the treatment, set the lid again, seal plates with surgical tape, and put the plants back to white light (7 W/m^2).
4. Sample aerial parts of seedlings for RT-PCR at appropriate time points. Immediately freeze harvested tissue in liquid N_2 and store them at −80 °C until use.

2.2.6 Methods for cold treatment

1. Set crashed ice in a styrofoam tray and put metal tray on the ice. Pour water into the metal tray to a depth of ~5 mm. Put the whole set into a growth chamber set at 4 °C.
2. Uncover plates containing precultured plants, cover with a cellophane sheet, and put the plates in the metal tray (with water) set in the growth chamber. Chilling plates with water is necessary for quick temperature shift.
3. Harvest aerial parts of seedlings for RT-PCR at appropriate time points. Immediately freeze harvested tissue in liquid N_2 and store them at −80 °C until use.

2.3. Extraction of total RNA

2.3.1 Reagents

Extraction buffer (see Appendix)
Liquid N_2 for grinding
2 *M* sodium acetate (RNase free) (see Appendix)
Phenol/chloroform/isoamyl alcohol mix (PCI) (see Appendix)
2-Isopropanol (RNase free)
75% Ethanol (RNase free)
RNase-free DNase I (5 U/μL, Takara Bio, Japan)
10 × DNase I buffer (Takara Bio, Japan)
RNase inhibitor (40 U/μL, Takara Bio, Japan)

2.3.2 Equipments

Latex gloves
Masher (EARTH MAN Drill & Driver, Takagi, Japan)
Microfuge tubes
High-speed microfuge (himac CF 15R, Hitachi Koki, Japan)
Vortex mixture (Delta mixer Se-08, TAITEC, Japan)
Spectrophotometer (Nano-Vue, GE Healthcare)

2.3.3 Methods

The method is modified from Suzuki, Kawazu, and Koyama (2004).

1. Take frozen tissue sample (up to 500 mg) in 1.5-mL microfuge tube. Grind in liquid N_2 using the masher.
2. Add 610 μL of extraction buffers (including 2-mercaptoethanol) and vortex for 30 s.
3. Centrifuge at 15,000 rpm at RT for 5 min.
4. Transfer supernatant (~600 μL) to a new tube containing 60 μL of 2 *M* sodium acetate.
5. Add equal volume (~660 μL) of PCI. Vortex for 1 min, and centrifuged at 15,000 rpm at RT for 10 min.
6. Transfer upper aqueous phase (~600 μL) to a new microfuge tube.
7. Add equal volume of 2-isopropanol.
8. Mix gently and centrifuge for 20 min at 4 °C at 15,000 rpm.
9. Discard supernatant by decantation and add 500 μL of 75% ethanol (RT).
10. Centrifuge for 5 min at 15,000 rpm at RT.
11. Discard supernatant by decantation, spin down again, and remove the remaining supernatant by pipetting.
12. Dry samples by leaving the lid open for ~5–30 min.
13. Add 25 μL of RNase-free water and dissolve RNA pellet.
14. Prepare 2 × DNase I digestion mixture. In a new tube, mix the followings per sample: 17 μL of RNase-free H_2O, 5 μL of 10 × DNase I Buffer, 1 μL of RNase inhibitor, and 2 μL of DNase I. Total 25 μL per sample.
15. Add 25 μL of 2 × DNase I digestion mixture and incubate at 37 °C for 30 min.
16. Add 250 μL of RNase-free water, and 300 μL of PCI.
17. Vortex for 1 min and centrifuge at 15,000 rpm for 5 min at RT.
18. Transfer ~250 μL of upper aqueous phase to a new microfuge tube.
19. Add 1/10 volume of 2 *M* sodium acetate (25 μL) and equal volume of 2-isopropanol (275 μL) and centrifuge at 15,000 rpm at 4 °C for 20 min.
20. Discard supernatant by decantation and add 200 μL of 75% ethanol (RT).
21. Centrifuge for 5 min at 15,000 rpm at RT.
22. Discard supernatant by decantation, spin down again, and remove the remaining supernatant by pipetting.
23. Dry samples by leaving the lid open for ~5–30 min.
24. Dissolve the pellet with 20 μL of RNase-free water.
25. Measure RNA concentration using spectrophotometer (e.g., NanoVue). Typically, ~20 μg of total RNA is obtained.

2.4. Quantitative real-time RT-PCR

2.4.1 Reagents

5 μ*M* Oligo d(T)$_{20}$ Primer
dNTP mix (10 m*M* each)
5 × RT buffer (TOYOBO, Japan)
RTase (RT Ace, 100 U/μL, TOYOBO, Japan)
RNase inhibitor (40 U/μL, Takara Bio, Japan)
2 × Fast SYBR® Green Master Mix (Applied Biosystems, USA)
cDNA Template
Primers of selected genes

2.4.2 Equipments

MicroAmp™ Fast Optical 96-Well Reaction Plate
MicroAmp™ Optical Adhesive Film
MicroAmp™ 96-Well Tray for VeriFlex™ Blocks
MicroAmp™ 96-Well Support Base
Step one plus Real-Time™ PCR system (Applied Biosystems, USA)

2.4.3 Methods

1. Put 1 μg of total RNA in a PCR tube and fill up to 10.5 μL with RNase-free water. Add 0.5 μL (20 units) of RNase inhibitor.
2. Incubate using a thermal cycler at 70 °C for 10 min, and then 4 °C for 1 min (denaturation of RNA).
3. Prepare RT mix. In a new tube, add the followings per sample: 4 μL of 5 × RT Buffer, 2 μL of oligo d(T) primer, 2 μL of 10 m*M* dNTP mix, 0.5 μL of RNase inhibitor, 0.5 μL of RTase. Total 9 μL per sample.
4. Add 9 μL of RT mix to the denatured RNA sample.
5. Incubate using a thermal cycler at 42 °C for 1 h, 99 °C for 5 min and then chill at 4 °C.
6. Add 80 μL of ×1/4 TE. Store at −20 °C until use.
7. Use 2.0 μL of cDNA for real-time PCR analysis using Fast SYBR® Green Master Mix following the manufacturer's instructions. For each primer pair, prepare a serial dilution of a cDNA sample (×1, ×1/5, ×1/10, ×1/50, ×1/100) which is thought to show the highest expression among the samples to make a standard curve.
8. After cycling reactions, calculate the relative signal intensity with standard curve from the threshold cycle (Ct) values according to the manufacturer's software, normalized by actin cDNA, and expressed as relative to control (0 h).

3. EXAMPLE OF ANALYSIS

Following the above protocols, a test experiment was conducted to see some H_2O_2 responses those are involved in abiotic stress response.

3.1. Selection of primers of test marker genes for real-time PCR

Four H_2O_2 responsive genes were selected for analysis (genes with an asterisk, Table 12.1) after survey of literatures. The selected primer (see Table 12.3) pairs have the following characteristics: $T_m = 60 \pm 2$ °C, primer lengths = 20–24 nucleotides, and PCR amplicon length = ~100 bp. Sequences of the used primers and PCR conditions are shown in Table 12.3. Making T_m as the same degree among the markers enables simultaneous PCRs, facilitating PCR analysis. Purity of PCR product was confirmed by analysis of melting curve (data not shown). Negative control was included every time during real-time PCR by adding water instead of the template to check the primer secondary structures. No amplifications were found in the negative controls (data not shown).

3.2. Expression of test marker genes to H_2O_2 treatment

To observe kinetic profile of the selected marker genes, seedlings were harvested at several time points after H_2O_2 treatment and subjected to real-time qRT-PCR analysis (Fig. 12.2). While *SAG21* showed quick induction after the treatment, *HSP17.6C-CI* responded more slowly, suggesting heterogeneity of the H_2O_2 response. Remaining times of the response were also different, response of *SAG21* and *HSP70* disappeared after 3 h, *HSP101* did after 6 h, and *HSP17.6C-CI* remained even after 24 h.

In this experiment, we used very low concentration of H_2O_2 (50 m*M*), which is 16 times lower than the concentration used by Yamamoto et al. (2004) (Table 12.1). With this difference in H_2O_2 treatment, results are somewhat different. While response of *HSP101* was reduced from 46.5- to 3.1-fold accompanying the reduction of H_2O_2 concentration, response of *SAG21* was rather enhanced from 1.5- to 2.5-fold. This difference might be due to the heterogeneity of the dose–response of the marker genes.

Another difference of the four marker genes is dependence of the expression profile on a peroxisomal catalase (*CAT2*). While *HSP17.7C-CI* and *HSP101* showed 86- and 31-fold enhancement by *CAT2* suppression

Table 12.3 List of primers used in real-time qRT-PCR

Gene	Primer (5′–3′)	Cycle conditions	References
SAG21	F CTC CAA TGC TAT CTT CCG ACG	2 steps (95 °C × 3 s and 62 °C × 30 s) × 40 cycles	Hundertmark and Hincha (2008)
	R TTC ATC ACA GCC GAA GCA AC		
HSP17.6C-CI	AAG AAT GAC AAG TGG CAC CGT G		Nishizawa et al. (2006)
	R GGC TTT GAT TTC CTC CAT CTT AGC		
HSP101	F GGT CGA TGG ATG CAG CTA AT		Yoshida et al. (2011)
	R CTT CAA GCG TTG TAG CAC CA		
HSP70	F AGG AGC TCG AGT CTC TTT GC		Lohmann et al. (2004)
	R AGG TGT GTC GTC ATC CAT TC		
ACT2	F GGC AAG TCA TCA CGA TTG G		Gao et al. (2011)
	R CAG CTT CCA TTC CCA CAA AC		

ACT2 (actin) can be used for normalization of each template amount.

under HL treatment, respectively, *HSP70* had little difference (0.7-fold) (Vandenabeele et al., 2004; Yamamoto et al., 2004).

3.3. Analysis of abiotic stress response

Subsequently, response of the marker genes to UV-B, HL, and cold stresses was analyzed. As shown in Fig. 12.3, all the markers showed positive response to UV-B, HL, and cold stresses ranging from 5- to 210-fold, demonstrating successful application of the molecular markers to analysis of these stress responses. Closely looking at the induction profiles, *SAG21* and *HSP17.6C-CI* showed faster responses than *HSP101* and *HSP70* in the UV-B and cold responses.

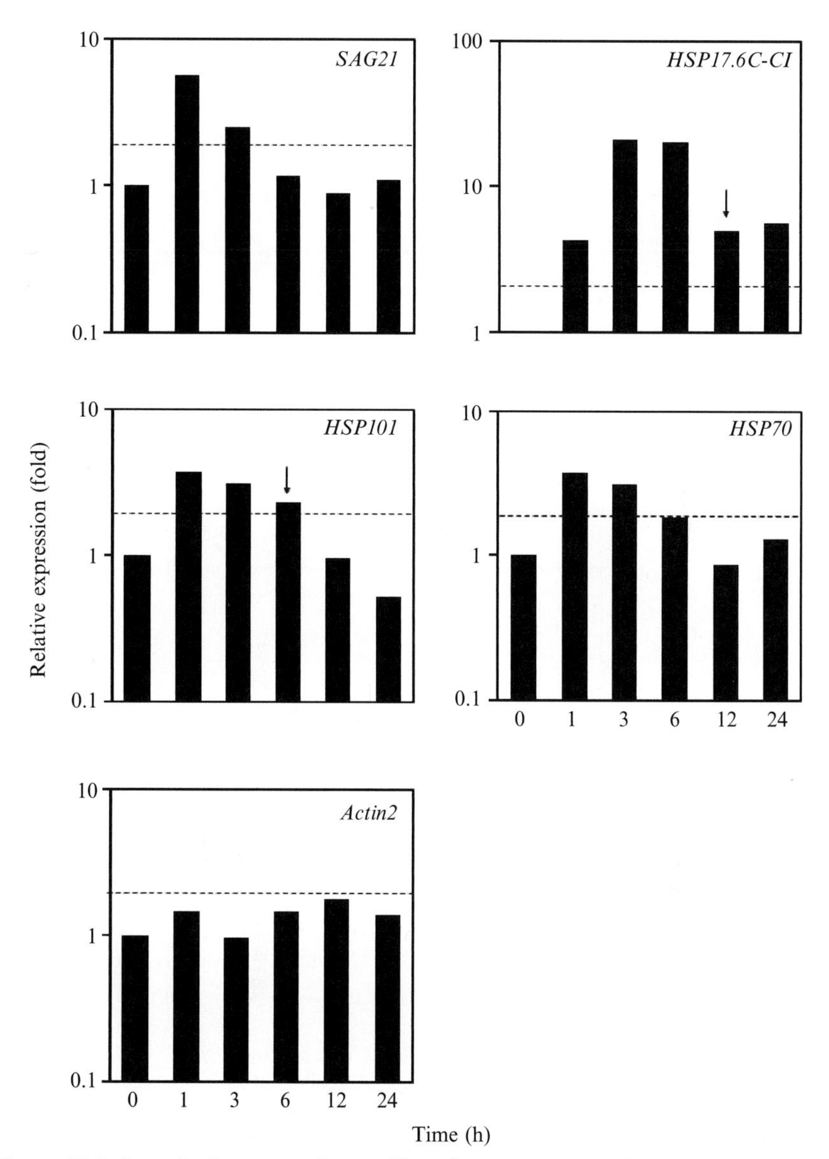

Figure 12.2 Quantitative expression profiles of test genes to H_2O_2 treatment. Quantitative real-time RT-PCR was done to check the expression profile of test genes to H_2O_2 treatment (50 m*M*). Vertical axis is expressed in a logarithmic scale of the fold change of cDNA levels of the treatments compared with control (0 h). Dotted line shows the level of twofold induction. Complementary DNA prepared from 10 days old about 60 plants in each of two replicated experiments. Data shown with an arrow means one replication.

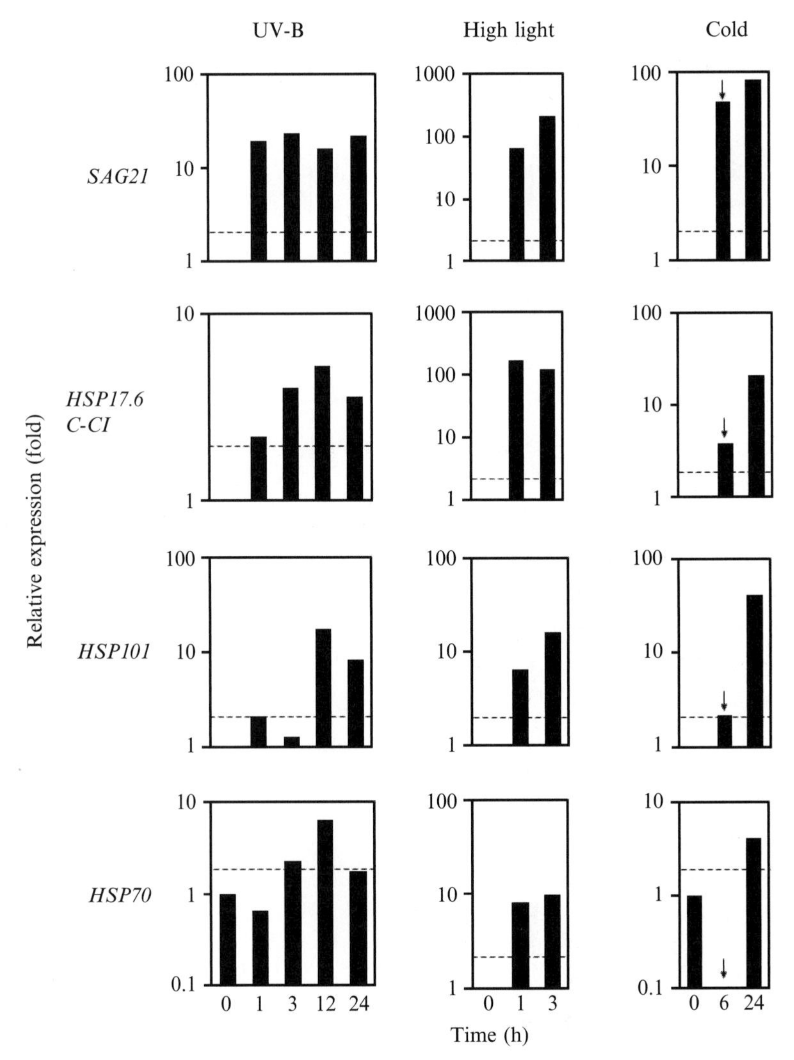

Figure 12.3 Expression profiles of test genes to different abiotic stress treatments. Quantitative expression profile of test genes to different abiotic stress treatments analyzed by real-time qRT-PCR. Fold change relative to the expression level at 0 h is shown. Dotted line shows the level of twofold induction. Experiments were repeated twice and the average is shown. Data shown with an arrow means one replication.

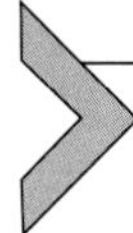

APPENDIX: RECIPES OF STOCK SOLUTIONS, BUFFERS, AND MEDIA

Germination medium (GM) (Valvekens, van Montagu, & van Lijsebettens, 1988)

Murashige and Skoog salt mix 4.43 g
Sucrose 10 g
MES buffer (2-*N*-morpholinolethanesulfonic acid) 0.5 g
$\times$1000 Gamborg B5 vitamin mix 1 mL
Bacto agar 8 g
Distilled water (DW) up to 1 L
Mix all. Adjust pH to 5.6 using potassium hydroxide (KOH) and Autoclave.

MGRL (Fujiwara, Hirai, Chino, Komeda, & Naito, 1992)

$\times$100 Sodium phosphate buffer ($NaPO_4$) (175 m*M*, pH 5.8)
$Na_2HPO_4 \cdot 12H_2O$ 15.67 g/500 mL
$NaH_2PO_4 \cdot 2H_2O$ 6.83 g/500 mL
Adjust pH to 5.8 with 1 N HCl.
$\times$100 $MgSO_4$ (150 μ*M*)
$MgSO_4 \cdot 7H_2O$ 18.49 g/500 mL
$\times$100 $MnSO_4$ (103 μ*M*)
$MnSO_4 \cdot 5H_2O$ 119.5 mg/500 mL
$\times$100 $FeSO_4$ (860 μ*M*)
$FeSO_4 \cdot 7H_2O$ 119.5 mg/500 mL
$\times$100 $ZnSO_4$ (100 μ*M*)
$ZnSO_4 \cdot 7H_2O$ 14.38 mg/500 mL
$\times$100 $CuSO_4$ (100 μ*M*)
$CuSO_4 \cdot 5H_2O$ 12.49 mg/500 mL
$\times$100 $Ca(NO_3)_2$ (200 m*M*)
$Ca(NO_3)_2 \cdot 4H_2O$ 23.62 mg/500 mL
$\times$100 KNO_3 (300 m*M*)
KNO_3 15.17 mg/500 mL
$\times$100 H_3BO_3 (3 m*M*)
H_3BO_3 92.7 mg/500 mL
$\times$100 $(NH_4)6MoO_{24}$ (2.4 μ*M*)
$(NH_4)6MoO_{24} \cdot 4H_2O$ 1.48 mg/500 mL
$\times$100 $CoC1_2$ (13 μ*M*)

$CoC1_2 \cdot 6H_2O$ 1.55 mg/500 mL
×100 EDTA (6.7 m*M*)
Na_2EDTA 1.25 g/500 mL

Mix all the stock solutions and prepare 2× MGRL. Adjust pH to 5.6 with 0.5 N HCl. Dilute to 0.2× MGRL.

Extraction Buffer

Guanidine Thiocyanate (GTC) 20 g
N-Lauroylsarcosine sodium salt 0.2 g
Tri-sodium citrate dehydrate 0.2 g
RNase-free water fill up to 40 mL

Mix well (incubate at 37 °C overnight to dissolve) and store at room temperature.

When use, add 10 μL of 2-mercaptoethanol per 600 μL of the above stock buffer.

Phenol/chloroform/isoamyl alcohol (PCI)

Mix [a]Phenol and [b]CIA (1:1, v/v) in a colored bottle, shake well, and keep overnight to separate layers of phenol and chloroform clearly. Use phenol from the lower layer and store at 4 °C in the dark.

[a]Water-saturated phenol:

Crystal phenol 500 g
SDW 150 mL
8-Hydroxyquinoline 0.5 g (0.1%, w/w)

Warm crystal phenol with hot water to melt in a colored glass bottle. Add DW and 8-hydroxyquinoline and mix well. Store at 4 °C in the dark.

[b]CIA (chloroform/isoamyl alcohol): Mix chloroform and isoamyl alcohol (Chl:Iso = 24:1, v/v) in a colored glass bottle and keep at 4 °C in the dark.

2 *M* sodium acetate

NaAcetate 54.43 g
RNase-free water fill up to 200 mL.
Adjust pH to 4.0 with acetic acid. Store at room temperature.

REFERENCES

Bechtold, U., Richard, O., Zamboni, A., Gapper, C., Geisler, M., Pogson, B., et al. (2008). Impact of chloroplastic- and extracellular-sourced ROS on high light-responsive gene expression in *Arabidopsis*. *Journal of Experimental Botany*, *59*(2), 121–133.

Desikan, R., Neill, S. J., & Hancock, J. T. (2000). Hydrogen peroxide-induced gene expression in *Arabidopsis thaliana*. *Free Radical Biology & Medicine*, *28*(5), 773–778.

Fujiwara, T., Hirai, M. Y., Chino, M., Komeda, Y., & Naito, S. (1992). Effects of sulfur nutrition on expression of the soybean seed storage protein genes in transgenic petunia. *Plant Physiology*, *99*, 263–268.

Gao, Y., Nishikawa, H., Badejo, A. A., Shibata, H., Sawa, Y., Nakagawa, T., et al. (2011). Expression of aspartyl protease and C_3HC_4-type RING zinc finger genes are responsive to ascorbic acid in *Arabidopsis thaliana*. *Journal of Experimental Botany*, *62*(10), 3647–3657.

Hundertmark, M., & Hincha, D. K. (2008). LEA (Late Embryogenesis Abundant) proteins and their encoding genes in *Arabidopsis thaliana*. *BMC Genomics*, *9*(118).

Karpinski, S., Reynolds, H., Karpinska, B., Wingsle, G., Creissen, G., & Mullineaux, P. (1999). Systemic signaling and acclimation in response to excess excitation energy in *Arabidopsis*. *Science*, *284*, 654–657.

Kimura, M., Yoshizumi, T., Manabe, T., Yamamoto, Y. Y., & Matsui, M. (2001). *Arabidopsis* transcriptional regulation by light stress *via* hydrogen peroxide-dependent and -independent pathways. *Genes to Cells*, *6*, 607–617.

Lohmann, C., Eggers-Schumacher, G., Wunderlich, M., & Schöffl, F. (2004). Two different heat shock transcription factors regulate immediate early expression of stress genes in *Arabidopsis*. *Molecular Genetics and Genomics*, *271*(1), 11–21.

Mackerness, A., John, C. F., Jordan, B., & Thomas, B. (2001). Early signaling components in ultraviolet-B responses: Distinct roles for different reactive oxygen species and nitric oxide. *FEBS Letters*, *489*, 237–242.

Nishizawa, A., Yabuta, Y., Yoshida, E., Maruta, T., Yoshimura, K., & Shigeoka, S. (2006). *Arabidopsis* heat shock transcription factor A2 as a key regulator in response to several types of environmental stress. *The Plant Journal*, *48*(4), 535–547.

Orozco-Cardenas, M., & Ryan, C. A. (1999). Hydrogen peroxide is generated systemically in plant leaves by wounding and systemin via the octadecanoid pathway. *Proceedings of the National Academy of Sciences of the United States of America*, *96*, 6553–6557.

Quan, L. J., Zhang, B., Shi, W.-W., & Li, H.-Y. (2008). Hydrogen peroxide in plants: A versatile molecule of the reactive oxygen species network. *Journal of Integrative Plant Biology*, *50*, 2–18.

Rizhsky, L., Davletova, S., Liang, H., & Mittler, R. (2004). The zinc finger protein Zat12 is required for cytosolic ascorbate peroxidase 1 expression during oxidative stress in *Arabidopsis*. *The Journal of Biological Chemistry*, *279*, 11736–11743.

Ryals, J. A., Neuenschwander, U. H., Willits, M. G., Molina, A., Steiner, H., & Hunt, M. D. (1996). Systemic acquired resistance. *The Plant Cell*, *8*, 1809–1819.

Suzuki, Y., Kawazu, T., & Koyama, H. (2004). RNA isolation from siliques, dry seeds and other tissues of *Arabidopsis thaliana*. *BioTechniques*, *37*, 542–544.

Suzuki, K., Yano, A., & Shinshi, H. (1999). Slow and prolonged activation of the p47 protein kinase during hypersensitive cell death in cultured tobacco cells. *Plant Physiology*, *119*, 1465–1472.

TAIR: http://www.arabidopsis.org/portals/expression/microarray/ATGenExpress.jsp.

Toda, T., Koyama, H., & Hara, T. (1999). A simple hydroponic culture method for the development of a highly viable root system in *Arabidopsis thaliana*. *Bioscience, Biotechnology, and Biochemistry*, *63*, 210–212.

Valvekens, D., van Montagu, M., & van Lijsebettens, M. (1988). *Agrobacterium tumefaciens*-mediated transformation of *Arabidopsis thaliana* root explants by using kanamycin selection. *Proceedings of the National Academy of Sciences*, *85*, 5536–5540.

Vandenabeele, S., Vanderauwera, S., Vuylsteke, M., Rombauts, S., Langerbartels, C., Seidlitz, H. K., et al. (2004). Catalase deficiency drastically affects gene expression induced by high light in *Arabidopsis thaliana*. *The Plant Journal*, *39*, 45–58.

Yamamoto, Y. Y., Shimada, Y., Kimura, M., Manabe, K., Sekine, Y., Matsui, M., et al. (2004). Global classification of transcriptional responses to light stress in *Arabidopsis thaliana*. *Endocytobiosis Cell Research*, *15*, 438–452.

Yoshida, T., Ohama, N., Nakajima, J., Kidokoro, S., Mizoi, J., Nakashima, K., et al. (2011). *Arabidopsis* HsfA1 transcription factors function as the main positive regulators in heat shock-responsive gene expression. *Molecular Genetics and Genomics*, *286*(5–6), 321–332.

CHAPTER THIRTEEN

The Role of Plant Bax Inhibitor-1 in Suppressing H_2O_2-Induced Cell Death

Toshiki Ishikawa[*], **Hirofumi Uchimiya**[†], **Maki Kawai-Yamada**[*,†,1]

[*]Graduate School of Science and Engineering, Saitama University, Sakura-ku, Saitama City, Saitama, Japan
[†]Institute for Environmental Science and Technology, Saitama University, Sakura-ku, Saitama City, Saitama, Japan
[1]Corresponding author: e-mail address: mkawai@mail.saitama-u.ac.jp

Contents

1. Introduction 240
2. Morphological Changes of Mitochondria Under ROS Stress 242
   2.1 Chemicals 242
   2.2 Plant materials and treatment 243
   2.3 Ion leakage measurement 243
   2.4 Microscopic analysis 243
3. Assay for Inhibitory Effect of BI-1 on ROS Stress-Induced Cell Death Using Heterologous Expression System in Suspension Cultured Cells 246
   3.1 Plant materials 246
   3.2 Treatment of cultured cells with H_2O_2 and chemicals 247
   3.3 H_2O_2 measurement 247
   3.4 Assessment of H_2O_2-induced damage of intracellular components 248
   3.5 Characterization of tolerance to stress induced by H_2O_2 and other ROS and their metabolism in BI-1 overexpressing plant cells 249
4. Summary 253
References 253

Abstract

Hydrogen peroxide (H_2O_2) is known to be a typical endogenous signaling molecule that triggers programmed cell death in plants and metazoan. In this respect, they seem to share the mechanism of cell death caused by H_2O_2 and other reactive oxygen species (ROS). Bax inhibitor-1 (BI-1) is a well-conserved protein in plants and animals that serves as the inhibitor of mammalian proapoptotic proteins as well as plant ROS-induced cell death. As a target of H_2O_2, mitochondrion is considered to be an organelle of the primary ROS generation and perception. Thus, analysis of mitochondrial behavior in relation to functional roles of regulatory proteins (e.g., BI-1) will lead us to understand the core mechanisms of cell death regulation conserved in eukaryotes. In this chapter, we

Methods in Enzymology, Volume 527
ISSN 0076-6879
http://dx.doi.org/10.1016/B978-0-12-405882-8.00013-1

first introduce techniques of analyzing H_2O_2- (and ROS-) mediated changes in mitochondrial behavior. Next, we describe our understanding of the functions of plant BI-1 in regulation of ROS-induced cell death, with a technical basis for assessment of tolerance to ROS-mediated cell death in model plant systems.

1. INTRODUCTION

Reactive oxygen species (ROS) are generated upon exposure to various biotic and abiotic stresses and trigger programmed cell death (PCD). One well-characterized ROS-generating event in plants is the oxidative burst that occurs within minutes after incompatible pathogen infection, in which NADPH oxidase produces ROS to induce hypersensitive cell death and defense responses (Jabs, Dietrich, & Dangl, 1996; Lamb & Dixon, 1997; Levine, Tenhaken, Dixon, & Lamb, 1994). The NADPH oxidase inhibitor diphenylene iodonium partially blocks elicitor-induced cell death in Arabidopsis (Jabs et al., 1996) and rice cells (Matsumura et al., 2003). In other cases, ROS generation is also reported under abiotic stresses such as ozone (Rao & Davis, 1999), high light (Karpinski, Escobar, Karpinska, Creissen, & Mullineaux, 1997), heat shock (Vacca et al., 2004), and chilling (O'Kane, Gill, Boyd, & Burdon, 1996).

In mammalian cells, the mitochondria are a platform for various proapoptotic events. For instance, mitochondria are one of the major sources of ROS. Proapoptotic Bcl-2 family proteins are localized in mitochondria, leading to disruption of membrane potential and release of cytochrome *c*, which initiates apoptosis via activation of the caspase cascade (Danial & Korsmeyer, 2004). Mitochondria also play important roles in plant PCD induction. A mitochondrial oxidative burst and a following breakdown of mitochondrial membrane potential are observed in oat cultured cells treated with victorin, a host-specific toxin secreted by *Cochliobolus victoriae* (Yao et al., 2002). Cyclosporin A, an inhibitor of mitochondrial membrane permeability, prevents the H_2O_2-induced oxidative burst and induction of cell death in Arabidopsis suspension cells (Tiwari, Belenghi, & Levine, 2002). These observations indicate that H_2O_2 and ROS affect mitochondrial function and stimulate further ROS generation prior to PCD. Our microscopic approach using a visible marker for mitochondria in plant cells has demonstrated that morphological changes of the organelle can be apparent by H_2O_2 and other ROS stresses (Yoshinaga, Arimura, Niwa, et al., 2005). Such evidence may support that the role of mitochondria in inducing ROS-mediated cell death in plant as well as mammalian cells (Jendrach, Mai, Pohl, Vöth, & Bereiter-Hahn, 2008). In this chapter, we first describe methods for microscopic observation of mitochondrial morphological change under ROS-induced stress in Section 2.

The second topic in this chapter is the mechanism of inhibition of responses to ROS-induced stress in plant cells, particularly focused on Bax inhibitor-1 (BI-1), a cell death suppressor widely conserved in higher plants and animals. In mammalian cells, Bax is a Bcl-2 family proapoptotic protein that induces apoptosis via mitochondrial disruption. BI-1 was originally isolated from a mammalian cDNA library as a suppressive factor of lethality of mouse Bax expression using an ectopic expression system in yeast cells (Xu & Reed, 1998). In the case of plants, our lab isolated plant genes orthologous to mammalian BI-1. Surprisingly, plant BI-1 orthologs also inhibit lethality of mouse Bax expression in plant cells, although no counterparts of mammalian Bax have been found in plant genomes (Kawai, Pan, Reed, & Uchimiya, 1999; Kawai-Yamada, Jin, Yoshinaga, Hirata, & Uchimiya, 2001). This suggests that a core PCD pathway downstream of Bax action in mitochondria would be conserved in plants and animals. However, because BI-1 is localized at the endoplasmic reticulum (ER) membrane, plant BI-1 may suppress Bax-induced plant cell death in the downstream of mitochondrial dysfunction, whereas Bcl-2 directly inhibits Bax at the mitochondria (Kawai-Yamada et al., 2001; Oltvai, Milliman, & Korsmeyer, 1993). One of the key events found in the downstream of Bax is ROS generation. In fact, we demonstrated the evidence for ROS accumulation along with cell death by ectopic expression of Bax in plant cells (Kawai-Yamada, Ohori, & Uchimiya, 2004). In addition, overexpression of plant BI-1 leads to enhanced tolerance to not only Bax but also H_2O_2 and other ROS-generating chemicals such as salicylic acid (SA) and menadione (Ishikawa et al., 2009; Kawai-Yamada et al., 2004). Furthermore, cell death induced by a pathogen-derived elicitor, which triggers cellular ROS levels, was inhibited by plant BI-1 (Matsumura et al., 2003). Transcription of plant BI-1 is highly upregulated under stressful conditions, suggesting that plant BI-1 acts as a universal suppressor of ROS-mediated cell death under various environmental stresses (Ishikawa, Watanabe, Nagano, Kawai-Yamada, & Lam, 2011). In Section 3, we describe methods for assessment of ROS stress tolerance in plant cells expressing BI-1. Finally, in Section 4, we summarize the function of plant BI-1, which suppresses the cell death induced by ROS stresses.

2. MORPHOLOGICAL CHANGES OF MITOCHONDRIA UNDER ROS STRESS

Mitochondria are one major source of intracellular generation of ROS through the electron transport chain (Møller, 2001). In addition, studies on mammalian apoptosis have unraveled that mitochondria play an important role in PCD regulatory mechanisms by Bcl-2 family proteins (Tsujimoto & Shimizu, 2000). The Bcl-2 family includes pro- (e.g., Bax, Bak, Bid) and antiapoptotic (Bcl-2, Bcl-xl, Ced-9) proteins that control the initiation of apoptosis via mitochondria (Gross, McDonnell, & Korsmeyer, 1999). Following sensing of a death signal, Bax is translocated from the cytosol to the mitochondrial outer membrane by conformational changes, which induces mitochondrial dysfunction and release of cytochrome *c* to the cytosol, followed by triggering of apoptosis via the caspase pathway (Jürgensmeier et al., 1998; Liu, Kim, Yang, Jemmerson, & Wang, 1996). Ectopic expression of mammalian Bax causes a PCD-like phenotype in yeasts and plants, although orthologous counterparts of the proapoptotic protein are absent in these organisms (Aravind, Dixit, & Koonin, 1999; Jürgensmeier et al., 1997; Kawai-Yamada et al., 2001; Lacomme & Cruz, 1999; Zha et al., 1996). Ectopically expressed Bax is also localized at the mitochondrial membrane in yeast and plant cells (Yoshinaga, Arimura, Hirata, et al., 2005). This evidence strongly suggests the presence of PCD-inducing molecular mechanisms commonly conserved in plants and animals, in which mitochondria play important roles. In this chapter, we describe a microscopic technique to assess cell death-inducing stresses by H_2O_2 and other ROS using a mitochondrial visible marker.

2.1. Chemicals

H_2O_2, paraquat (Nacalai Tesque, Tokyo, Japan), and menadione are ROS stress-inducing agents. H_2O_2 is diluted with distilled water to 50–150 m*M*. Paraquat and menadione, which are ROS-generating chemicals in the chloroplast and mitochondria, respectively, are dissolved in DMSO at desired concentrations so that the DMSO concentration is never higher than 0.2% (v/v) during treatment. In our work, the final concentration is 0.2–0.6 μ*M* for paraquat and 30–100 μ*M* for menadione; therefore, the concentration of the stock solution should be above 0.3 and 50 m*M*, respectively.

2.2. Plant materials and treatment

Transgenic *Arabidopsis thaliana* expressing mtGFP, mitochondrially targeted GFP (S65T) driven by cauliflower mosaic virus (CaMV) 35S promoter (Niwa, Hirano, Yoshimoto, Shimizu, & Kobayashi, 1999), is grown at 23 °C under continuous light. Fully developed leaves of 3-week-old plants are punched out using a cork borer. Leaf discs (0.5 mm diameter) are floated on distilled water in 24-well plates (three discs per well). After addition of chemicals to discs, leaf samples were vacuum infiltrated for 5 min and incubated at 23 °C under continuous light for ion leakage measurement (Section 2.3) and/or microscopic observation (Section 2.4).

2.3. Ion leakage measurement

ROS-induced cell death leads to leakage of intracellular electrolytes through the injured plasma membrane, which is one of the quantitative indicators of ROS stress and induced PCD (Kawai-Yamada et al., 2004; Mitsuhara, Malik, Miura, & Ohashi, 1999). Electrolyte leakage from leaf discs exposed to stress is easily measured as an increase in conductivity of the treatment solution (Fig. 13.1A). If necessary, total electrolytes are determined after autoclaving the sample with the treatment solution, and electrolyte leakage can be expressed on a relative basis as percent of total electrolytes. To determine the optimal stress conditions for microscopic analysis, such as time and dose dependence, electrolyte leakage should be monitored temporarily after treatment with various concentrations of chemicals (prepared as in Section 2.2). For example, treatment of Arabidopsis leaf discs with 100 mM H_2O_2, 0.3 μM paraquat, and 60 μM menadione for 24–72 h is optimal under our usual conditions, in which 30–50% of total cellular electrolytes leak from leaves. A typical result of time- and dose-dependent electrolyte leakage from H_2O_2-treated leaves is shown in Fig. 13.1A.

2.4. Microscopic analysis

For microscopic observation of mitochondria under ROS-induced stress, 3-week-old mtGFP Arabidopsis leaves are treated with each chemical as described earlier. The leaves are then placed on glass slides and observed using a confocal laser scanning microscope equipped with an argon ion laser (488 nm). Figure 13.1B shows representative microscopic images obtained from mock-treated control and 100 mM H_2O_2-treated leaf discs (24 h after treatment). In the control cells, mitochondria retain their bacilliform shape. The steady state of the mitochondrial shape is kept for at least 3 days. On the

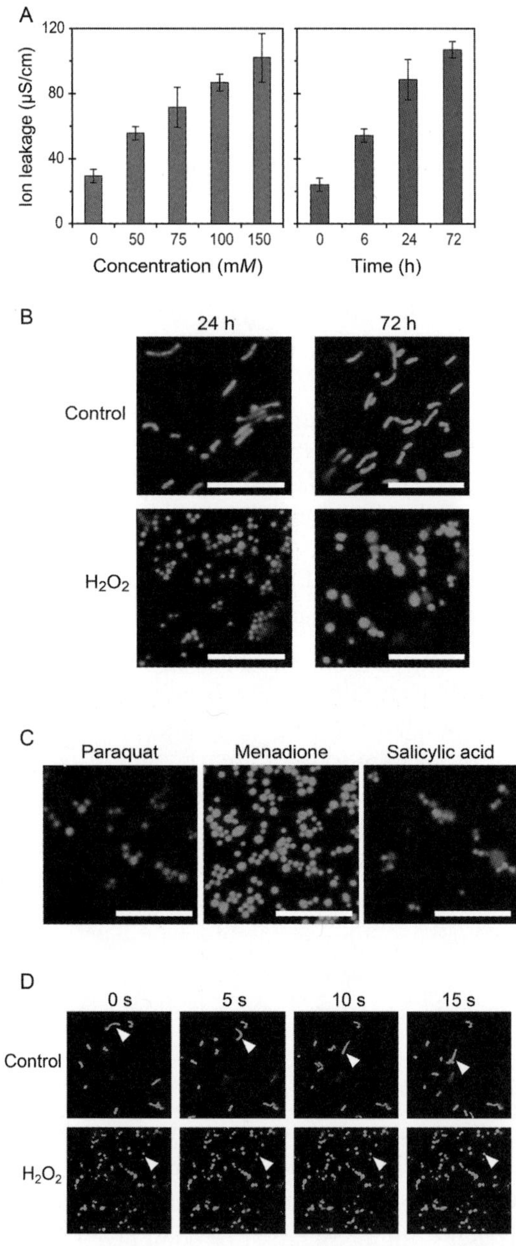

Figure 13.1 H_2O_2-induced cell death and morphological change of mitochondria in Arabidopsis leaves. (A) Leaf discs prepared from 3-week-old Arabidopsis plants were floated on distilled water and treated with H_2O_2 for 3 days at 23 °C under continuous light. Ion leakage is determined using various concentrations of H_2O_2 (0–150 m*M*) and treatment times (0–72 h at 100 m*M* H_2O_2). Data are means ± SD ($n = 3$). (B) Morphology of mitochondria

other hand, in H_2O_2-treated leaves, the shape of mitochondria changed from bacilliform to entirely round, resulting in a smaller cross-sectional area than those in control leaves (~50% in the case of 100 m*M* H_2O_2 for 24 h as in Fig. 13.1B). The morphology of the round-shaped mitochondria in leaves treated with H_2O_2 and the other chemicals changed further by swelling 3 days after treatment.

Similar mitochondrial morphological changes are observed in leaves treated with ROS-generating chemicals such as paraquat and menadione (Yoshinaga, Arimura, Niwa, et al., 2005) as well as SA (Fig. 13.1C). Paraquat is a nonselective herbicide that disturbs proton translocation through the thylakoid membrane, leading to the production of ROS in chloroplasts (Babbs, Pham, & Coolbaugh, 1989). Menadione is a redox-active quinone that generates superoxide, mainly in the inner mitochondrial membrane (Cadenas, Boveris, Ragan, & Stoppani, 1977), followed by accumulation of intracellular H_2O_2 (Ishikawa et al., 2009). Treatment of plant cells with SA also enhances H_2O_2 levels and the oxidative burst prior to induction of cell death and defense responses (Draper, 1997; Rao & Davis, 1999; Shirasu, Nakajima, Rajashekar, Dixon, & Lamb, 1997). Thus, the morphological change of mitochondria is a common feature independent of the various ROS stresses (ROS themselves or ROS-generating chemicals) and their source (generation in the mitochondria or chloroplasts, or exogenous treatment). These results indicate that morphological change of mitochondria can be considered a general indicator of ROS stress and induced cell death, supporting the utility of observation of this phenomenon in microscopic approach. For this purpose, fluorescent indicators of mitochondria, such as MitoTracker dye (Invitrogen), are useful in combination with other dyes or fluorescently tagged targeted proteins.

---

in leaves incubated with 100 m*M* H_2O_2 for 24 and 72 h. (C) Morphology of mitochondria incubated with 0.3 μ*M* paraquat (24 h), 60 μ*M* menadione (72 h), or 300 μ*M* salicylic acid (24 h). Epidermal cells of leaf discs obtained from transgenic mtGFP Arabidopsis were examined using a confocal laser scanning microscope at 488 nm excitation wavelength to detect mtGFP. Controls were treated with water (0 m*M* H_2O_2). (D) Arrest of mitochondrial movement by exogenous treatment with H_2O_2. Movement of mitochondria in epidermal cells of Arabidopsis leaves expressing mtGFP was observed after treatment with 100 m*M* H_2O_2 for 72 h. Images were taken at 5-s intervals. Arrowheads indicate traces of one mitochondrial particle, actively moving in control cells, but remaining motionless in H_2O_2-treated cells. Scale bars = 10 μm. *Modified from Yoshinaga, Arimura, Niwa, et al. (2005).* (See Color Insert.)

In the ROS-treated cells, mitochondrial movement is also blocked in parallel with the morphological change (Fig. 13.1D). This phenomenon seems to be closely associated with PCD induction, supported by the observation that treatment of Arabidopsis leaves with butanedione monoxime, which is an inhibitor of myosin ATPase, causes not only cessation of cytoplasmic movement but also cell death and morphological changes in mitochondria, as observed in H_2O_2-treated leaves. The mitochondrial morphological change is observed only 1 h after treatment with the inhibitor, suggesting that the cessation of mitochondrial movement precedes and triggers cell death.

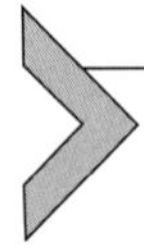

3. ASSAY FOR INHIBITORY EFFECT OF BI-1 ON ROS STRESS-INDUCED CELL DEATH USING HETEROLOGOUS EXPRESSION SYSTEM IN SUSPENSION CULTURED CELLS

Heterologous expression of mammalian Bax in plant cells causes PCD-like cell death, implying the presence of a common cell death-inducing mechanism in plant and animal cells. Our previous study revealed that Bax also causes accumulation of ROS in plant cells; Bax is death-triggering molecule that acts commonly in plants and animals. To address the involvement of plant BI-1 in stress of H_2O_2 and other ROS, we developed model assay systems using cell cultures to assess the role of BI-1 in ROS stress. For example, BI-1 overexpression lines were established in tobacco BY-2 cells and rice suspension cultured cells, which are both very useful for analysis of cell death inhibition, stress tolerance, and maintenance of metabolism of ROS and cellular components. In this section, we describe experimental approach and summarize BI-1 behavior on ROS stress in plant cells.

3.1. Plant materials

Suspension cultures of tobacco (*Nicotiana tabacum*) Bright Yellow 2 (BY-2) cells are one of the most convenient systems for assessment of stress tolerance mediated by BI-1 expression (Kawai-Yamada et al., 2004). For example, transgenic BY-2 cells harboring a cassette of CaMV 35S promoter-driven AtBI–GFP or GFP (vector control) can be obtained by Agrobacterium-mediated transformation. These cell lines are maintained in Linsmaier and Skoog (1965) medium supplemented with 0.2 mg/L 2,4-D by transferring

2 mL of a 1-week-old culture into 50 mL of fresh medium and culturing on a rotating shaker at 100 rpm at 27 °C in the dark.

Rice (*Oryza sativa*) suspension cultures are also a good system in which AtBI-1 overexpression results in drastic elevation of tolerance to ROS-mediated cell death (Ishikawa et al., 2009; Matsumura et al., 2003). Suspension cells are induced from sterilized seeds of wild-type and AtBI-1 overexpressors in Chu's N6 medium (Wako, Tokyo, Japan) supplemented with 3% sucrose and 2 mg/L 2,4-D, with rotation at 130 rpm at 27 °C in the dark (Ishikawa et al., 2009). Three to four weeks after sowing, emerging suspension cells are refreshed with new medium. After further culture for 4 days, media with stress-inducing chemicals are added and used for assays.

3.2. Treatment of cultured cells with H_2O_2 and chemicals

The condition of cultured cells (e.g., culture duration after addition of refreshing medium) is one of the most important factors that should be optimized to obtain good data. For example, in our experiments using AtBI-1 overexpressing cells and ROS treatment, most notable and reproducible results are obtained when cells are treated just after refreshing media for BY-2 cells and 4 days after refreshing media for rice cells.

After treatment with chemicals, cells are stained with 0.05% (w/v) Evans blue, and the number of unstained living cells and Evans blue-stained dead cells are counted under a microscope to determine the cell viability or percent of dead cells. Instead, living cells can be stained with fluorescein diacetate (Matsumura et al., 2003) and then counted using a fluorescence microscope. When using rice suspension cells, it is often difficult to count cells since cultured rice cells form larger tissue aggregates compared to BY-2 cells. In this case, uptake of Evans blue dye is determined by extracting it from cells with 1% SDS in 50% methanol for 2 h at 50 °C. The absorbance at 600 nm of the extract is measured, and the degree of cell death is expressed as dye uptake relative to the dead control cells, which are prepared by freeze-thawing them twice.

3.3. H_2O_2 measurement

The intracellular H_2O_2 level after treatment with ROS-generating chemicals, such as menadione, is determined by a peroxidase photometric assay essentially as described by Okuda, Matsuda, Yamanaka, and Sagisaka (1991) with some modifications. Sample (~40 mg) is frozen and ground to a fine powder in liquid N_2. The powder is extracted with 400 μL 0.2 *M* perchloric acid and centrifuged at 15,000 × *g* for 3 min at 4 °C. The supernatant (320 μL) is neutralized with 16 μL 4 N KOH and centrifuged again to remove excess perchloric acid as its insoluble potassium salt. The supernatant is mixed with 200 μL AG-1 strong anion exchange resin to remove cellular anions inhibiting measurement (e.g., ascorbate). After centrifugation, 50 μL of the supernatant is added to 100 μL reaction mixture (150 m*M* sodium phosphate, pH 6.5, 0.8 mg/mL 3-dimethylaminobenzoic acid, 24 μg/mL 3-methyl-2-benzothiazolinone hydrazine, 25 mU horseradish peroxidase). After incubation at 30 °C for 10 min, the absorbance at 590 nm is measured. Absolute H_2O_2 amounts can be determined using a standard curve, but there are both background absorbance and inhibitory effects on the enzymatic colorimetric assay derived from sample matrices, and these are variable according to sample type, particularly under stressed conditions. Thus, we usually perform blank and spike experiments to correct for background and inhibitory effects for each analyte. For the blank assay, the same volume of extract is added with 0.5 μL catalase (10 μg/mL) and incubated for 10 min at 30 °C. For the spike experiment, a known concentration of H_2O_2 (e.g., 50 μ*M*) is added to each analyte. These pretreated analytes are similarly assayed and used for calculation of H_2O_2 concentration.

In vivo antioxidant activity of suspension cells is assayed by measuring the decay of 1 m*M* H_2O_2 overtime (Tiwari et al., 2002). The media of cells treated for various time intervals (0–10 min) with 1 m*M* H_2O_2 are mixed with 0.1 mL of titanium sulfate and incubated for 15 min at room temperature. The oxidation of titanium sulfate is recorded by reading the absorbance at 410 nm. These values are converted to corresponding concentrations using a standard calibration plot of known concentrations of H_2O_2.

3.4. Assessment of H_2O_2-induced damage of intracellular components

The degree of intracellular oxidative damage from H_2O_2 and other ROS can be estimated by assay of ROS-sensitive enzymes. Aconitase (aconitate hydratase; EC 4.2.1.3), which catalyzes isomerization of citrate to isocitrate via *cis*-aconitate in the TCA cycle, is an enzyme highly sensitive to irreversible oxidative inactivation by H_2O_2 (Verniquet, Gaillard, Neuburger, & Douce, 1991). Thus, aconitase activity is used as an indicator of levels of intracellular H_2O_2 and its oxidative damage to cellular components. For example, substantially decreased aconitase activity is observed in parallel with accumulation of H_2O_2 prior to initiation of PCD in menadione-treated rice suspension cells (Ishikawa et al., 2009). A method for assay of aconitase is briefly described below. The plant sample (~50 mg) is homogenized in 10 volumes of ice-cold extraction buffer (100 m*M* Tris–HCl, pH 7.5, 1 m*M* EDTA, 1 m*M* DTT, 10% glycerol), and crude extract is recovered by centrifugation (15,000 × *g*, 5 min, 4 °C) to remove cell debris. Reaction mixture (0.8 volume Tris–HCl, pH 8.0, and 0.1 volume crude extract) is prepared in a quartz cell for reading in a spectrophotometer, and the reaction is started by adding 0.1 volume 200 m*M* isocitrate as substrate. Formation of *cis*-aconitate is monitored as increase in absorbance at 240 nm for 1 min at room temperature, and enzyme activity is determined using the extinction coefficient 3.6 mM^{-1}. Protein concentration of the crude extract is determined by a standard Bradford assay. Similar experiments can be conducted using other ROS-sensitive cellular enzymes, such as pyruvate dehydrogenase, 2-oxoglutarate dehydrogenase, and NAD-dependent malic enzyme (Millar & Leaver, 2000).

3.5. Characterization of tolerance to stress induced by H_2O_2 and other ROS and their metabolism in BI-1 overexpressing plant cells

One of our recent interests is how plant BI-1 inhibits cell death induced by H_2O_2 and other stresses. For this purpose, we established stable lines overexpressing AtBI-1 in suspension cell cultures of tobacco and rice. In BY-2 cells of tobacco, treatment with H_2O_2 causes cell death in a dose-dependent manner within 18 h, as determined by counting dead cells as described in Section 3.2; these cells show shrunken cytoplasm when stained with Evans blue (Fig. 13.2A). When treated with 3 m*M* H_2O_2, almost 100% cells died within 18 h after treatment (Fig. 13.2B). On the other hand, transgenic cells overexpressing AtBI-1 exhibit attenuated dye uptake, showing enhanced tolerance to H_2O_2-induced cell death. Oxidative stress tolerance mediated by AtBI-1 overexpression is also clearly observed in rice suspension cells. Treatment of rice suspension cells with menadione causes a drastic increase in intracellular H_2O_2 levels within 2 h (Fig. 13.2B), followed by induction of cell death as judged by Evans blue uptake ~8 h after menadione treatment (Fig. 13.2C). In AtBI-1 overexpressing cell lines, Evans blue uptake 24 h after menadione treatment is significantly decreased compared to wild-type cells (Fig. 13.2D). These results support the idea that the suppressive effect of AtBI-1 on H_2O_2-induced cell death is finely expressed in various plants and that model systems using suspension cell cultures, such as BY-2 cells and rice

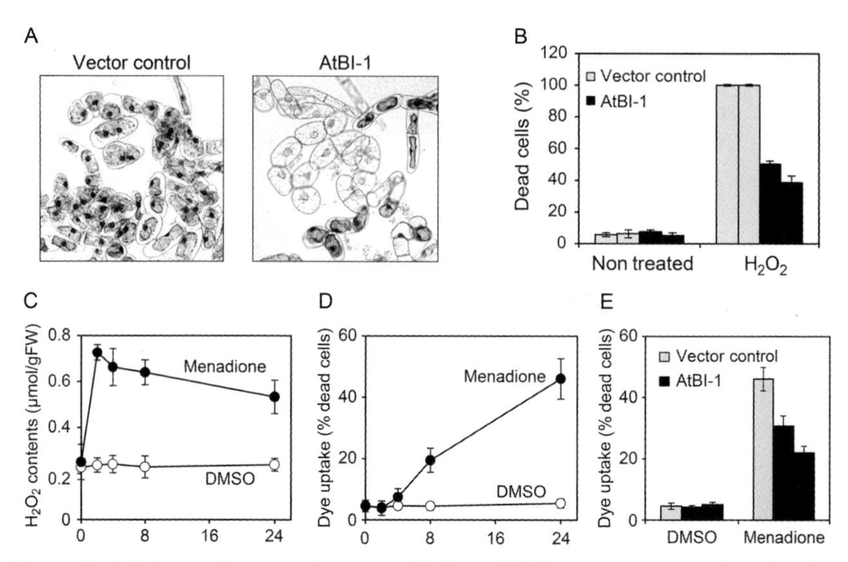

Figure 13.2 Enhanced stress tolerance to H_2O_2 and ROS-generator menadione in plant suspension cultured cells overexpressing Arabidopsis BI-1. (A) Comparison of tobacco BY-2 cells expressing AtBI-1 and vector control cells stained with Evans blue at 18 h after treatment with 3 m*M* H_2O_2. Dead cells show shrunken blue cytoplasm. (B) Cell death as a result of exogenously supplied H_2O_2. Dead cells as in (A) were counted 18 h after treatment with 3 m*M* H_2O_2 and expressed as percent of total cell number. Two-independent transgenic lines, for vector control and AtBI-1 overexpressor, were analyzed. (C and D) Intracellular H_2O_2 level (C) and induced cell death (D) were measured in rice suspension cells (wild type, cv. Sasanishiki) treated with 0.4 m*M* menadione for 24 h. H_2O_2 content was determined by a colorimetric assay and expressed as absolute amount per gram fresh weight (gFW) using a standard curve. Dead cells were stained by Evans blue, and the absorbance at 600 nm of the dye in SDS/methanol extract was measured. Dead cells were prepared by twice repeating a freezing and thawing treatment and were defined as 100% dead cells. (E) Vector control and two-independent AtBI-1 overexpressing rice cell lines were treated with 0.4 m*M* menadione or DMSO, and induced cell death was monitored by Evans blue uptake. Data are means $\pm$ SE from three or four independent experiments. (For color version of this figure, the reader is referred to the online version of this chapter.)

cells, are useful to investigate the functions of AtBI-1 and related factors. For example, catabolic activity of exogenous H_2O_2 and intracellular H_2O_2 levels are analyzed using the cell culture system by methods described in Section 3.3. As a result, AtBI-1 overexpressing BY-2 cells show similar activity in causing decay of exogenously added H_2O_2 (Fig. 13.3A). In addition, AtBI-1 overexpression did not affect the increase in intracellular H_2O_2 levels after treatment of rice suspension cells with menadione (Fig. 13.3B). These results strongly suggest that inhibition of H_2O_2-induced cell death

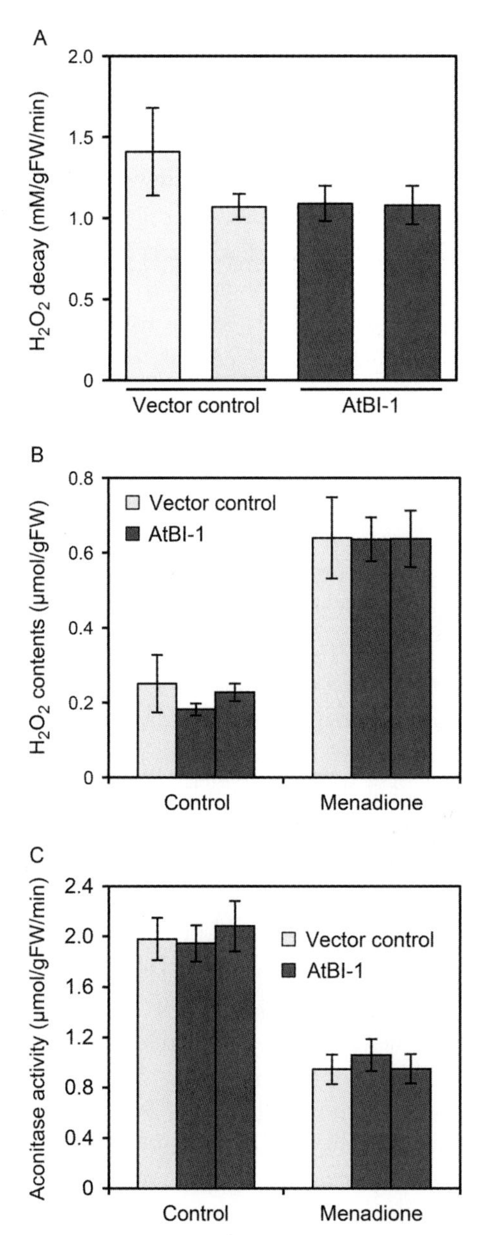

Figure 13.3 AtBI-1 does not affect H_2O_2 detoxification and its oxidative damage of cellular components. (A) Turnover of H_2O_2 after exogenous addition to vector control and AtBI-1 overexpressing BY-2 cell lines. H_2O_2 (1 m*M*) was added to each cell line, and the H_2O_2 concentration of the medium was measured after 5 min. (B) Intracellular content of H_2O_2 after treatment of rice cells with menadione. Vector control and AtBI-1 overexpressing rice cells (cv. Sasanishiki) were treated with 0.4 m*M* menadione for 8 h,

mediated by AtBI-1 overexpression does not depend on detoxification of ROS. This is also supported by our previous result that AtBI-1 overexpression does not attenuate accumulation of ROS in leaves that express mammalian Bax despite the significant inhibition of Bax-induced cell death observed in AtBI-1 overexpressing Arabidopsis (Kawai-Yamada et al., 2004). To estimate oxidative damage to cellular components, a ROS-sensitive enzyme such as aconitase is analyzed as described in Section 3.4. In menadione-treated rice cells, aconitase activity is substantially reduced, reflecting an accumulation of H_2O_2 in the treated cells (Fig. 13.3C). In AtBI-1 overexpressing cells, as expected from H_2O_2 measurement, decreased aconitase activity is observed at a level similar to that in wild-type cells (Fig. 13.3C). Moreover, our metabolomic analysis described elsewhere demonstrated that BI-1 overexpression does not directly affect the redox state of ascorbate and glutathione pools (Ishikawa et al., 2009). These insights lead to the conclusion that plant BI-1 does not regulate generation and catabolism of H_2O_2 and other ROS under stress conditions, resulting in comparable oxidative inactivation of cellular components such as ROS-sensitive enzymes. Therefore, BI-1 seems to inhibit the cell death induction pathway downstream of ROS signaling (Ishikawa et al., 2011). This is consistent with observations of mammalian BI-1. Mammalian BI-1 is also localized at the ER but blocks Bax-induced cell death downstream of the action of Bax, which is resident at the mitochondria, whereas Bcl-2 directly antagonizes Bax action by physical interaction at the mitochondria (Xu & Reed, 1998). Although orthologs of Bax and Bcl-2 are not present in plants, downstream mechanisms for PCD induction through the ER might be conserved in plants and animals, in which BI-1 acts as a key factor to regulate the suicide program. One of our recent interests in BI-1 function at the ER is the involvement of lipid metabolism: BI-1 overexpressing plants accumulate sphingolipid components, which are associated with oxidative stress tolerance (T. Ishikawa et al., unpublished data; Nagano et al., 2009, 2012). Sphingolipids are involved in regulation of PCD and apoptosis in plants as well as animals, although we have no insights that sphingolipid metabolism under BI-1 regulation is conserved in both plants and animals. Further

and intracellular H_2O_2 contents were determined colorimetrically. (C) Aconitase activity in rice cells treated with menadione as in (B) was determined as conversion of isocitrate to *cis*-aconitate, which was monitored as an increase in the absorbance at 240 nm. Enzyme activity is represented as millimolar *cis*-aconitate produced per gFW per minute. Data are means $\pm$ SE from at least three independent experiments.

studies of the mechanism of cell death suppression of BI-1 will help our understanding of the core regulation of cell death regulation common to plants and animals.

4. SUMMARY

This chapter presents basic analytical methods for H_2O_2-induced oxidative stress and BI-1 function, which confers tolerance to ROS-induced cell death in plant cells. Morphological changes in mitochondria are a useful tool to visualize the degree of oxidative stress caused by H_2O_2 in living cells, and it is applicable to not only H_2O_2 but also intracellular ROS generation induced by paraquat, menadione, SA, etc. Since our experiments described earlier employed GFP-tagging of mitochondria, this method can be flexibly modified using other fluorescent proteins or commercially available mitochondrion-targeting fluorescent dyes (e.g., MitoTracker) for multiple color fluorescence imaging. For example, we have succeeded in covisualization of mitochondrial Bax localization and active mitochondria, maintaining their transmembrane potential during Bax-induced cell death using GFP-tagged Bax and MitoTracker Red staining (Yoshinaga, Arimura, Hirata, et al., 2005); this method will help in understanding the behavior of particular molecules in plant cells in association with stress-induced cell death at the subcellular level.

The last half of this chapter introduces analytical methods for assessment of cell death inhibition and H_2O_2 metabolism in BI-1 overexpressing plant cells. Analysis based on these methods has evidenced that plant BI-1 effectively suppresses cell death induced by mammalian Bax as well as various sources of oxidative stress, such as SA and pathogen elicitor (Ishikawa et al., 2009; Kawai-Yamada et al., 2004; Matsumura et al., 2003). In view of the molecular function of BI-1, it is important that enhanced stress tolerance does not depend on detoxifying endogenous and exogenous H_2O_2 and on preventing intracellular molecules from oxidative damage, such as inactivation of aconitase (Fig. 13.3). We hope that analytical methods presented in this chapter are applicable to the study of other cell death-regulating factors.

REFERENCES

Aravind, L., Dixit, V. M., & Koonin, E. V. (1999). The domains of death: Evolution of the apoptosis machinery. *Trends in Biochemical Sciences*, *24*, 47–53.

Babbs, C. F., Pham, J. A., & Coolbaugh, R. C. (1989). Lethal hydroxyl radical production in paraquat-treated plants. *Plant Physiology*, *90*, 1267–1270.

Cadenas, E., Boveris, A., Ragan, C. I., & Stoppani, A. O. (1977). Production of superoxide radicals and hydrogen peroxide by NADH-ubiquinone reductase and ubiquinol-cytochrome *c* reductase from beef-heart mitochondria. *Archives of Biochemistry and Biophysics, 180*, 248–257.

Danial, N. N., & Korsmeyer, S. J. (2004). Cell death: Critical control points. *Cell, 116*, 205–219.

Draper, J. (1997). Salicylate, superoxide synthesis and cell suicide in plant defense. *Trends in Plant Science, 2*, 162–165.

Gross, A., McDonnell, J. M., & Korsmeyer, S. J. (1999). BCL-2 family members and the mitochondria in apoptosis. *Genes & Development, 13*, 1899–1911.

Ishikawa, T., Takahara, K., Hirabayashi, T., Matsumura, H., Fujisawa, S., Terauchi, R., et al. (2009). Metabolome analysis of response to oxidative stress in rice suspension cells over-expressingcell death suppressor Bax inhibitor-1. *Plant & Cell Physiology, 51*, 9–20.

Ishikawa, T., Watanabe, N., Nagano, M., Kawai-Yamada, M., & Lam, E. (2011). Bax inhibitor-1: A highly conserved endoplasmic eticulum-resident cell death suppressor. *Cell Death and Differentiation, 18*, 1271–1278.

Jabs, T., Dietrich, R. A., & Dangl, J. L. (1996). Initiation of runaway cell death in an *Arabidopsis* mutant by extracellular superoxide. *Science, 273*, 1853–1856.

Jendrach, M., Mai, S., Pohl, S., Vöth, M., & Bereiter-Hahn, J. (2008). Short- and long-term alterations of mitochondrial morphology, dynamics and mtDNA after transient oxidative stress. *Mitochondrion, 8*, 293–304.

Jürgensmeier, J. M., Krajewski, S., Armstrong, R. C., Wilson, G. M., Oltersdorf, T., Fritz, L. C., et al. (1997). Bax- and Bak-induced cell death in the fission yeast *Schizosaccharomyces pombe. Molecular and Cellular Biology, 8*, 325–339.

Jürgensmeier, J. M., Xie, Z., Deveraux, Q., Ellerby, L., Bredesen, D., & Reed, J. C. (1998). Bax directly induces release of cytochrome *c* from isolated mitochondria. *Proceedings of the National Academy of Sciences of the United States of America, 95*, 4997–5002.

Karpinski, S., Escobar, C., Karpinska, B., Creissen, G., & Mullineaux, P. M. (1997). Photosynthetic electron transport regulates the expression of cytosolic ascorbate peroxidase genes in *Arabidopsis* during excess light stress. *The Plant Cell, 9*, 627–640.

Kawai, M., Pan, L., Reed, J. C., & Uchimiya, H. (1999). Evolutionally conserved plant homologue of the Bax inhibitor-1 (BI-1) gene capable of suppressing Bax-induced cell death in yeast. *FEBS Letters, 464*, 143–147.

Kawai-Yamada, M., Jin, L., Yoshinaga, K., Hirata, A., & Uchimiya, H. (2001). Mammalian Bax-induced plant cell death can be down-regulated by overexpression of *Arabidopsis Bax Inhibitor-1. Proceedings of the National Academy of Sciences of the United States of America, 98*, 12295–12300.

Kawai-Yamada, M., Ohori, Y., & Uchimiya, H. (2004). Dissection of *Arabidopsis* Bax inhibitor-1 suppressing Bax, hydrogen peroxide and salicylic acid-induced cell death. *The Plant Cell, 16*, 21–32.

Lacomme, C., & Cruz, S. S. (1999). Bax-induced cell death in tobacco is similar to the hypersensitive response. *Proceedings of the National Academy of Sciences of the United States of America, 96*, 7956–7961.

Lamb, C. J., & Dixon, R. A. (1997). The oxidative burst in plant disease resistance. *Annual Review of Plant Physiology and Plant Molecular Biology, 48*, 251–275.

Levine, A., Tenhaken, R., Dixon, R., & Lamb, C. (1994). H_2O_2 from the oxidative burst orchestrates the plant hypersensitive disease resistance response. *Cell, 79*, 583–593.

Linsmaier, E. M., & Skoog, F. (1965). Organic growth factor requirements of tobacco tissue cultures. *Physiologia Plantarum, 18*, 100–127.

Liu, X., Kim, C. N., Yang, J., Jemmerson, R., & Wang, X. (1996). Induction of apoptotic program in cell-free extracts: Requirement for dATP and cytochrome *c. Cell, 86*, 147–157.

Matsumura, H., Nirasawa, S., Kiba, A., Urasaki, N., Saitoh, H., Ito, M., et al. (2003). Overexpression of Bax inhibitor suppresses the fungal elicitor-induced cell death in rice (*Oryza sativa* L.) cells. *The Plant Journal, 33*, 425–434.
Millar, A. H., & Leaver, C. J. (2000). The cytotoxic lipid peroxidation product, 4-hydroxy-2-nonenal, specifically inhibits decarboxylating dehydrogenases in the matrix of plant mitochondria. *FEBS Letters, 481*, 117–121.
Mitsuhara, I., Malik, K. A., Miura, M., & Ohashi, Y. (1999). Animal cell-death suppressors Bcl-x(L) and Ced-9 inhibit cell death in tobacco plants. *Current Biology, 9*, 775–778.
Møller, I. M. (2001). Plant mitochondria and oxidative stress: Electron transport, NADPH turnover, and metabolism of reactive oxygen species. *Annual Review of Plant Physiology and Plant Molecular Biology, 52*, 561–591.
Nagano, M., Ihara-Ohori, Y., Imai, H., Inada, N., Fujimoto, M., Tsutsumi, N., et al. (2009). Functional association of cell death suppressor, Arabidopsis Bax inhibitor-1, with fatty acid 2-hydroxylation through cytochrome b_5. *The Plant Journal, 58*, 122–134.
Nagano, M., Takahara, K., Fujimoto, M., Tsutsumi, N., Uchimiya, H., & Kawai-Yamada, M. (2012). Arabidopsis sphingolipid fatty acid 2-hydroxylases (AtFAH1 and AtFAH2) are functionally differentiated in fatty acid 2-hydroxylation and stress responses. *The Plant Journal, 159*, 1138–1149.
Niwa, Y., Hirano, T., Yoshimoto, K., Shimizu, M., & Kobayashi, H. (1999). Non-invasive quantitative detection and applications of non-toxic, S65T-type green fluorescent protein in living plants. *The Plant Journal, 18*, 455–463.
O'Kane, D., Gill, V., Boyd, P., & Burdon, R. (1996). Chilling, oxidative stress and antioxidant responses in *Arabidopsis thaliana* callus. *Planta, 198*, 371–377.
Okuda, T., Matsuda, Y., Yamanaka, A., & Sagisaka, S. (1991). Abrupt increase in the level of hydrogen peroxide in leaves of winter wheat is caused by cold treatment. *Plant Physiology, 97*, 1265–1267.
Oltvai, Z. N., Milliman, C. L., & Korsmeyer, S. J. (1993). Bcl-2 heterodimerizes in vivo with a conserved homolog, Bax, that accelerates programed cell death. *Cell, 74*, 609–619.
Rao, M. V., & Davis, K. R. (1999). Ozone-induced cell death occurs via two distinct mechanisms in *Arabidopsis*: The role of salicylic acid. *The Plant Journal, 17*, 603–614.
Shirasu, K., Nakajima, H., Rajashekar, K., Dixon, R. A., & Lamb, C. (1997). Salicylic acid potentiates an agonist-dependent gain control that amplifies pathogen signals in the activation of defense mechanisms. *The Plant Cell, 9*, 261–270.
Tiwari, B. S., Belenghi, B., & Levine, A. (2002). Oxidative stress increased respiration and generation of reactive oxygen species, resulting in ATP depletion, opening of mitochondrial permeability transition, and programmed cell death. *Plant Physiology, 128*, 1271–1281.
Tsujimoto, Y., & Shimizu, S. (2000). Bcl-2 family: Life-or-death switch. *FEBS Letters, 466*, 6–10.
Vacca, R. A., de Pinto, M. C., Valenti, D., Passarella, S., Marra, E., & De Gara, L. (2004). Production of reactive oxygen species, alteration of cytosolic ascorbate peroxidase, and impairment of mitochondrial metabolism are early events in heat shock-induced programmed cell death in tobacco Bright-Yellow 2 cells. *Plant Physiology, 134*, 1100–1112.
Verniquet, F., Gaillard, J., Neuburger, M., & Douce, R. (1991). Rapid inactivation of plant aconitase by hydrogen peroxide. *The Biochemical Journal, 276*, 643–648.
Xu, Q., & Reed, J. C. (1998). Bax inhibitor-1, a mammalian apoptosis suppressor identified by functional screening in yeast. *Molecular Cell, 1*, 337–346.
Yao, N., Tada, Y., Sakamoto, M., Nakayashiki, H., Park, P., Tosa, Y., et al. (2002). Mitochondrial oxidative burst involved in apoptotic response in oats. *The Plant Journal, 30*, 567–579.

Yoshinaga, K., Arimura, S., Hirata, A., Niwa, Y., Yun, D. J., Tsutsumi, N., et al. (2005). Mammalian Bax initiates plant cell death through organelle destruction. *Plant Cell Reports, 24*, 408–417.

Yoshinaga, K., Arimura, S., Niwa, Y., Tsutsumi, N., Uchimiya, H., & Kawai-Yamada, M. (2005). Mitochondrial behaviour in the early stages of ROS stress leading to cell death in *Arabidopsis thaliana*. *Annals of Botany, 96*, 337–342.

Zha, H., Fisk, H. A., Yaffe, M. P., Mahajan, N., Herman, B., & Reed, J. C. (1996). Structure-function comparisons of the proapoptotic protein Bax in yeast and mammalian cells. *Molecular and Cellular Biology, 16*, 6494–6508.

CHAPTER FOURTEEN

Comparative Analysis of Cyanobacterial and Plant Peroxiredoxins and Their Electron Donors: Peroxidase Activity and Susceptibility to Overoxidation

Marika Lindahl, Francisco Javier Cejudo[1]

Instituto de Bioquímica Vegetal y Fotosíntesis, Universidad de Sevilla, CSIC IBVF(CSIC/US), Seville, Spain
[1]Corresponding author: e-mail address: fjcejudo@us.es

Contents

1. Introduction 258
2. Expression and Purification of Recombinant Prxs and Thioredoxins 259
   2.1 Cloning and expression of Prxs and thioredoxins 259
   2.2 Purification of Prxs and thioredoxins 260
3. Prx Activity Assays *In Vitro* 260
   3.1 Peroxide decomposition as measured by a colorimetric assay 260
   3.2 The coupled NTRC/2-Cys Prx assay 262
4. Peroxide Decomposition in Cyanobacteria *In Vivo* 264
   4.1 Growth of cyanobacterial cultures under standard conditions 264
   4.2 Chlorophyll measurement of intact cyanobacterial cells 265
   4.3 Quantification of peroxides in cyanobacterial cultures 265
5. Overoxidation of Plant and Cyanobacterial 2-Cys Prx 267
   5.1 Immunological detection of overoxidized 2-Cys Prx *in vitro* 267
   5.2 Overoxidation of cyanobacterial 2-Cys Prx *in vivo* as detected by nonreducing SDS-PAGE and Western blot 268
   5.3 Overoxidation of cyanobacterial 2-Cys Prx as detected by two-dimensional isoelectric focusing/SDS-PAGE and Western blot 270
6. Concluding Remarks 272
References 272

Abstract

Peroxiredoxins (Prxs) are peroxidases that use thiol-based catalytic mechanisms implying redox-active cysteines. The different Prx families have homologs in all photosynthetic organisms, including plants, algae, and cyanobacteria. However, recent studies show that the physiological reduction systems that provide Prxs with reducing equivalents to sustain their activities differ considerably between cyanobacterial strains.

Methods in Enzymology, Volume 527
ISSN 0076-6879
http://dx.doi.org/10.1016/B978-0-12-405882-8.00014-3

Thus, for example, the filamentous cyanobacterium *Anabaena* sp. PCC 7120 is similar to the chloroplast in that it possesses an abundant 2-Cys Prx, which receives electrons from the NADPH-dependent thioredoxin reductase C (NTRC). In contrast, the unicellular cyanobacterium *Synechocystis* sp. PCC 6803, which lacks NTRC, has little 2-Cys Prx but high amounts of PrxII and 1-Cys Prx. The characterization of cyanobacterial Prxs and their electron donors relies on straightforward enzymatic assays and tools to study the physiological relevance of these systems. Here, we present methods to measure peroxidase activities *in vitro* and peroxide decomposition *in vivo*. Several approaches to detect overoxidation of the active site cysteine in cyanobacterial 2-Cys Prxs are also described.

1. INTRODUCTION

Cyanobacteria are photosynthetic prokaryotes that share a common ancestor with plant and algal chloroplasts. Whereas reactive oxygen species (ROS) are generated during normal aerobic metabolism in most organisms, the oxygenic photosynthesis of cyanobacteria and chloroplasts leads to additional production of ROS, such as superoxide anion radicals and hydrogen peroxide. The ROS accumulated may cause oxidative damage to macromolecules and are therefore potentially harmful to the cell. However, some ROS, particularly hydrogen peroxide, have important functions as signaling molecules. Therefore, the intracellular levels of ROS are controlled by various enzymatic systems, some of which are common to prokaryotic and eukaryotic photosynthetic organisms. Peroxiredoxins (Prxs) represent a class of thiol-dependent peroxidases that are divided into four principal groups based on phylogeny, 2-cysteine peroxiredoxin (2-Cys Prx), 1-cysteine peroxiredoxin (1-Cys Prx), PrxII, and PrxQ. Cyanobacteria encode Prxs from all four groups (Bernroitner, Zamocky, Furtmüller, Peschek, & Obinger, 2009), whereas chloroplasts contain all but the 1-Cys Prx (Dietz, 2011; Dietz et al., 2006), which in plants is localized to the nucleus (Pulido, Cazalis, & Cejudo, 2009; Stacy, Nordeng, Culiáñez-Macià, & Aalen, 1999). Since the catalytic mechanism of Prxs implies oxidation of cysteines to cystine bridges, these enzymes must be regenerated through disulfide reduction. Chloroplast Prxs have been found to use a variety of endogenous electron donors, such as thioredoxins, glutaredoxins, or cyclophilin, to sustain their activities (Dietz, 2011). The ubiquitous chloroplast 2-Cys Prx together with its reductant, the NADPH thioredoxin reductase C (NTRC), constitutes one of the most efficient plant systems for peroxide detoxification reported to date (Pérez-Ruiz et al., 2006; Pulido et al., 2010). While all cyanobacterial species examined contain 2-Cys Prx, NTRC is present in some but not all cyanobacteria. For instance, NTRC is present in the filamentous nitrogen-fixing cyanobacterium *Anabaena* sp. PCC 7120, but not in the unicellular cyanobacterium *Synechocystis* sp. PCC 6803 (Pascual, Mata-Cabana, Florencio, Lindahl, & Cejudo, 2010). However, the five different *Synechocystis* Prxs may receive reducing equivalents from three of the *Synechocystis* thioredoxins (Pérez-Pérez, Mata-Cabana, Sánchez-Riego, Lindahl, & Florencio, 2009). This is just one example of the heterogeneity among the cyanobacterial Prx systems. This chapter describes some of the most useful tools for characterization of the cyanobacterial Prxs and their reductants, aimed at comparative studies with the plant chloroplast enzymes.

2. EXPRESSION AND PURIFICATION OF RECOMBINANT Prxs AND THIOREDOXINS

2.1. Cloning and expression of Prxs and thioredoxins

Cyanobacterial gene and protein sequences are obtained from the CyanoBase (http://genome.kazusa.or.jp/cyanobase) and plant gene and protein sequences may be retrieved from PlantGDB (http://www.plantgdb.org). DNA fragments corresponding to the entire coding region of each cyanobacterial gene of interest are amplified by PCR and cloned into the expression vectors, whereas for chloroplast proteins only the sequence encoding the mature protein, without the transit peptide, is amplified. If there are doubts regarding the length of the chloroplast transit peptides and the position of the cleavage sites, it might be useful to consult the ChloroP server (http://www.cbs.dtu.dk/services/ChloroP) or PSORT (http://www.psort.org) for a prediction of mature protein sequences. We recommend cloning of the PCR products into vectors that add a histidine tag to the protein in order to enable purification by nickel affinity chromatography. The histidine tag should be added to the N-terminus of the Prx proteins (Pérez-Pérez et al., 2009) in order to avoid interference with the C-terminal region that is closer to the catalytic site (Pascual, Mata-Cabana, Florencio, Lindahl, & Cejudo, 2011; Wood, Poole, & Karplus, 2003). Suitable vectors are, for example, pQE-30 (Qiagen) and pET28a (Novagen®, EMD Millipore). For protein production, *Escherichia coli* (e.g., strain BL21) cells are transformed with the resulting plasmids, grown at 37 °C and expression is induced by adding 1 m*M* isopropyl-L-D-thiogalactose at an optical density at 600 nm of about 0.4. Prxs and thioredoxins are usually well tolerated by the *E. coli* cells and expressed as soluble proteins at high levels.

2.2. Purification of Prxs and thioredoxins

Ni-NTA (nickel-nitrilotriacetic acid) agarose resin (Qiagen), His bind® Resin (Novagen®, EMD Millipore), or HisTrap columns (GE Healthcare) may be used for purification of the expressed proteins. If additional purification would be necessary, we recommend exclusion chromatography (i.e., gel filtration) using, for example, a HiLoad 16/600 Superdex 75 column (GE Healthcare). Before carrying out this step, the pooled fractions containing the protein of interest eluted from the nickel affinity chromatography should be incubated with 20 m*M* DTT on ice for 1 h. This ensures that all disulfide bonds are broken prior to the second chromatography and that the protein shows a more consistent migration behavior in the gel filtration column. Exclusion chromatography also has the advantage that the high amounts of imidazole and NaCl, which are present in concentrations of up to 0.5 *M* in the eluates from nickel affinity chromatography, are eliminated along with the DTT. Furthermore, it offers the opportunity to change the buffer in accordance with requirements for subsequent analyses. In general terms, Prxs and thioredoxins from a wide variety of sources are straightforward to purify and give yields of about 10 mg of pure protein per liter of *E. coli* culture.

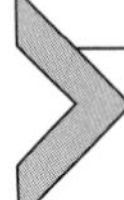

3. Prx ACTIVITY ASSAYS *IN VITRO*

3.1. Peroxide decomposition as measured by a colorimetric assay

Reduction of peroxides catalyzed by Prxs may be conveniently measured using the ferrous ion oxidation (FOX) assay in the presence of xylenol orange, as described by Wolff (1994). In dilute acid hydroperoxides oxidize selectively ferrous to ferric ions, which can be determined using ferric-sensitive dyes, such as xylenol orange. The so-called FOX1 reagent is composed of 100 μ*M* xylenol orange, 250 μ*M* ammonium ferrous sulfate $(NH_4^+)_2Fe(SO_4)_2$, 100 m*M* sorbitol and 25 m*M* H_2SO_4. Xylenol orange binds the ferric ion to produce a colored complex with an extinction coefficient of $1.5 \times 10^4\ M^{-1}\ cm^{-1}$ at 560 nm, the absorbance maximum. Hence, the rate of H_2O_2 decomposition catalyzed by Prx can be monitored through changes in the H_2O_2 concentration as determined by the colorimetric FOX assay.

For measurement of Prx activity the reaction mixture usually contains a buffer at pH 7.0. This buffer may be 50 m*M* Hepes–NaOH (pH 7.0) as in

Pérez-Pérez et al. (2009) or 100 m*M* sodium phosphate buffer (pH 7.0) as in Pérez-Ruiz et al. (2006). The initial concentration of H_2O_2 or alkyl hydroperoxides used in this kind of assays reported in the literature ranges from 100 to 500 μ*M*. However, taking into account the susceptibility of some classes of Prx to hyperoxidation and concomitant inactivation at 500 μ*M* H_2O_2 (Pascual et al., 2010), we advice the use of the lower peroxide concentration, 100 μ*M*, in the reaction mixture. In order to facilitate the calculation of initial reaction rates and kinetic constants, the concentration of the enzyme, Prx, should be adjusted to decompose about half of the peroxide content in the first 2 min of reaction when saturated with reducing agent. This concentration is usually found between 5 and 15 μ*M* Prx, depending on the efficiency of the enzyme (Pérez-Pérez et al., 2009; Pérez-Ruiz et al., 2006).

Various kinds of electron donors may be tested as reducing agents for each Prx. When using Trxs, these need to be kept reduced throughout the assay by including a low concentration of DTT (0.2–0.5 m*M*) (Pérez-Pérez et al., 2009; Pérez-Ruiz et al., 2006). The concentration of DTT chosen should be sufficiently low as not to allow for direct reduction of the Prx, since this *per se* would sustain its peroxidase activity. It should also be noted that the ability to receive electrons from DTT varies considerably between different Prxs. For instance, the 2-Cys Prx from the cyanobacterium *Anabaena* sp. PCC 7120, much like the plant enzyme, is reduced to 15% by a 10-min incubation with 0.5 m*M* DTT, whereas the *Synechocystis* sp. PCC 6803 2-Cys Prx is not reduced under these conditions (Pascual et al., 2010). A series of reactions with increasing Trx concentrations will be needed for determination of kinetic parameters. A range of concentrations between 0.1 and 25 μ*M* Trx is suitable, since the apparent K_m values of the five Prxs from *Synechocystis* sp. PCC 6803 for three different Trxs as substrates were found in this range (Pérez-Pérez et al., 2009). Reaction volumes of 250–500 μL are recommended because larger volumes would consume unnecessarily high amounts of purified enzymes. The final volume of the reaction in the protocol detailed below is 400 μL. Prior to the assays, the enzymes should be diluted to stock solutions of 2 mg mL^{-1} for Prx and 1 mg mL^{-1} for Trx. The reactions are performed at 25 °C.

1. Pipette 200 μL 100 m*M* Hepes–NaOH (pH 7.0) into a 1.5-mL microfuge tube.
2. Add 40 μL 2 mg mL^{-1} Prx, which yields 0.2 mg mL^{-1} final concentration. This would correspond to 8 μ*M* for a Prx with a molecular mass of 25 kDa.

3. Add 10 μL of 8 m*M* DTT, which gives a final concentration of 0.2 m*M*.
4. Add 125 μL 1 mg mL^{-1} Trx. For an average Trx of 12 kDa, this corresponds to approximately 25 μ*M* final concentration.
5. The reaction is started with the addition of 25 μL 1.6 m*M* H_2O_2, yielding 100 μ*M* initial concentration, followed by vigorous mixing.
6. At time points 0, 1, 2, 4,. . ., 10 min, 50 μL-aliquots of the reaction mixture are withdrawn and mixed with 950 μL of the FOX1 reagent. Color development takes about 30 min at room temperature and is thereafter stable for some hours.

The absorbance is read at 560 nm against a blank consisting of 950 μL FOX1 reagent mixed with 50 μL H_2O. Obviously, the maximal H_2O_2 concentration in the mixture between 950 μL FOX1 reagent and a 50 μL-aliquot of reaction withdrawn at time point 0 never exceeds 5 μ*M*, when the initial H_2O_2 concentration in the reaction is 100 μ*M*. This value is within the linear response range of the FOX assay (Wolff, 1994), which we have also confirmed independently. Therefore, the absorbance is directly proportional to the H_2O_2 content. For example, if the initial absorbance at time point 0 is 1.382 AU, then an absorbance of 0.846 AU after 1 min of reaction means that there remains 61% of 100 μ*M* H_2O_2, that is, 61 μ*M*, in the reaction mixture. The concentrations of Prx and Trx suggested here should be used as a guide to try out the best conditions for each couple of enzymes to be examined. If the aim of the assay is to determine the K_m value of a given Prx for a particular Trx as substrate, serial dilutions of the Trx stock solution should be performed in order to obtain progressively lower Trx concentrations. A graph representing the result of a typical assay of this kind is displayed in Fig. 14.1. As may be appreciated, longer times of reaction are needed to accurately determine the reaction rates at the lowest Trx concentrations, whereas 1 or 2 min are sufficient to determine the rates at the highest Trx concentrations.

3.2. The coupled NTRC/2-Cys Prx assay

The plant NTRC comprises a NADPH thioredoxin reductase (NTR)/ thioredoxin system in a single polypeptide chain (Pérez-Ruiz et al., 2006; Serrato, Pérez-Ruiz, Spínola, & Cejudo, 2004). The functional quaternary structure of NTRC is a homodimer (Pérez-Ruiz, González, Spínola, Sandalio, & Cejudo, 2009), and there is evidence that electrons are transferred from the NTR domain of one subunit to the thioredoxin domain

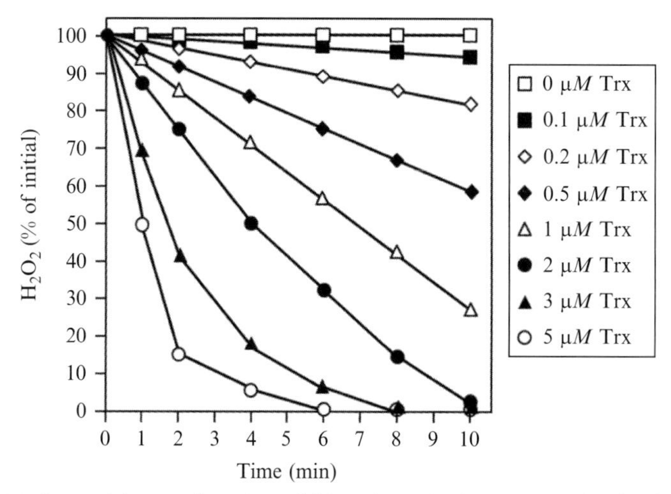

Figure 14.1 Prx activity as a function of thioredoxin (Trx) concentration for determination of kinetic constants. The graph represents an example of H_2O_2 reduction catalyzed by a Prx *in vitro* using a Trx as electron donor. Each sample contains a low concentration of DTT, which does not *per se* sustain the Prx activity (0 μ*M* Trx, open squares). A series of Trx concentrations is required to accurately determine the kinetic parameters.

of the other subunit (Bernal-Bayard, Hervás, Cejudo, & Navarro, 2012; Cejudo, Ferrández, Cano, Puerto-Galán, & Guinea, 2012; Pérez-Ruiz & Cejudo, 2009). The chloroplast 2-Cys Prx receives reducing equivalents from NTRC *in vivo* (Kirchsteiger, Pulido, González, & Cejudo, 2009) and *in vitro* (Pérez-Ruiz et al., 2006). Thus, the activity of the NTRC/2-Cys Prx system that catalyzes the transfer of electrons from NADPH to H_2O_2 can be measured as consumption of NADPH, which absorbs light at 340 nm with an extinction coefficient of 6220 M^{-1} cm^{-1}. The following protocol has been used for the measurement of activity of NTRC from rice (Pérez-Ruiz et al., 2006) and from the cyanobacterium *Anabaena* sp. PCC 7120 (Pascual et al., 2011). The assay may be performed directly in a 500 μL quartz cuvette at room temperature. In order not to waste the purified enzymes, the reaction volume in this protocol is 200 μL. However, it might be necessary to adjust the reaction volume depending on the height of the light beam in the spectrophotometer. The purified NTRC and 2-Cys Prx should be diluted to stock solutions of 2 mg mL^{-1} and kept on ice.

1. Pipette 100 μL 200 m*M* sodium phosphate buffer into the cuvette.
2. Add 31 μL of water. (This volume should be adjusted to give a 200 μL final reaction volume.)

3. Add 21 μL 2 mg mL^{-1} NTRC. This corresponds to about 4 μ*M* of the 53 kDa NTRC protein.
4. Add 18 μL 2 mg mL^{-1} 2-Cys Prx. This corresponds to about 8 μ*M* of the 23 kDa 2-Cys Prx protein.
5. Add 10 μL 5 m*M* NADPH to yield a 0.25 m*M* final concentration within the cuvette and mix well. This maximal concentration of NADPH gives an absorbance at 340 nm that is in the upper part of the linear range.
6. The reaction is started by adding 20 μL 1 m*M* H_2O_2, which results in a 100 μ*M* initial concentration.

The absorbance should be read continuously at 340 nm during 5 min against a blank containing all components except NADPH. Note that 4 μ*M* is a suitable concentration for the *Anabaena* sp. PCC 7120 NTRC (Pascual et al., 2011) but may be too high when measuring the activity of the plant chloroplast NTRC. Previous measurements of the rice NTRC activity have been performed using 2 μ*M* concentration (Pascual et al., 2011; Pérez-Ruiz et al., 2006). Since NTRC presents a low diaphorase activity (Pérez-Ruiz et al., 2006), a sample containing all components except Prx should also be prepared and the measured activity should be subtracted from each activity measured in the presence of Prxs. If the K_m value of NTRC for a particular Prx as substrate is to be determined, serial dilutions of the Prx stock solution should be performed in order to obtain a range of Prx concentrations between 0.1 and 20 μ*M*.

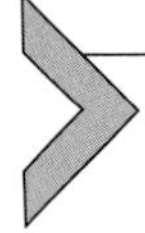

4. PEROXIDE DECOMPOSITION IN CYANOBACTERIA *IN VIVO*

Levels of endogenously produced peroxides in cyanobacteria are usually too low to be detected by colorimetric assays. However, the decomposition of peroxides exogenously added to cyanobacterial cultures may be measured using the FOX reagent. H_2O_2 readily diffuses through the cell wall and lipid membranes and is degraded within the cell. It must be remembered that the observed activity is the sum of all peroxidase and/or catalase activities of the cyanobacterium.

4.1. Growth of cyanobacterial cultures under standard conditions

Synechocystis sp. PCC 6803 and *Anabaena* sp. PCC 7120, as well as other strains, are grown in BG-11 medium (Rippka, Deruelles, Waterbury,

Herdman, & Stanier, 1979) supplemented with 12 m*M* $NaHCO_3$ as carbon source. Cultures are bubbled with a stream of 1% (v/v) CO_2 in air under continuous illumination at a light intensity of 50 μmol photons m^{-2} s^{-1} and a temperature of 30 °C.

4.2. Chlorophyll measurement of intact cyanobacterial cells

The growth is monitored by measuring the chlorophyll concentration of the culture. To this end, 1 mL of cyanobacterial culture is centrifuged for 2 min at 12,000 × *g* at room temperature. Nine hundred microliter of the supernatant is discarded and the pellet containing the cells is resuspended in the remaining volume of medium by vortexing. Nine hundred microliter of methanol is added and mixed with the cell suspension by vigorous vortexing for 30 s. The sample is centrifuged for 2 min at 12,000 × *g*, and the absorbance of the supernatant is read at 665 nm. According to Mackinney (1941), the extinction coefficient of chlorophyll *a* at 665 nm is 74.46 mM^{-1} cm^{-1} when using this method of extraction. For all practical purposes, the absorbance at 665 nm multiplied by 13.43 equals the amount of chlorophyll (μg) in the sample. For example, an absorbance of 0.361 AU × 13.43 yields an amount of 4.8 μg of chlorophyll, that is, the chlorophyll concentration in the culture is 4.8 μg mL^{-1}. An exponentially growing culture has a chlorophyll concentration ranging from 3 to 5 μg mL^{-1}.

4.3. Quantification of peroxides in cyanobacterial cultures

When comparing peroxide reduction rates between cyanobacterial strains and mutants, exponentially growing cultures should first be diluted to equal chlorophyll concentrations. The following protocol was used to compare peroxide decomposition in *Synechocystis* sp. PCC 6803 and *Anabaena* sp. PCC 7120 (Pascual et al., 2010). Since the cyanobacterial catalase-peroxidase KatG has been reported to be specifically inhibited by hydroxylamine (NH_2OH) (Miller, Hunter, O'Leary, & Hart, 2000), peroxide decomposition may be analyzed with and without 100 μ*M* NH_2OH to determine the proportion of activity due to KatG (Pascual et al., 2010).

1. Aliquots of 20 mL cyanobacterial cultures at a concentration of 3.5 μg chlorophyll per mL are pipetted into 100-mL E-flasks and placed on an orbital shaker at 100 rpm in order to avoid sedimentation of cells.
2. The temperature should be kept at 30 °C and the light intensity at 50 μmol photons m^{-2} s^{-1}.

3. Add 4 μL of 0.5 *M* NH_2OH to the aliquots, which are to contain 100 μ*M* NH_2OH for inhibition of KatG.
4. Each reaction is started with the addition of 100 μL 100 m*M* H_2O_2, yielding a 0.5 m*M* initial concentration. It is important to mix rapidly the peroxide with the culture through a circular movement of the E-flask and to withdraw instantaneously the time 0 aliquot, since some strains display very high rates of peroxide decomposition.
5. At regular time intervals up to 15 min, 20 μL-aliquots of the reaction mixture are withdrawn and mixed with 1.98 mL of the FOX1 reagent in 2-mL microfuge tubes, to obtain a maximal H_2O_2 concentration of 5 μ*M*.
6. After at least 30 min color development at room temperature, the absorbance is read at 560 nm against a blank consisting of 1.98 mL FOX1 reagent mixed with 20 μL H_2O.

The outcome of a typical experiment comparing peroxide decomposition rates between *Synechocystis* sp. PCC 6803 and *Anabaena* sp. PCC 7120 in the presence and in the absence of NH_2OH is shown in Fig. 14.2. The *Synechocystis* cells decompose 80% of the H_2O_2 within the first 2 min of reaction in the absence of NH_2OH, but less than 5% in the presence of NH_2OH (Fig. 14.2). This is in agreement with a highly active and/or abundant KatG

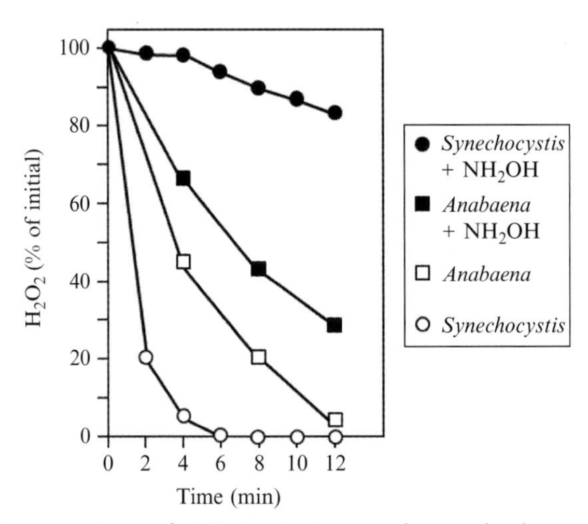

Figure 14.2 Decomposition of H_2O_2 *in vivo* in cyanobacterial cultures. Exponentially, growing *Synechocystis* sp. PCC 6803 and *Anabaena* sp. PCC 7120 decompose 0.5 m*M* exogenously added H_2O_2 within minutes. The strong inhibition of peroxide elimination in *Synechocystis* by 100 μ*M* NH_2OH shows that its high peroxide detoxifying activity is largely due to the catalase-peroxidase KatG.

enzyme that is responsible for most of the peroxide detoxification in this organism, as has previously been reported in a study of a *Synechocystis* KatG deletion mutant (Tichy & Vermaas, 1999). In contrast, the *Anabaena* cells display considerably lower rates of peroxide decomposition but are much less affected by NH_2OH (Fig. 14.2).

5. OVEROXIDATION OF PLANT AND CYANOBACTERIAL 2-Cys Prx

Eukaryotic and some prokaryotic 2-Cys Prxs, for example, the cyanobacterial 2-Cys Prxs, are sensitive to overoxidation, also referred to as "hyperoxidation," at elevated peroxide concentrations (Pascual et al., 2010). This means that one of the catalytic cysteines is oxidized by the peroxide substrate to a sulfinic acid, which renders the enzyme inactive (Wood et al., 2003). However, the active form of 2-Cys Prx may be restored through the action of sulfiredoxin (Biteau, Labarre, & Toledano, 2003), which has also been identified in plant chloroplasts (Iglesias-Baena et al., 2010; Rey et al., 2007) and in the cyanobacterium *Anabaena* sp. PCC 7120 (Boileau et al., 2011). The physiological importance of Prx hyperoxidation in eukaryotes and cyanobacteria has been debated (Jeong, Bae, Toledano, & Rhee, 2012; Pascual et al., 2010). There are several methods to detect the overoxidized form of a 2-Cys Prx.

5.1. Immunological detection of overoxidized 2-Cys Prx *in vitro*

Some years ago, polyclonal antibodies were raised in rabbit against a peptide comprising 10 amino acids from the active site of a human 2-Cys Prx carrying the catalytic cysteine in the sulfonic acid form (Woo et al., 2003). These antibodies, which are commercially available (LabFrontier, Seoul, South Korea), recognize the sulfinic and sulfonic acid forms of the enzyme, but not the thiol-, disulfide-, or sulfenic acid forms (Woo et al., 2003). Since the antigenic peptide is completely conserved in plant 2-Cys Prx, a protocol for Western blot analysis was developed for detection of overoxidized *Arabidopsis thaliana* 2-Cys Prx *in vivo* (Iglesias-Baena et al., 2010). To this end, 10 μg of total soluble leaf protein is loaded on the SDS-PAGE gels and the antibodies against overoxidized 2-Cys Prx are used at a dilution of 1:2000 (Iglesias-Baena et al., 2010). Western blot has also been used to detect overoxidation *in vitro* of plant and cyanobacterial 2-Cys Prx (Pascual et al., 2010).

In this protocol 25 μg each of purified 2-Cys Prx from rice, *Synechocystis* sp. PCC 6803 and *Anabaena* sp. PCC 7120 are first incubated for at least 15 min with 10 m*M* DTT and 5 m*M* H_2O_2 at room temperature. The presence of DTT is absolutely necessary, since the disulfide-linked dimer of 2-Cys Prx is inert to peroxide treatment, and the enzyme must be in the reduced form to be susceptible to overoxidation. Thereafter, the proteins are subjected to normal SDS-PAGE gels under reducing conditions, electroblotted onto nitrocellulose membranes and immunodetected using the aforementioned antibodies at a dilution of 1:2000 (Pascual et al., 2010). It should be noted that, although the antigenic decapeptide is conserved also in cyanobacteria, the detection of the overoxidized cyanobacterial 2-Cys Prx is less efficient than the detection of the overoxidized rice 2-Cys Prx using these antibodies. Therefore, 5 μg of the cyanobacterial proteins are loaded per lane, whereas 0.3 μg of the rice protein is loaded per lane, in order to obtain strong signals (Pascual et al., 2010). Moreover, this is one of the reasons why the immunological method has limited utility for the analysis of 2-Cys Prx overoxidation in cyanobacteria *in vivo*. The other reason is that the abundance of 2-Cys Prx varies substantially between cyanobacterial species. The *Anabaena* 2-Cys Prx is an abundant protein that accounts for about 1% of the total cytosolic protein (Pascual et al., 2010), which is similar to the plant 2-Cys Prx that constitutes about 0.6% of the total chloroplast protein (Dietz et al., 2006). In contrast, the *Synechocystis* 2-Cys Prx is at least 20 times less abundant (Pascual et al., 2010). It is possible to load 50 μg of total protein per lane in a small (9×8 cm^2) SDS-PAGE gel without loss of resolution and this would correspond to 0.5 μg of *Anabaena* 2-Cys Prx and 0.3 μg of plant chloroplast 2-Cys Prx. However, since 50 μg of *Synechocystis* cytosolic protein corresponds to just 25 ng 2-Cys Prx, this would be below the detection level, even when the protein is fully overoxidized.

5.2. Overoxidation of cyanobacterial 2-Cys Prx *in vivo* as detected by nonreducing SDS-PAGE and Western blot

A simple way to assess the degree of overoxidation *in vivo* is to resolve cytosolic proteins from cyanobacteria using one-dimensional SDS-PAGE under nonreducing conditions combined with Western blot analysis. The overoxidized sulfinic acid form of 2-Cys Prx is unable to form disulfide-linked dimers and, therefore, migrates as a monomer also under nonreducing conditions. In contrast, the pools of 2-Cys Prx that at the moment of isolation were found in the thiol- or sulfenic acid forms will rapidly form disulfides and migrate as dimers. Thus, the monomeric pool of the enzyme detected by Western blot corresponds exclusively to overoxidized 2-Cys Prx.

The susceptibility to overoxidation *in vivo* differs between 2-Cys Prxs from different cyanobacterial species. The *Anabaena* sp. PCC 7120 2-Cys Prx is highly sensitive to overoxidation induced by illumination at high light intensity or by addition of H_2O_2 at high concentrations, whereas the *Synechocystis* sp. PCC 6803 2-Cys Prx is largely inert to these treatments (Pascual et al., 2010). The following protocol describes the detection of *Anabaena* 2-Cys Prx overoxidation by nonreducing SDS-PAGE and Western blot following exposure of cultures to high light intensities. Since maximal overoxidation of the *Anabaena* 2-Cys Prx is observed after 15–30 min under high light conditions (Pascual et al., 2010), the duration of the light exposures in a trial experiment should be 0 min (control), 15, 30, and 60 min.

5.2.1 High light treatment of cyanobacterial cultures

1. *Anabaena* sp. PCC 7120 cultures are grown under standard conditions to mid-exponential phase (3–5 μg chlorophyll per mL).
2. Aliquots of 40 mL culture diluted to 2.5 μg chlorophyll per mL are poured into 50-mL glass tubes and placed in a transparent water bath kept at 30 °C. Cultures are continuously bubbled with a stream of 1% (v/v) CO_2 in air, which also avoids sedimentation of cells during the light treatment.
3. A light source emitting strong white light is placed directly in front of the water bath and the intensity should be measured inside a glass beaker placed in the bath at the same distance from the light as the tubes containing the cultures. The intensity for high light treatment should be at least 500 μmol photons m^{-2} s^{-1} and, hence, the temperature of the water bath must be controlled through a cooling system to maintain 30 °C.

5.2.2 Isolation of cytosolic extract from cyanobacterial cells

1. At each time point after the onset of high light exposure, cells are harvested by centrifugation at 14,000 × *g* for 30 min at 4 °C.
2. The pellets are resuspended in 350 μL of Buffer A consisting of 25 m*M* Hepes–NaOH (pH 7.6), 15% (v/v) glycerol, and supplemented with 1 m*M* phenylmethylsulfonyl fluoride. These suspensions are transferred to 1.5-mL microfuge tubes and kept on ice.
3. Add to each tube 0.5 mL glass beads with a diameter of 212–300 μm.

4. Cells are broken through vigorous shaking on a vortex at maximum speed. Thirty seconds of vortexing should be followed by 30 s of cooling on ice to avoid heating of the material. Repeat this procedure six times in order to achieve the best possible lysis.
5. Pipette the lysate to a fresh tube, rinse the glass beads with 150 μL Buffer A, and pool this volume with the rest of the lysate.
6. Centrifuge at 2300 × *g* for 5 min at 4 °C, to remove unbroken cells. Thereafter, centrifuge the supernatant at 16,000 × *g* for 30 min at 4 °C to pellet the total cellular membranes, that is, thylakoid and plasma membranes. The remaining supernatant corresponds to the cytosolic extract.

5.2.3 One-dimensional nonreducing SDS-PAGE and Western blot analysis of 2-Cys Prx

Five microgram of *Anabaena* cytosolic proteins are loaded per lane on 15% acrylamide gels. The sample buffer must be devoid of reductants, such as DTT and β-mercaptoethanol, in order to leave disulfides intact. Gels are electroblotted onto nitrocellulose membranes and the antibodies raised against rice 2-Cys Prx (Pérez-Ruiz et al., 2006) are used at a dilution of 1:1500. If *Synechocystis* sp. PCC 6803 2-Cys Prx is to be analyzed, 25 μg of *Synechocystis* cytosolic proteins should be loaded per lane and the antibodies raised against *Synechocystis* 2-Cys Prx (Pascual et al., 2010) are used at a dilution of 1:1000. It should be remarked that overoxidation of *Synechocystis* 2-Cys Prx has not been previously observed after illumination at high light intensities, though addition of H_2O_2 at m*M* concentrations to *Synechocystis* cultures does lead to some overoxidation of 2-Cys Prx (Pascual et al., 2010). Figure 14.3A illustrates a typical experiment demonstrating *Anabaena* 2-Cys Prx overoxidation following high light treatment.

5.3. Overoxidation of cyanobacterial 2-Cys Prx as detected by two-dimensional isoelectric focusing/SDS-PAGE and Western blot

In this protocol, both isoelectric focusing and SDS-PAGE are performed under denaturing conditions in the presence of high concentrations of DTT. Therefore, the entire pool of 2-Cys Prx remains monomeric throughout the procedure. The sulfinic acid-containing 2-Cys Prx is distinguished by a slightly lower pI value and, thus, the overoxidized form of the enzyme is found in a spot shifted toward the acidic side of the gel (Fig. 14.3B). The pI values of unmodified 2-Cys Prxs from most cyanobacteria and plants are close to 5.

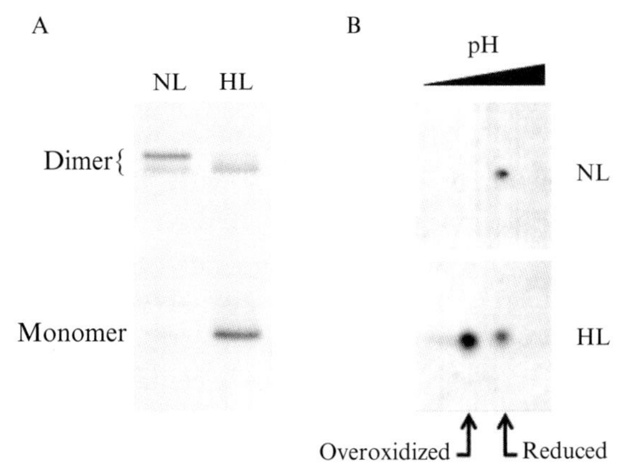

Figure 14.3 Overoxidation of the *Anabaena* sp. PCC 7120 2-Cys Prx. *Anabaena* cultures are kept at normal light intensity (NL; 50 μmol photons $m^{-2} s^{-1}$) or exposed to high light intensity (HL; 800 μmol photons $m^{-2} s^{-1}$) for 15 min. The cytosolic extracts are thereafter analyzed by one-dimensional SDS-PAGE under nonreducing conditions (A) and two-dimensional isoelectric focusing/SDS-PAGE (B) combined with Western blot. The overoxidised 2-Cys Prx is characterized by its migration as a monomer in (A) and by the pI shift toward the acidic side in (B).

1. *Anabaena* cytosolic extracts containing a minimum of 15 μg of protein are precipitated with 5% trichloroacetic acid for 30 min on ice and centrifuged at 16,000 × *g* for 20 min at 4 °C.
2. The supernatant is discarded and the pellet is washed twice with 200 μL acetone and centrifuged as above. The acetone is removed and the pellet should be well dried in air.
3. Add 100 μL rehydration buffer (e.g., ReadyPrep™, Bio-Rad) containing 8 *M* urea, 2% CHAPS, 50 m*M* DTT, and 0.2% ampholytes. Do not attempt to resuspend the pellet immediately, but leave it overnight at −20 °C to absorb slowly the buffer.
4. Mix the proteins solubilized in dehydration buffer using the pipette tip and centrifuge the sample at 16,000 × *g* for 5 min to remove non-solubilized material.
5. Resolve the proteins by isoelectric focusing with an immobilized pH gradient (pH 4–7) of 11 cm, using for instance ReadyStrip™ IPG strips (Bio-Rad).
6. The second dimension SDS-PAGE should be performed on 16-cm long 15% acrylamide gels. After electrophoresis, the proteins are transferred onto nitrocellulose membranes and probed with the antibody against rice 2-Cys Prx (Pérez-Ruiz et al., 2006) at a dilution of 1:1500.

6. CONCLUDING REMARKS

The precise roles of the different families of Prxs in peroxide detoxification and signaling of plants and cyanobacteria still remain to be established (Dietz, 2011). This chapter should serve as a practical guide in the effort to explore the Prx systems of cyanobacteria and chloroplasts. The methods describing measurement of activity may be used to test the substrate specificities of various Prxs with respect to electron donors as well as peroxides and to examine the importance of particular residues or domains for Prx activity in site-directed mutant versions. The protocols for *in vivo* measurements of peroxide decomposition should be useful to assess the detoxifying capacity of knockout mutants lacking one or more Prxs. Finally, the degree of overoxidation of 2-Cys Prx *in vivo* under different physiological conditions might be informative for future studies on peroxide-mediated signaling in photosynthetic organisms.

REFERENCES

Bernal-Bayard, P., Hervás, M., Cejudo, F. J., & Navarro, J. A. (2012). Electron transfer pathways and dynamics of chloroplast NADPH-dependent thioredoxin reductase C (NTRC). *The Journal of Biological Chemistry*, *287*, 33865–33872.

Bernroitner, M., Zamocky, M., Furtmüller, P. G., Peschek, G. A., & Obinger, O. (2009). Occurrence, phylogeny, structure and function of catalases and peroxidases in cyanobacteria. *Journal of Experimental Botany*, *60*, 423–440.

Biteau, B., Labarre, J., & Toledano, M. B. (2003). ATP-dependent reduction of cysteine-sulphinic acid by *S. cerevisiae* sulphiredoxin. *Nature*, *425*, 980–984.

Boileau, C., Eme, L., Brochier-Armanet, C., Janicki, A., Zhang, C.-C., & Latifi, A. (2011). A eukaryotic-like sulfiredoxin involved in oxidative stress responses and in the reduction of the sulfinic form of 2-Cys peroxiredoxin in the cyanobacterium *Anabaena* PCC 7120. *The New Phytologist*, *191*, 1108–1118.

Cejudo, F. J., Ferrández, J., Cano, B., Puerto-Galán, L., & Guinea, M. (2012). The function of the NADPH thioredoxin reductase C—2-Cys peroxiredoxin system in plastid redox regulation and signaling. *FEBS Letters*, *586*, 2974–2980.

Dietz, K.-J. (2011). Peroxiredoxins in plants and cyanobacteria. *Antioxidants & Redox Signaling*, *15*, 1129–1159.

Dietz, K.-J., Jacob, S., Oelze, M.-L., Laxa, M., Tognetti, V., Nunes de Miranda, S. M., et al. (2006). The function of peroxiredoxins in plant organelle redox metabolism. *Journal of Experimental Botany*, *57*, 1697–1709.

Iglesias-Baena, I., Barranco-Medina, S., Lázaro-Payo, A., López-Jaramillo, F. J., Sevilla, F., & Lázaro, J.-J. (2010). Characterization of plant sulfiredoxin and role of sulphinic form of 2-Cys peroxiredoxin. *Journal of Experimental Botany*, *61*, 1509–1521.

Jeong, W., Bae, S. H., Toledano, M. B., & Rhee, S. G. (2012). Role of sulfiredoxin as a regulator of peroxiredoxin function and regulation of its expression. *Free Radical Biology & Medicine*, *53*, 447–456.

Kirchsteiger, K., Pulido, P., González, M., & Cejudo, F. J. (2009). NADPH thioredoxin reductase C controls the redox status of chloroplast 2-Cys peroxiredoxins in *Arabidopsis thaliana*. *Molecular Plant*, *2*, 298–307.

Mackinney, G. (1941). Absorption of light by chlorophyll solutions. *The Journal of Biological Chemistry*, *140*, 315–322.

Miller, A. G., Hunter, K. J., O'Leary, S. J. B., & Hart, L. J. (2000). The photoreduction of H_2O_2 by *Synechococcus* sp. PCC 7942 and UTEX 625. *Plant Physiology*, *123*, 625–636.

Pascual, M. B., Mata-Cabana, A., Florencio, F. J., Lindahl, M., & Cejudo, F. J. (2010). Overoxidation of 2-Cys peroxiredoxin in prokaryotes. Cyanobacterial 2-Cys peroxiredoxin sensitive to oxidative stress. *The Journal of Biological Chemistry*, *285*, 34485–34492.

Pascual, M. B., Mata-Cabana, A., Florencio, F. J., Lindahl, M., & Cejudo, F. J. (2011). A comparative analysis of the NADPH thioredoxin reductase C—2-Cys peroxiredoxin system from plants and cyanobacteria. *Plant Physiology*, *155*, 1806–1816.

Pérez-Pérez, M. E., Mata-Cabana, A., Sánchez-Riego, A. M., Lindahl, M., & Florencio, F. J. (2009). A comprehensive analysis of the peroxiredoxin reduction system in the cyanobacterium *Synechocystis* sp. strain PCC 6803 reveals that all five peroxiredoxins are thioredoxin dependent. *Journal of Bacteriology*, *191*, 7477–7489.

Pérez-Ruiz, J. M., & Cejudo, F. J. (2009). A proposed reaction mechanism for rice NADPH thioredoxin reductase C, an enzyme with protein disulfide reductase activity. *FEBS Letters*, *583*, 1399–1402.

Pérez-Ruiz, J. M., González, M., Spínola, M. C., Sandalio, M. L., & Cejudo, F. J. (2009). The quaternary structure of NADPH thioredoxin reductase C is redox-sensitive. *Molecular Plant*, *2*, 457–467.

Pérez-Ruiz, J. M., Spínola, M. C., Kirchsteiger, K., Moreno, J., Sahrawy, M., & Cejudo, F. J. (2006). Rice NTRC is a high-efficiency redox system for chloroplast protection against oxidative damage. *The Plant Cell*, *18*, 2356–2368.

Pulido, P., Cazalis, R., & Cejudo, F. J. (2009). An antioxidant redox system in the nucleus of wheat seed cells suffering oxidative stress. *The Plant Journal*, *57*, 132–145.

Pulido, P., Spínola, M. C., Kirchsteiger, K., Guinea, M., Pascual, M. B., Sahrawy, M., et al. (2010). Functional analysis of the pathways for 2-Cys peroxiredoxin reduction in *Arabidopsis thaliana* chloroplasts. *Journal of Experimental Botany*, *61*, 4043–4054.

Rey, P., Bécuwe, N., Barrault, M.-B., Rumeau, D., Havaux, M., Biteau, B., et al. (2007). The *Arabidopsis thaliana* sulfiredoxin is a plastidic cysteine-sulfinic acid reductase involved in the photooxidative stress response. *The Plant Journal*, *49*, 505–514.

Rippka, R., Deruelles, J., Waterbury, J. B., Herdman, M., & Stanier, R. Y. (1979). Generic assignments, strain histories and properties of pure cultures of cyanobacteria. *Journal of General Microbiology*, *111*, 1–61.

Serrato, A. J., Pérez-Ruiz, J. M., Spínola, M. C., & Cejudo, F. J. (2004). A novel NADPH thioredoxin reductase, localized in the chloroplast, which deficiency causes hypersensitivity to abiotic stress in *Arabidopsis thaliana*. *The Journal of Biological Chemistry*, *279*, 43821–43827.

Stacy, R. A., Nordeng, T. W., Culiáñez-Macià, F. A., & Aalen, R. B. (1999). The dormancy-related peroxiredoxin anti-oxidant, PER1, is localized to the nucleus of barley embryo and aleurone cells. *The Plant Journal*, *19*, 1–8.

Tichy, M., & Vermaas, W. (1999). *In vivo* role of catalase-peroxidase in *Synechocystis* sp. strain PCC 6803. *Journal of Bacteriology*, *181*, 1875–1882.

Wolff, S. P. (1994). Ferrous ion oxidation in presence of ferric ion indicator xylenol orange for measurement of hydroperoxides. *Methods in Enzymology*, *233*, 182–189.

Woo, H. A., Kang, S. W., Kim, H. K., Yang, K.-S., Chae, H. Z., & Rhee, S. G. (2003). Reversible oxidation of the active site cysteine of peroxiredoxins to cysteine sulfinic acid. Immunoblot detection with antibodies specific for the hyperoxidized cysteine-containing sequence. *The Journal of Biological Chemistry*, *278*, 47361–47364.

Wood, Z. A., Poole, L. B., & Karplus, P. A. (2003). Peroxiredoxin evolution and the regulation of hydrogen peroxide signaling. *Science*, *300*, 650–653.

CHAPTER FIFTEEN

Using Hyper as a Molecular Probe to Visualize Hydrogen Peroxide in Living Plant Cells: A Method with Virtually Unlimited Potential in Plant Biology

Alejandra Hernández-Barrera[*], Carmen Quinto[*], Eric A. Johnson[†,‡], Hen-Ming Wu[†,‡], Alice Y. Cheung[†,‡,§], Luis Cárdenas[*,1]

[*]Instituto de Biotecnología, Universidad Nacional Autónoma de México, Morelos, Mexico
[†]Department of Biochemistry and Molecular Biology, University of Massachusetts, Amherst, Massachusetts, USA
[‡]Molecular Cell Biology Program, University of Massachusetts, Amherst, Massachusetts, USA
[§]Plant Biology Program, University of Massachusetts, Amherst, Massachusetts, USA
[1]Corresponding author: e-mail address: luisc@ibt.unam.mx

Contents

1. Introduction 276
2. NADPH Oxidase in Plant Cells 277
3. Plant Cells Respond to External and Internal Stimuli 279
4. Visualizing Hydrogen Peroxide in Living Plant Cells 280
5. Hyper as a New Genetically Encoded Probe 281
6. Vector Description and Plant Transformation 282
7. Preparation and Sterilization of Modified Petri Dishes for Growing *Arabidopsis* Plants for Microscopy Analysis 284
8. Seeds Sterilization and Stratification 284
9. Growth Conditions 285
10. Image Acquisition and Processing 286

Acknowledgments 288
References 288

Abstract

Reactive oxygen species (ROS) are highly reactive reduced oxygen molecules that play a myriad of roles in animal and plant cells. In plant cells, the production of ROS occurs as a result of aerobic metabolism during respiration and photosynthesis. Therefore mitochondria, chloroplasts, and peroxisomes constitute an important source of ROS. However, they can be produced in response to many physiological stimuli such as pathogen attack, hormone signaling, abiotic stresses, or during cell wall organization and plant

Methods in Enzymology, Volume 527
ISSN 0076-6879
http://dx.doi.org/10.1016/B978-0-12-405882-8.00015-5

morphogenesis. Monitoring ROS in plant cells has been limited to biochemical assays and use of fluorescent probes, however, the irreversible oxidation of the fluorescent dyes make it impossible to visualize dynamic changes of ROS. Hyper is a recently developed live cell probe for H_2O_2 and consists of a circularly permutated YFP (cpYFP) inserted into the regulatory domain of the *Escherichia coli* hydrogen peroxide (H_2O_2) sensor OxyR rendering it a H_2O_2 specific ratiometric, and therefore quantitative probe. Herein, we describe a protocol for using Hyper as a dynamic probe for H_2O_2 in *Arabidopsis* with virtually unlimited potential to detect H_2O_2 throughout the plant and under a broad range of developmental and environmental conditions.

1. INTRODUCTION

In plant cells, ROS are continuously produced as a by-product of various metabolic pathways that occur in several organelles such as mitochondria, chloroplasts, and peroxisomes. ROS might generally be considered to have a deleterious effect on plant cells, as all different ROS types have been associated with their capacity to cause oxidative damage to several macromolecules such as proteins, DNA, and lipids. However, ROS can be generated in a regulated way by other enzymatic pathways in response to different stimuli and may be used as a signal. For instance, many responses in plant cells depend on ROS, due to its capacity to activate calcium channels and receptors involved in signaling processes and metabolism.

In plant cells, ROS accumulation has been involved in several processes including development, hypersensitive response to pathogen attack, hormonal perception, gravitropism, and stress response (Mittler & Berkowitz, 2001; Tsukagoshi, Busch, & Benfey, 2010). In guard cells from *Vicia faba*, ROS regulate the opening of stomata and more recently they have been shown to have an important role in root hair growth in *Arabidopsis*. In the polarized root hair cells, ROS levels generate and maintain an apical calcium gradient to support apical growth (Cárdenas & Quinto, 2008a; Foreman et al., 2003; Mori & Schroeder, 2004; Pei et al., 2000). The key role of ROS in supporting polar growth is also thought to play a similar role in pollen tubes (Potocky, Jones, Bezvoda, Smirnoff, & Zarsky, 2007; Cárdenas et al., 2008b) but they may also be involved in self-incompatibility induced death of pollen tubes (Wilkins et al., 2011).

The ability of ROS to regulate ion channels indicates another order of complexity when working as signaling molecules. Furthermore, ROS can play a key role as second messengers that potentially regulate protein conformation and activity (Mori & Schroeder, 2004). Under physiological conditions, ROS production and removal should be tightly regulated and, as they are continuously produced, cells have a robust scavenging system to maintain a balance between the two processes. This scavenging system involves several antioxidative defense mechanisms that respond to cellular demands to either decrease antioxidative capacity or increase the activity of enzymes generating ROS, resulting in a net increase in ROS production. However, this balance can be perturbed by several conditions such as high light, drought, low and high temperature, pathogen attack or herbivorous wound, and mechanical stress, leading to oxidative responses.

It is important to point out that among the active oxygen species, superoxide is not very mobile and it is not permeable to the plasma membrane due to its negative charge. Therefore, a more stable and diffusible derivative is required for cell to cell signaling. It has been widely accepted that superoxide can be spontaneously dismutated to hydrogen superoxide or by the activity of the superoxide dismutase (SOD). These reactions generate hydrogen peroxide, which is considerably more stable and able to cross the plasma membrane, and constitutes a long range cell to cell signal (Allan & Fluhr, 1997).

2. NADPH OXIDASE IN PLANT CELLS

In animal cells, Nicotinamide Adenine Dinucleotide Phosphate (NADPH) oxidases (NOXs) are enzymes involved in the production of ROS such as superoxide anion and hydrogen peroxide, through single-electron reduction of molecular oxygen with NADPH as the electron donor. This family is composed of NOX1-5, DUOX1, and DUOX2, and has been defined on the basis of its members' structural homology with gp91^{PHOX} (now renamed NOX2), the catalytic core of the phagocyte NOX. The homologs are now referred to as the NOX family of NADPH oxidases. These enzymes share the capacity to transport electrons across the plasma membrane to generate superoxide, which serves as a starting point for other reactive oxidants. On the other hand, plant NADPH oxidases are referred to as RBOHs (*r*espiratory *b*urst *o*xidase *h*omologs), due to their similarity with proteins identified in mammals as being responsible for the respiratory burst in activated neutrophils (Keller et al., 1998). Plant RBOHs are membrane proteins whose activity is highly regulated, in fact evidence suggests that they might be located in specific microdomains, or lipid rafts, membrane regions enriched in cholesterol and gangliosides able to cluster proteins for specific functions (Liu et al., 2009; Vilhardt & van Deurs, 2004). RBOHs contain in their

N-terminus well described EF-hands domains, which are calcium-binding motifs and therefore their activity is highly dependent on calcium, and can also be regulated by phosphorylation of serine residues or small GTPase binding (Kobayashi et al., 2007; Sirichandra et al., 2009).

In the model plant *Arabidopsis*, the genome encodes a small family of 10 RBOH proteins, which are differentially expressed and involved in several roles (Sagi & Fluhr, 2006). The zinc-finger protein Zat12 responds to a large number of biotic and abiotic stresses and therefore has been widely used as an indicator of oxidative stress; however, its mode of action and regulation are largely unknown. Using a fusion between the Zat12 promoter and the reporter gene luciferase (Zat12p::LUC), it has been elegantly demonstrated in *Arabidopsis* that expression of Zat12 depends on NADPH oxidase activity and ROS production (Miller et al., 2009) as diphenyleneiodonium (DPI) and inhibitor of the NADPH oxidase inhibit the propagation of the Zat12p::LUC expression.

NADPH enzymes are not the only enzymes that generate ROS in plant cells and other alternative mechanisms have been proposed. For instance, H_2O_2 can also be generated by enzymes such as apoplastic oxalate oxidase, diamine oxidase, or class III peroxidase (Caliskan & Cuming, 1998; Elstner & Heupel, 1976; Federico & Angelini, 1986). For instance, many peroxidases located in the apoplastic space can be linked to cell wall polymers through ionic or covalent bonds, contributing to some very interesting regulatory mechanisms of ROS production at the cell wall space that are important for wall properties. In the presence of H_2O_2 and phenolic substrates, peroxidases operate in the peroxidatic cycle and are involved in the synthesis of lignin and other phenolic polymers. However, in the presence of NADPH there is a mechanism for generating H_2O_2 through a NADH-oxidase activity of peroxidases (Chen & Schopfer, 1999). In fact, it has been suggested that cell wall peroxidases in contact with superoxide and H_2O_2 could be involved in originating hydroxyl radical in the cell wall. This free radical in plant cells has a very special role for other physiological responses including a controlled breakdown of structural polymers during rearrangement of cell wall in roots, hypocotyls or coleoptiles (Ogawa, Kanematsu, & Asada, 1997; Vianello & Macri, 1991). On the other hand, H_2O_2 in the presence of transition metal such as iron can catalyze the Fenton's reaction to generate hydroxyl radical. Therefore the cells have evolved to have several strategies to keep superoxide, H_2O_2, and metals such as Fe^{2+} and Cu^{2+} under tight control (Apel & Hirt, 2004).

3. PLANT CELLS RESPOND TO EXTERNAL AND INTERNAL STIMULI

Plants present a sessile lifestyle and have evolved sophisticated mechanisms to cope with several stresses in their environment. These can be activated locally in tissues that initially interact with the threat, but also can be transmitted systemically to tissues that are not directly challenged. The activation of defense or acclimation mechanisms in systemic tissues is often termed systemic acquired resistance (SAR) or systemic acquired acclimation (SAA), respectively, and serve important roles in preventing further infection or damage to the entire plant during stress (Jung, Tschaplinski, Wang, Glazebrook, & Greenberg, 2009). The most striking finding in plant cells is that many signals can travel long distances, for instance signal from the stress-responsive Zat12p::LUC can move at a rate of up to 8.4 cm min^{-1} and is dependent on the RBOHD enzyme, which produces superoxide that is rapidly transformed to H_2O_2 and able to mediate a rapid systemic signaling (Miller et al., 2009).

One of the most interesting responses of plant cells to external stimuli is the response to pathogen attack generating the so called oxidative burst, which constitute the production of a high level of superoxide that is rapidly transformed to H_2O_2 at the site of infection (Apostol, Heinstein, & Low, 1989; Doke et al., 1996). During this response a massive oxidative burst is induced, causing local cell death at the infection site and stop pathogen propagation to other tissues. This response has been widely explored to depict the role of the NADPH oxidases involved in this process and how it is regulated. In *Arabidopsis*, roots under mechanical stimuli respond with a cell wall localized ROS production that is dependent on NADPH oxidase (Monshausen, Bibikova, Weisenseel, & Gilroy, 2009). Furthermore, there is a clear role of H_2O_2 as a second messenger for wound-induced response in tomato (Orozco-Cardenas, Narvaez-Vasquez, & Ryan, 2001) or during symbiotic interaction (Marino et al., 2011; Montiel, Arthikala, & Quinto, 2013). On the other hand, for the polarized root hair cell, it is well established that ROS oscillations play a key role during tip growth process (Monshausen, Bibikova, Messerli, Shi, & Gilroy, 2007) which requires the activity of the NADPH oxidase since a loss of function mutant in this enzyme is unable to support root hair development (Foreman et al., 2003). Furthermore, the finding that calcium channel activity increased in response to ROS opens the venue for a tight regulation of ROS in a calcium dependent manner.

4. VISUALIZING HYDROGEN PEROXIDE IN LIVING PLANT CELLS

Changes in ROS are recognized as specific signatures of plant responses to biotic and abiotic responses. The general idea is very similar to what has been described for calcium signatures in that not only the spatial but also the temporal dynamic of ROS could be crucial for a given response. Therefore, a change in ROS level at the cell wall induces a response that differs from a similar change in ROS in the cytoplasm or a given organelle.

Biochemical assays and fluorescent markers are widely used for analyzing H_2O_2 production. 3,3'-Diaminobenzidine (DAB) is a widely used compound to visualize H_2O_2 production, resulting in a dark brown precipitate as a result of DAB polymerization (Driever, Fryer, Mullineaux, & Baker, 2009). In fact, DAB has been used as much as Nitro Blue Tetrazolium (NBT) for imaging the superoxide oxidative burst during plant–pathogen interaction, and several other biotic and abiotic responses (Swanson, Choi, Chanoca, & Gilroy, 2011). Therefore DAB is useful for rapid *in situ* detection of H_2O_2 production, but not for dynamic analysis. Furthermore, similar to what has been described for NBT, the experimental design should be carefully taken when using DAB due to possible toxic and photodegradation properties (Driever et al., 2009).

Fluorescein derived compounds such as dihydrodichlorofluorescein diacetate (H_2DCF-DA) represent the most widely used fluorescent probes for visualizing H_2O_2 in living cells. H_2DCF-DA is a nonfluorescent compound that can be ester loaded in living cells and thereafter rendered nonpermeable by the activity of intracellular esterases and thus remain sequestered in the cells. As the dye can be easily loaded and imaged (basically use the same filters required for GFP), this probe is very popular for visualizing H_2O_2 in plant cells during several responses. However, H_2DCF-DA is nonselective and thus unspecific. Furthermore, the strongest drawback of H_2DCF-DA is probably its instability in living cells, being easily photooxidized and photobleached (or highly sensitive to photooxidation and photobleaching). These two main disadvantages have been carefully addressed by Simon Gilroy's lab (Swanson et al., 2011), describing that the normal exposure to excitation light is enough for inducing photooxidation, or even photobleaching. Finally, the main problem with H_2DCF-DA is very similar to DAB and NBT, once the H_2DCF-DA reacts with ROS it becomes irreversibly fluorescent, therefore decline in ROS level will not be reported by this probe under normal conditions. A combination of several approaches and conditions, for instance in root hairs where ROS production occurs in a limited region and cytoplasmic diffusion helps to remove the oxidized dye from the tip or in other cases when the dye is added to the medium, a dynamic analysis has been achieved in the cell wall (Cárdenas & Quinto, 2008; Monshausen et al., 2007).

5. HYPER AS A NEW GENETICALLY ENCODED PROBE

Hyper consists of a circularly permutated YFP (cpYFP) inserted into the regulatory domain of the *Escherichia coli* hydrogen peroxide sensor OxyR. The key function of this regulatory domain lies in two cysteines residues (Cys199 and Cys208). When the OxyR domain is exposed to H_2O_2, which promotes disulfide bond formation, an intramolecular change occurs shifting the excitation peak of the attached cpYFP from 420 to 500 while the emission peak remains the same (516 nm). This H_2O_2 -induced change in the excitation wavelength of Hyper allows its use as a ratiometric probe, similar to other ratiometric probes such as fura-2 for calcium. However, there are other properties of Hyper that make it a powerful probe for monitoring H_2O_2 distribution in living cells.

With its oxidoreductive sensitive conformational change being reversible, Hyper is able to register both increases and decreases in H_2O_2 levels. As it is a molecular probe, it can be genetically engineered and targeted to specific cells and organelles such as peroxisomes, mitochondria, chloroplasts, etc. (Costa et al., 2010). Hyper could be developed into a universal H_2O_2-specific biosensor to be explored in plant cells. For instance, the described oscillation in extracellular ROS levels registered in root hair tip growth (Monshausen et al., 2007) could be nicely correlated with intracellular H_2O_2 changes, or oxidative bursts could be elegantly illustrated in specific compartments, such as the cell wall, or specific tissues and cells such as the meristems and pollen tubes by targeting Hyper expression and localization specifically.

It has been described recently that the Casparian strip, a lignin-based structure serving as a diffusion barrier between cells is formed by NADPH oxidase and peroxidases co-recruited by specific membrane domains at the plasma membrane for site-specific lignin deposition. Therefore localized production of ROS could a broadly used mechanism for cell wall modifications that require subcellular precision in plant (Lee, Rubio, Alassimone, & Geldner, 2013; Roppolo & Geldner, 2012) and animal

(Ushio-Fukai, 2006) cells alike. Hyper can conceivably be a powerful reporter for these important cellular events in future studies.

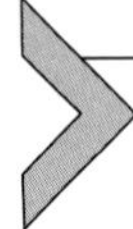

6. VECTOR DESCRIPTION AND PLANT TRANSFORMATION

The HyPer gene was removed from its mammalian expression vector (Evrogen, Moscow, Russia) and cloned into a Ti plasmid-based vector (Rodermel, Abbott, & Bogorad, 1988) behind either the constitutively active CaMV35S promoter (Benfey & Chua, 1990) of the pollen specific promoter LAT52 (Twell et al., 1990). The constructs 35s::Hyper and LAT52::Hyper are shown in Fig. 15.1. The HyPer gene coding region can be removed through digestion with BamH1 and Sal1 for introduction behind other promoters of choice.

The floral dip method (Clough & Bent, 1998) was used to transform *Arabidopsis*. It is described briefly as follows.

1. Transformation: *Arabidopsis thaliana* Col-1 plants were grown in a growth chamber (Percival) at 21 °C and under 14 h illumination. Robustly flowering plants were used. Log phase *35s:HyPer* Agrobacterium culture was used for floral dip transformation (Clough and Bent, 1998). As *Arabidopsis* flowering is not synchronized, siliques (i.e. already bearing seeds) were removed prior to dipping. After application of the *35s:HyPer* Agrobacterium, dipped plants were left to grow to seed-set. Once the siliques were dried, about 3 weeks after dipping, seeds were harvested. In general, ~4 to 6 pots of *Arabidopsis* (5–6 flowering plants per pot) should produce enough seeds to have a reasonable recovery of transformed seedlings.
2. Screening for transformants: Seeds were surface sterilized and germinated on tissue culture medium (Gamborg B5, 1% Sucrose, 5 mM MES pH 5.7, 0.7% agar, and supplemented with kanamycine @ 50 mg l^{-1}). Kanamycin-resistant seedlings (T1 plants) would emerge

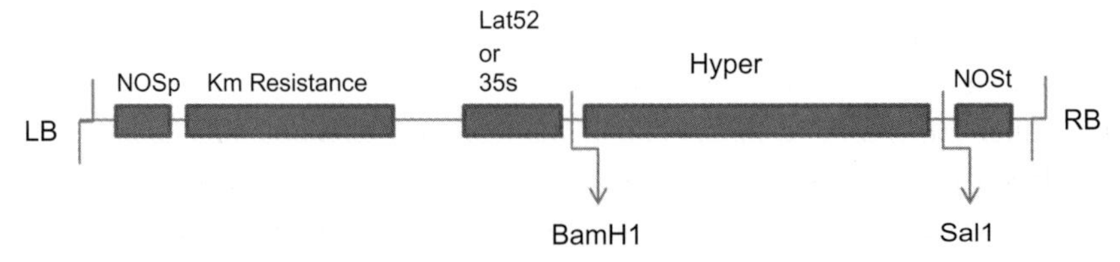

Figure 15.1 A restriction map for the CaMV 35s:Hyper construct. (See Color Insert.)

within 2 weeks in the midst of a lawn of dead (kanamycin-sensitive) seedlings. These T1 plants were transferred to soil till flowering and self-fertilization to produce T2 seeds.

3. These T2 seeds were germinated on Gamborg B5 with 1% sucrose, 0.7% agar. Roots from these T2 seedlings (~1 week) were examined for YFG fluorescence and for sensitivity to H_2O_2-induced shift in excitation maximum.

Some of the YFP positive seedlings were transferred to soil again and grown to maturity for seed collection. These seeds (i.e. T3 seeds) and seeds from future generations are ready for experimental use.

The *Agrobacterium*-mediated leaf disc transformation method (Delebrese et al., 1986) was used to obtain transformed tobacco plants. It is described briefly here.

1. Transformation: Leaf discs were prepared from tissue culture-grown *Nicotiana tabacum* SR1 and placed in liquid callus-inducing medium (CIM) (Murashige and Skoog (MS), with 3% sucrose, 5 mM MES, pH 5.7, 0.1 mg ml^{-1} naphthalene acetic acid, NAA) in a Petri dish. A log phase culture of LAT52::Hyper was added (100 μl per 10 ml medium with about 15 leaf discs each).
2. After 2 days of cocultivation at ~28 °C in the dark, leaf discs were removed from the medium and passed through three washes, each with 10 ml of CIM.
3. These leaf discs were plated onto shoot-inducing medium (SIM) medium (same as CIM except with 1 mg ml^{-1} 6-benzylaminopurine and the antibiotics cefotaxamine @ 250 μg ml^{-1} and kanamycin @ 50 μg ml^{-1}, 0.7% agar added), and left in the dark for about 5–7 days. Cefotaxamine prevents *Agrobacterium* growth, kanamycin would select for transformed cells.
4. Leaf discs were transferred to a fresh Petri dish of SIM, and they were kept under 16 h/8 h light/dark cycle.
5. Tissue callusing and plantlet regeneration would occur in the next 2–4 weeks. An additional change of fresh medium would be recommended every two weeks.
6. Plantlets were excised from the leaf discs and planted on to basal ½ MS, 1% sucrose medium, supplemented with antibiotics. Transformed plants (T1) produced roots, and were transferred to soil and grown to flowering.
7. Pollen from these transformed plants were screened for YFP. Seeds from these YFP positive T1 plants were collected and pollen from their progeny was used for experiments. Pollen could be kept at −80 °C for long term storage.

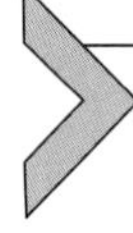

7. PREPARATION AND STERILIZATION OF MODIFIED PETRI DISHES FOR GROWING *ARABIDOPSIS* PLANTS FOR MICROSCOPY ANALYSIS

1. Commercial Petri dishes were perforated in the center with a hot cutter. The cutter is heated with the Bunsen burner until the knife of the cutter becomes red.
2. Draw with an ink pencil a square in the bottom part of the Petri dish according to the size of the coverslip that will be used (we recommend 66×48 mm, Gold Seal).
3. Use the heated cutter to melt down the line describing the square of the bottom part of the Petri dish.
4. When the hole is made, polish the edges of the square and put a line of silicon glue along all the edge of the square. Then set the coverslip carefully, the binding should be perfect to avoid bubbles between the glass and the Petri dish. Leave them to dry out overnight.
5. The modified Petri dishes are ready for sterilization.
6. Modified Petri dishes were sterilized by vapor phase method used to sterilize *Arabidopsis* seeds (Clough & Bent, 1998) and exposing to UV light.
7. Modified Petri dishes were placed into a plastic chamber or a desiccator jar, placed in a fume hood.
8. A beaker containing 100 ml commercial bleach was placed into the plastic chamber and 3 ml hydrochloric acid 36% (w/v) was carefully added to the bleach before the plastic chamber was completely sealed.
9. Leave the modified Petri dishes in the chamber for 2 h for complete sterilization.
10. Remove all the Petri dishes and close immediately.
11. We recommend a second sterilization procedure by exposing to UV light in the flow chamber for 30 min.
12. After the two sterilization procedures, the modified Petri dishes are ready to use (Fig. 15.2).

8. SEEDS STERILIZATION AND STRATIFICATION

Arabidopsis seeds were surfaced sterilized by vapor phase method (Clough & Bent, 1998).

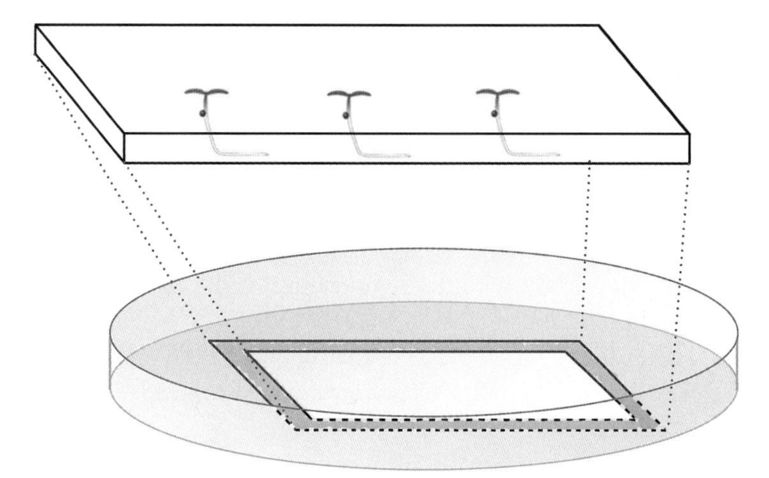

Figure 15.2 Purpose modified Petri dish to growth *Arabidopsis* seedling for microscopy analysis. The Petri dish with a hole in the middle is replaced with glass coverslip silicon glued to the bottom. Upper part represents the seedlings growing in the solid medium. (For color version of this figure, the reader is referred to the online version of this chapter.)

1. Seeds were placed in an open Eppendorf tube for vapor phase sterilization as described previously for Petri dish sterilization.
2. Eppendorf was marked with sufficiently chlorine resistant ink such as black Staedtler permanent Lumocolor.
3. Leave the *Arabidopsis* seeds in the chamber for surface sterilization vapor for 2 h.
4. Eppendorf tubes were closed immediately after opening the sealed sterilization chamber.
5. Sterilized surface seeds were resuspended in 1 ml of sterilized water and were stratified by incubation at 4 °C for 2 days, in the dark. Stratification improves the rate and synchrony of germination.
6. Seeds are ready for germination.

9. GROWTH CONDITIONS

Medium used were Linsmaier & Skoog Basal Medium (L477, Phyto Technology Laboratories, Lenexa, KS, USA) which presents a similar composition to Murashige & Skoog basal salts, supplemented with Linsmaier and Skoog vitamines (thiamine·HCl and myo-inositol). Linsmaier & Skoog medium was pH adjusted (5.25–6.25) and buffered with MES.

1. Seeds were germinated and grown in 0.2× Linsmaier & Skoog Basal Medium supplemented with vitamins (0.1 mg l^{-1} piridoxine, 0.1 mg l^{-1} nicotinic acid), 1% (w/v) sucrose, 0.8% (w/v) agar, and pH was adjusted to 5.7 with KOH 1 *M*.
2. Medium was autoclaved at 120 °C for 20 min.
3. Autoclaved medium was cooled (to about 55 °C) and distributed into the vapor-sterilized Petri dishes described above (approximately 30 ml per plate). Be careful with the temperature, if the medium is too hot, the coverslip will break down.
4. Sterilized surface seeds were set in the modified Petri dishes containing the solid medium and each seed was lay down around 5–7 mm deep into the medium by using a 200-μl micropipette.
5. Plates were wrapped with Parafilm and placed horizontal for germination and growth at 21 °C, in a photoperiod 16 h light/8 h dark, light intensity of 105 μmol photons m^{-2} s^{-1}.
6. After 3 days of germination and when the primary root reaches the bottom of the chamber (coverslip) we set the Petri dishes vertically in order to keep the main root growing over the surface of the glass.
7. Usually by about 6 days after germination, when the main root is over 1 cm in length they are in good moment for microscopy analysis.

10. IMAGE ACQUISITION AND PROCESSING

All images were acquired with a CCD camera (Sensys, Roper Scientific) attached to a Nikon TE300 inverted microscope coupled to a xenon illumination source (DG-4, Sutter Instruments), which contains a 175-W ozone-free xenon lamp (330–700 nm) and a galvanometer-driven wavelength switcher. We have incorporated a uniblitz shutter (Vincent Associates) that allows the acquisition of the transmitted light for every ratio image. In order to make a ratio analysis, a filter wheel should be incorporated to switch between the two different excitation wavelengths. Otherwise a monochromatic UV illumination source with a galvanometer incorporated can make the switching automatically. Nowadays, we have also LED illumination sources that can incorporate several wavelengths such as the XCITE LED (Lumen Dynamics, Canada). All the system is operated by MetaMorph/MetaFluor software (Universal Imaging, Molecular Devices) (Fig. 15.3).

1. Seedlings presenting a root more that 1 cm in length were set into the microscope stage and visualized under the 20×/0.8 or 40×/1 NA water immersion objective lens.

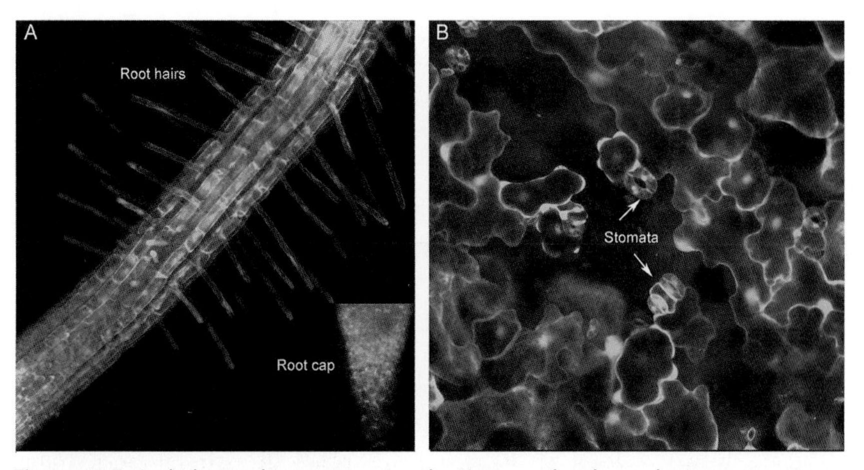

Figure 15.3 *Arabidopsis* plants expressing the Hyper molecular probe in transgenic stable lines. (A) Images from *Arabidopsis* roots under the confocal microscope, note that epidermal cell and root hairs present a good expression of hyper. (B) A region of the leave showing the pavement cells and stomata showing Hyper expression. Inset, shows the cap of the root tip with Hyper expression of the same root. (See Color Insert.)

2. Open the MetaFluor software for ratio image analysis
3. Select three windows for the experiment
4. For window 1 select the excitation wavelength (495 nm) and the emission at 530 nm (band pass 20).
5. For window 2, select the excitation wavelength (420 nm) and the emission at 530 nm (band pass 20).
6. For window 3, select the transmitted illumination source. This will acquire the transmitted image.
7. For windows 1 and 2, select a good exposure time that allows the best signal to noise ratio. It is important that both windows acquire a very similar image regarding the intensity level, the ratio values should come out in general 1. It is fine even if increasing the exposure time for one or the other window is necessary to achieve this.
8. Once you have a good ratio pair, you are ready to start acquiring long sequences for looking at dynamic changes or single ratio pairs for spatial distributions.
9. Selected images or sequences with good results can be chosen for analysis.
10. For image analysis open the ratio pair images, select background subtraction and apply.
11. MetaFluor make the ratio analysis automatically, dividing the windows 1 (excited at 495 nm) with the windows 1 (excited at 420 nm).

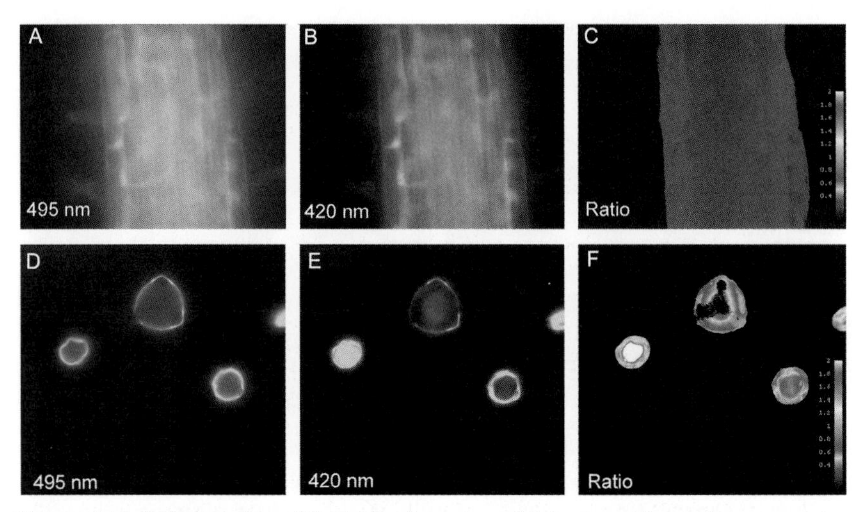

Figure 15.4 Ratio imaging of *Arabidopsis* roots and tobacco pollen tubes expressing Hyper. Panels A, B, and C represent the same region excited at 495 and 420 nm and the corresponding ratio image, respectively. Emission was collected at 530 nm with a band pass of 30 nm. Panels D, E, and F represent another ratio image experiment using the transgenic tobacco pollen tube. Note that the red color indicates a higher ROS concentration and blue a low level. (See Color Insert.)

12. Edited images can be scaled to represent the ratio image with the scale bar representing the ratio values. This is very useful for visualizing and comparing changes between different regions of the same cell or tissue (Fig. 15.4).
13. Images can be exported as a movie (quick time or avi or single images as tiff).

ACKNOWLEDGMENTS

This work was supported by awards from the National Science Foundation (NSF-0544222; 11-46941) and from Dirección General de Asuntos del Personal Académico (DGAP) IN-204409 and Consejo Nacional de Ciencia y tecnología (CONACyT) 58323 and 132155 to LC. E. J. was partially supported by an American Society of Plant Biology SURF award for this work. We also thank the NSF RCN on Integrative Pollen Biology for supporting meeting forums where discussions on this collaborative project were conducted. We thank Noreide Nava, Olivia Santana, and Yuridia Cruz for technical help.

REFERENCES

Allan, A. C., & Fluhr, R. (1997). Two distinct sources of elicited reactive oxygen species in tobacco epidermal cells. *The Plant Cell*, *9*, 1559–1572.

Apel, K., & Hirt, H. (2004). Reactive oxygen species: Metabolism, oxidative stress, and signal transduction. *Annual Review of Plant Biology*, *55*, 373–399.

Apostol, I., Heinstein, P. F., & Low, P. S. (1989). Rapid stimulation of an oxidative burst during elicitation of cultured plant cells: Role in defense and signal transduction. *Plant Physiology*, *90*, 109–116.

Benfey, P. N., & Chua, N. H. (1990). The cauliflower mosaic virus 35S promoter: Combinatorial regulation of transcription in plants. *Science*, *250*, 959–966.

Caliskan, M., & Cuming, A. C. (1998). Spatial specificity of H_2O_2-generating oxalate oxidase gene expression during wheat embryo germination. *The Plant Journal*, *15*, 165–171.

Cárdenas, L., Martinez, A., Sanchez, F., & Quinto, C. (2008a). Fast, transient and specific intracellular ROS changes in living root hair cells responding to Nod factors (NFs). *The Plant Journal*, *56*, 802–813.

Cárdenas, L., & Quinto, C. (2008b). Reactive oxygen species (ROS) as early signals in root hair cells responding to rhizobial nodulation factors. *Plant Signaling & Behavior*, *3*, 1–3.

Chen, S. X., & Schopfer, P. (1999). Hydroxil radical production in physiological reaction. A novel function of peroxidase. *European Journal of Biochemistry*, *260*, 726–735.

Clough, S. J., & Bent, A. F. (1998). Floral dip: A simplified method for agrobacterium-mediated transformation of *Arabidopsis thaliana*. *The Plant Journal*, *16*, 735–743.

Costa, A., Drago, I., Behera, S., Zottini, M., Pizzo, P., Schroeder, J. I., et al. (2010). H_2O_2 in plant peroxisomes: An in vivo analysis uncovers a Ca(2+)-dependent scavenging system. *The Plant Journal*, *62*, 760–772.

Delebrese, R., Reynaert, A., Hofte, H., Hernalsteens, J. P., Leemans, J., & Van Montagu, M. (1986). Vectors for cloning in plant cells. *Methods in Enzymology*, *153*, 277–290.

Doke, N., Miura, Y., Sanchez, L. M., Park, H. J., Noritake, T., Yoshioka, H., et al. (1996). The oxidative burst protects plants against pathogen attack: Mechanism and role as an emergency signal for plant bio-defence—A review. *Gene*, *179*, 45–51.

Driever, S. M., Fryer, M. J., Mullineaux, P. M., & Baker, N. R. (2009). Imaging of reactive oxygen species in vivo. *Methods in Molecular Biology*, *479*, 109–116.

Elstner, E. F., & Heupel, A. L. (1976). Formation of hydrogen peroxide by isolated cell walls from horseradish (*Armoracia lapathifolia Gilib*). *Planta*, *130*, 175–180.

Federico, R., & Angelini, R. (1986). Occurrence of diamine oxidase in the apoplast of pea epicotyls. *Planta*, *167*, 300–302.

Foreman, J., Demidchik, V., Bothwell, J. H., Mylona, P., Miedema, H., Torres, M. A., et al. (2003). Reactive oxygen species produced by NADPH oxidase regulate plant cell growth. *Nature*, *422*, 442–446.

Jung, H. W., Tschaplinski, T. J., Wang, L., Glazebrook, J., & Greenberg, J. T. (2009). Priming in systemic plant immunity. *Science*, *324*, 89–91.

Keller, T., Damude, H. G., Werner, D., Doerner, P., Dixon, R. A., & Lamb, C. (1998). A plant homolog of the neutrophil NADPH oxidase gp91phox subunit gene encodes a plasma membrane protein with Ca2+ binding motifs. *The Plant Cell*, *10*, 255–266.

Kobayashi, M., Ohura, I., Kawakita, K., Yokota, N., Fujiwara, M., Shimamoto, K., et al. (2007). Calcium-dependent protein kinases regulate the production of reactive oxygen species by potato NADPH oxidase. *The Plant Cell*, *19*, 1065–1080.

Lee, Y., Rubio, M. C., Alassimone, J., & Geldner, N. (2013). A mechanism for localized lignin deposition in the endodermis. *Cell*, *153*(2), 402–412.

Liu, P., Li, R. L., Zhang, L., Wang, Q. L., Niehaus, K., Baluska, F., et al. (2009). Lipid microdomain polarization is required for NADPH oxidase-dependent ROS signaling in Picea meyeri pollen tube tip growth. *The Plant Journal*, *60*, 303–313.

Marino, D., Andrio, E., Danchin, E. G., Oger, E., Gucciardo, S., Lambert, A., et al. (2011). A Medicago truncatula NADPH oxidase is involved in symbiotic nodule functioning. *The New Phytologist*, *189*, 580–592.

Miller, G., Schlauch, K., Tam, R., Cortes, D., Torres, M. A., Shulaev, V., et al. (2009). The plant NADPH oxidase RBOHD mediates rapid systemic signaling in response to diverse stimuli. *Science Signaling*, *2*, ra45.

Mittler, R., & Berkowitz, G. (2001). Hydrogen peroxide, a messenger with too many roles? *Redox Report*, *6*, 69–72.

Monshausen, G. B., Bibikova, T. N., Messerli, M. A., Shi, C., & Gilroy, S. (2007). Oscillations in extracellular pH and reactive oxygen species modulate tip growth of *Arabidopsis* root hairs. *Proceedings of the National Academy of Sciences of the United States of America*, *104*, 20996–21001.

Monshausen, G. B., Bibikova, T. N., Weisenseel, M. H., & Gilroy, S. (2009). Ca^{2+} regulates reactive oxygen species production and pH during mechanosensing in *Arabidopsis* roots. *The Plant Cell*, *21*, 2341–2356.

Montiel, J., Arthikala, M., & Quinto, C. (2013). *Phaseolus vulgaris* RbohB functions in lateral root development. *Plant Signaling & Behavior*, *8*, 144–146.

Mori, I. C., & Schroeder, J. I. (2004). Reactive oxygen species activation of plant Ca^{2+} channels. A signaling mechanism in polar growth, hormone transduction, stress signaling, and hypothetically mechanotransduction. *Plant Physiology*, *135*, 702–708.

Ogawa, K., Kanematsu, S., & Asada, K. (1997). Generation of superoxide anion and localization of CuZn-superoxide dismutase in the vascular tissue of spinach hypocotyls: Their association with lignification. *Plant & Cell Physiology*, *38*, 1118–1126.

Orozco-Cardenas, M. L., Narvaez-Vasquez, J., & Ryan, C. A. (2001). Hydrogen peroxide acts as a second messenger for the induction of defense genes in tomato plants in response to wounding, systemin, and methyl jasmonate. *The Plant Cell*, *13*, 179–191.

Pei, Z. M., Murata, Y., Benning, G., Thomine, S., Klusener, B., Allen, G. J., et al. (2000). Calcium channels activated by hydrogen peroxide mediate abscisic acid signalling in guard cells. *Nature*, *406*, 731–734.

Potocky, M., Jones, M. A., Bezvoda, R., Smirnoff, N., & Zarsky, V. (2007). Reactive oxygen species produced by NADPH oxidase are involved in pollen tube growth. *The New Phytologist*, *174*, 742–751.

Rodermel, S. R., Abbott, M. S., & Bogorad, L. (1988). Nuclear-organelle interactions: Nuclear antisense gene inhibits ribulose bisphosphate carboxylase enzyme levels in transformed tobacco plants. *Cell*, *55*, 673–681.

Roppolo, D., & Geldner, N. (2012). Membrane and walls: Who is master, who is servant? *Current Opinion in Plant Biology*, *15*, 608–617.

Sagi, M., & Fluhr, R. (2006). Production of reactive oxygen species by plant NADPH oxidases. *Plant Physiology*, *141*, 336–340.

Sirichandra, C., Gu, D., Hu, H. C., Davanture, M., Lee, S., Djaoui, M., et al. (2009). Phosphorylation of the *Arabidopsis* AtrbohF NADPH oxidase by OST1 protein kinase. *FEBS Letters*, *583*, 2982–2986.

Swanson, S. J., Choi, W. G., Chanoca, A., & Gilroy, S. (2011). In vivo imaging of Ca2+, pH, and reactive oxygen species using fluorescent probes in plants. *Annual Review of Plant Biology*, *62*, 273–297.

Tsukagoshi, H., Busch, W., & Benfey, P. N. (2010). Transcriptional regulation of ROS controls transition from proliferation to differentiation in the root. *Cell*, *143*, 606–616.

Twell, D., Yamaguchi, J., & McCormick, S. (1990). Pollen-specific gene expression in transgenic plant: coordinate regulation of two different tomato gene promoters during microsporogenesis. *Development*, *109*, 705–713.

Ushio-Fukai, M. (2006). Localizing NADPH oxidase-derived ROS. *Science's STKE*, *8*, 349.

Vianello, A., & Macri, F. (1991). Generation of superoxide and hydrogen peroxide at the surface of plant cells. *Journal of Bioenergetics and Biomembranes*, *23*, 409–423.

Vilhardt, F., & van Deurs, B. (2004). The phagocyte NADPH oxidase depends on cholesterol-enriched membrane microdomains for assembly. *The EMBO Journal*, *23*, 739–748.

Wilkins, K. A., Bancroft, J., Bosch, M., Ings, J., Smirnoff, N., & Franklin-Tong, V. E. (2011). ROS and NO mediate actin reorganization and programmed cell death in the Self-Incompatibility response of Papaver. *Plant Physiology*, *156*, 404–416.

AUTHOR INDEX

Note: Page numbers followed by "*f*" indicate figures, and "*t*" indicate tables, and "np" indicate footnotes.

A

Aalen, R. B., 258–259
Abbas, K., 114–115, 115*f*, 116, 118*f*, 124*f*
Abbott, M. S., 282
Abell, A. N., 151–152
Abidi, P., 174
Abo, A., 146
Aburatani, H., 76–78
Adams, A. G., 153–154, 159–160
Addepalli, B., 208, 217
Aebersold, R., 118
Aebi, H. E., 15
Agisheva, N., 52
Ahicart, P., 123
Ahn, C. Y., 54–55
Ahn, J. H., 207
Ahn, Y., 54–55
Aimi, N., 105
Akazawa, H., 76–78
Alassimone, J., 281–282
Albrecht, H. L., 190–191
Alexander, J., 189–190
Allan, A. C., 277
Allen, G. J., 190–191, 276
Allen, R. A., 146
Almeida, A., 130, 131–132, 133, 137–138
Ambruso, D. R., 146, 148, 149, 151–152, 153, 154, 158*f*, 161*f*, 163–165, 163*f*, 164*f*
Ambruso, D. W., 148, 149, 157, 157*np*, 159*np*
Andersen, J. N., 50
Anderson, M. E., 6–8
Andres-Mateos, E., 45*t*
Andrio, E., 279
Angelini, R., 278
Anilkumar, N., 42
Antunes, F., 4–5, 5*f*, 6*t*, 14, 16–17, 47–48
Apel, K., 204–205, 278
Apostol, I., 279
Aravind, L., 242
Arias, A. A., 146–147
Arimura, S., 240–241, 242, 244*f*, 245, 253
Armit, L. J., 74, 76–78
Armstrong, R. C., 242
Arnault, I., 105
Arnér, E. S., 70–71, 78
Arnold, R. S., 42
Artemenko, I. P., 177
Arthikala, M., 279
Arthur, J. R., 88–89
Asada, K., 278
Asashima, M., 67, 74, 77*f*
Ascenzi, R., 191–192
Åslund, F., 38
Asok, M. L., 186–187
Ataliotis, P., 90
Atsriku, C., 50–51
Audi, R., 45*t*, 51–52
Auger, J., 105
Augusto, O., 45*t*, 51–52
Avery, A. M., 8
Avery, S. V., 8, 130
Avila, J., 78
Awasthi, Y. C., 133
Azrak, R. G., 74–75

B

Babbs, C. F., 245
Babior, B. M., 146–147
Babu, C. V., 116
Badger, M. R., 187–188
Bae, S. H., 23, 48, 114–116, 171, 172–174, 172*f*, 175, 177–179, 178*f*, 267
Bae, S. W., 46–47
Baek, J. Y., 23
Baines, I. C., 148, 155–156, 157
Baker, A., 123
Baker, L. M., 26, 27, 31*f*, 33
Baker, N. R., 186–188, 196, 280
Balazadeh, S., 206–207, 217
Ball, L., 187
Ballikaya, S., 55
Baluska, F., 277–278

Bancroft, J., 276
Banning, A., 66–67, 68–69, 70–71, 72, 75, 77*f*, 78–79, 89, 91, 93, 94, 102, 106–107, 108
Barelier, S., 50–51
Barger, J. L., 108
Barja, G., 140–141
Barker, N., 72
Barnes, K. M., 88–89, 108
Barnes, P. J., 123–125
Barranco-Medina, S., 267
Barrault, M.-B., 52, 54–55, 267
Barrett, W. C., 45*t*, 46*t*
Bartel, D. P., 153–154
Bartels, D., 217–218
Baudo, M., 186–187
Baxter, P., 116, 123, 173–174
Bechtold, E., 50–51, 50*t*
Bechtold, U., 186–187, 223*t*
Beckwith, J., 38, 130
Béclin, C., 190–191
Bécuwe, N., 267
Beeckman, T., 205
Behera, S., 186–188, 190, 191, 281
Behne, D., 88–89
Behring, J. B., 50–51
Bejaoui, I., 74, 76–78
Belenghi, B., 240–241, 248
Belousov, V. V., 123–125, 130–131, 189–190, 194
Benfey, P. N., 276, 282
Benina, M., 214
Benning, G., 276
Bensimon, A., 70–71
Bent, A. F., 191, 282–283, 284–285
Bereiter-Hahn, J., 240–241
Berkowitz, G., 276
Berlett, B. S., 45*t*
Bermano, G., 88–89
Bernal-Bayard, P., 262–264
Bernroitner, M., 258–259
Berry, M. J., 88–89
Bezvoda, R., 276
Bi, X., 74–75
Biard, D., 77*f*, 78
Bibikova, T. N., 279, 280–281
Bienert, G. P., 47–48
Bienz, M., 72–73
Bignon, J., 116, 124*f*
Bilan, D. S., 189–190, 194
Bilic, J., 72–73
Binder, A., 189–190
Birringer, M., 89, 105–106
Biteau, B., 23, 55, 66–67, 115–116, 170–171, 267
Björnstedt, M., 104
Blajecka, K., 186–187
Blake, S. M., 106–107
Blanchard-Fillion, B., 45*t*
Blázquez, M. A., 207
Block, E., 89
Bloom, D. A., 70*f*
Bo, S., 108
Bobo-Jimenez, V., 130, 131–132, 133, 137–138
Böcher, M., 68, 88–89
Bode, H. H., 176
Boersema, P. J., 74
Bogorad, L., 282
Bohme, E., 140, 141
Boileau, C., 115–116, 267
Bolanos, J. P., 130, 137
Bolscher, B. G., 146, 151–152
Boonefaes, T., 205
Borevitz, J. O., 207
Bosch, M., 276
Bosse, A. C., 108
Bothwell, J. H., 276, 279
Botti, H., 44–45, 50*t*, 51–52, 78–79
Bouget, F. Y., 115–116
Bounds, P. L., 43
Bouschet, T., 76–78
Bouton, C., 114–115, 116, 118*f*, 124*f*
Boveris, A., 4–5, 245
Bowers, R. R., 123
Bowler, R. P., 123–125
Boyd, P., 240
Boyes, D. C., 191–192
Boyhan, A., 146
Brach, T., 189–190
Branco, M. R., 6*t*, 47–48
Bredesen, D., 242
Brennan, J. P., 50–51
Breton, J., 115*f*, 116, 124*f*
Breton-Romero, R., 123
Bridges, R. J., 137

Brigelius-Flohé, R., 57–58, 66–67, 68–71, 72, 78–80, 88–89, 91, 93, 104–105
Britt, A., 206
Britton, D. J., 50–51
Brochier-Armanet, C., 115–116, 267
Bromberg, Y., 146–147
Bronson, R., 54, 55
Brown, J. D., 55
Brunoud, G., 189–190
Bubenik, J. L., 107
Budde, H., 45*t*
Budiman, M. E., 107
Burdon, R., 240
Burk, R. F., 66, 88–89, 104–105, 106–107
Burrow, A. H., 189–190
Busch, W., 276
Bush, D. F., 209–210
Butterfield, D. A., 138
Byrne, D. W., 104–105, 106–107

C

Cadenas, E., 5*f*, 6*t*, 14, 16–17, 245
Cadigan, K. M., 74
Cai, W., 76–78
Caliskan, M., 278
Calvisi, D. F., 79–80
Cannon, M. B., 130–131
Cano, B., 262–264
Cao, G., 54
Cao, J., 54, 55
Cao, Z., 42
Capecchi, M. R., 78
Carballal, S., 50*t*
Cárdenas, L., 276, 280–281
Carlberg, I., 133
Carlson, B. A., 67, 68–69, 71, 79–80, 88
Carnac, G., 76–78
Casagrande, S., 56
Cash, A., 107
Castranova, V., 140–141
Caudy, A. A., 153–154
Cazalis, R., 258–259
Cejudo, F. J., 114–115, 258–259, 260–264, 265–266, 267, 268, 269, 270, 271
Cetin-Atalay, R., 79–80
Cha, Y. Y., 163–165
Chae, H. Z., 22–23, 27, 52, 55, 118, 147–148, 150, 155–156, 170–171, 267
Chalker, D. K., 88
Chan, E. C., 137–138
Chance, B., 4–5
Chang, C. J., 42, 123–125
Chang, T. S., 55, 56–57, 130–131, 132, 139, 140–141, 147–148, 170–171
Chanoca, A., 280–281
Charette, S. J., 54
Chaudhuri, B., 190–191
Chavan, A. J., 146–147
Chavez-Tapia, N. C., 123–125
Chen, J. W., 149, 187–188
Chen, R. H., 74, 76–78
Chen, S. X., 278
Chen, W., 116
Chen, X., 67, 148, 149
Chen, Y., 44
Cheng, G., 42
Cherradi, N., 171–172
Chi, Y. H., 55–56, 114–115
Chino, M., 235
Chiu, D. T. Y., 6–8
Cho, C. S., 46–47, 52
Cho, D. H., 56, 116
Cho, E. J., 116, 175
Cho, H. Y., 70–71
Cho, S. H., 54–55, 189–190
Chock, P. B., 45*t*, 46*t*, 56
Choi, E. J., 52
Choi, H. I., 55–56, 189–190
Choi, S. W., 163–165
Choi, S. Y., 56–57
Choi, W.-G., 190, 280–281
Chomczynski, 93–94
Chow, J., 88
Chow, W. S., 217–218
Chowdhry, S., 76–78, 77*f*
Chowdhury, G., 45*t*
Christensen, S. K., 207
Chu, C. T., 54
Chu, F. F., 68–69
Chu, P. G., 68–69
Chu, S. P., 190–191
Chua, N. H., 282
Chung, S. J., 22–23, 147, 155–156
Church, G., 147–148
Clark, B. J., 174
Clark, J. B., 137–138

Clark, L. C., 88
Clark, R. A., 146
Clayton, C. C., 146–147
Clevers, H., 72
Clippe, A., 45*t*, 46*t*, 50–52, 50*t*
Clough, S. J., 191, 282–283, 284–285
Cnudde, F., 209–210
Cochrane, C. G., 146
Cohen, P., 76
Colell, A., 138
Colton, C., 114
Combs, G. F. Jr., 88
Comhair, S. A., 123–125
Coolbaugh, R. C., 245
Cooma, I., 74–75
Cooper, A. J., 104–105
Coort, S. L., 68–69, 78–79, 89, 106–107, 108
Corellou, F., 115–116
Cornelis, R., 105
Cortes, D., 278, 279
Cortez, Y., 174
Costa, A., 186–188, 190, 191, 281
Cotte, A., 77*f*
Cotter, T. G., 42
Couvelard, L., 116
Cox, A. G., 45, 51–52
Cox, D. L., 33–34
Crapo, J. D., 123–125
Creissen, G., 186–187, 222, 240
Cross, A. R., 146–147
Cross, J. V., 69–70
Crowdus, C. A., 108
Cruciat, C. M., 72–73
Cruz, S. S., 242
Cuadrado, A., 76–78, 77*f*
Cuadros, R., 78
Culiáñez-Macià, F. A., 258–259
Cullinan, S. B., 69–70
Culmsee, C., 78–79
Cuming, A. C., 278
Cummings, A. H., 44–45
Curnutte, J. T., 146–147
Cusack, N., 151–152
Cyrne, L., 4–5, 6*t*, 16–17, 47–48

D

Dakin, K. A., 116
Damude, H. G., 277–278
Danchin, E. G., 279
Danenberg, E., 74, 76–78
Dang, P. M., 146–147
Dangl, J. L., 204–205, 240
Danial, N. N., 240–241
Daniel, V., 70–71
Danielson, S. R., 50–51
Dao, D. D., 133
Dat, J. F., 205
D'Autreaux, B., 130
Davanture, M., 277–278
Davidson, G., 72–73
Davis, K. R., 191–192, 240, 245
Davletova, S., 222
Davydov, D. R., 171–172
Dawes, I. W., 130
Dawson, K. A., 108
Day, A. M., 42, 55
De Duve, C., 4
De Fay, E., 43
De Gara, L., 240
de la Cal, C., 176
de Lau, W., 74, 76–78
De Laurentiis, A., 176
de Mendez, I., 153–154, 159–160
de Pinto, M. C., 240
de Sagarra, R. M., 76
Declercq, J. P., 45*t*, 46*t*, 50*t*, 51–52
DeGnore, J. P., 45*t*, 46*t*
Delaunay, A., 52, 54–55, 116
Delebrese, R., 283–284
DeLeo, F. R., 146–147
Demidchik, V., 276, 279
Demol, H., 56
Denev, I., 204–206, 211
Deng, Q., 123–125
DenHerder, G., 189–190
Denicola, A., 44–45, 45*t*, 46*t*, 51–52, 78–79
Deruelles, J., 264–265
Desagher, S., 130–131, 140–141
Desai, D., 74–75
Desikan, R., 223*t*
Desrosiers, D. C., 33–34
Deubel, S., 66–67, 68–69, 70–71, 78–79, 93
Deuschle, K., 190–191
Deveraux, Q., 242
Dick, T. P., 123–125
Dickinson, B. C., 42, 123–125

Die, L., 76–78
Diehl, J. A., 69–70
Diet, A., 114–115, 116, 118*f*
Dietrich, R. A., 240
Dietz, K.-J., 258–259, 268, 272
Dijkwel, P. P., 214
Dinakar, C., 217–218
Dinauer, M. C., 146–147, 159–160
Ding, V. W., 74, 76–78
Dixit, V. M., 242
Dixon, L. E., 115–116
Dixon, R. A., 240, 245, 277–278
Djaoui, M., 277–278
Dobrowolski, R., 73–74
Dodia, C., 148, 149
Doerner, P., 277–278
Doherty, C. J., 214
Doke, N., 279
Dong, H. F., 79–80
Douce, R., 248–249
Drago, I., 186–188, 190, 191, 281
Drake, J., 138
Draper, J., 245
Drapier, J. C., 115*f*, 116–117
Driever, S. M., 280
Dringen, R., 130–131, 137, 138, 139–141
Driscoll, D. M., 107
Du, Z., 217–218
Duan, H., 171–172
Dubois, L. G., 44–45
Duchen, M. R., 190
Dukiandjiev, S., 205–207
Dumont, E., 105
Dunoyer, P., 190–191
Duprez, J., 130–131
Duran, R., 50*t*
Dyson, M. T., 171–172, 177

E

Eaton, P., 50–51
Eberini, I., 56
Eckhaus, M., 79–80
Edgar, R. S., 23–24, 115–116
Egea, J., 76
Eichele, E., 46*t*, 51–52
Eken, A., 123–125
Elbashir, S. M., 153–154
El-Bayoumy, K., 74–75
Elger, W., 88–89
Ellerby, L., 242
Ellis, H. R., 26
Ellison, M. A., 148, 149, 152, 153, 154, 158*f*, 161*f*, 163–165, 163*f*, 164*f*
Elmayan, T., 190–191
Elsner, A., 104–105
Elstner, E. F., 278
Elzi, D., 151–152, 155*f*, 156*f*
Eme, L., 115–116, 267
Epperlein, M. M., 114
Erdjument-Bromage, H., 57
Erzurum, S. C., 123–125
Esaki, N., 102–104
Escobar, C., 240
Escobar, J., 130, 131–132, 133, 137–138
Espey, M. G., 114
Esworthy, R. S., 68–69
Eva, H., 188–189
Evelo, C. T., 68–69, 71, 72, 75, 77*f*, 78–79, 89, 94, 102, 106–107, 108
Evenson, J. K., 88–89, 106–107, 108
Everaerts, F. M., 47–48
Evrard-Todeschi, N., 77*f*

F

Facciuolo, T., 208, 217
Factor, V. M., 79–80
Fang, J., 56, 116
Fang, W., 74–75
Fankhauser, C., 207
Faulkner, M. J., 130
Federico, R., 278
Feinstein, S. I., 148, 149
Felix, K., 79–80
Ferhan, A., 188–189
Fernandes, A. P., 104
Fernandez, A., 76–78, 138
Fernandez, M., 44–45
Fernandez-Checa, J. C., 138
Fernandez-Fernandez, S., 130, 131–132, 133, 137–138
Fernandez-Solari, J., 176
Ferrández, J., 262–264
Ferrer-Sueta, G., 44–45, 45*t*, 46*t*, 50*t*, 51–52, 78–79, 114–115
Ferwerda, M., 206–207, 217
Feuerbach, F., 190–191

Finley, J. W., 70–71
Fischer, S., 108
Fisher, A. B., 147, 155–156
Fisk, H. A., 242
Flohé, L., 4–5, 6–8, 45*t*, 46*t*, 51–52, 57–58, 66, 67, 68, 69–70, 78, 79–80, 88–89, 114–115, 130, 138
Florencio, F. J., 114–115, 258–259, 260–264, 265–266, 267, 268, 269, 270
Florian, S., 67, 70–71, 72, 78, 91
Fluhr, R., 277, 278
Folmer, V., 6*t*
Fomenko, D. E., 52, 71, 79–80
Foreman, J., 276, 279
Forman, H. J., 78–79
Forrester, M. T., 116
Förster, B., 186–187
Forster, H., 78–79
Forstrom, J. W., 6–8, 14
Foster, M. W., 56, 116
Foster, S. J., 105
Fourquet, S., 77*f*, 78, 114–115, 116, 118*f*
Fradkov, A. F., 123–125, 130–131, 189–190, 194
Frame, S., 76
Francis, F. J., 217
Franklin-Tong, V. E., 276
Fratelli, M., 56
Freeman, B. A., 141
Freeman, M. L., 69–70, 76, 77*f*
Fricker, M. D., 190
Fridovich, I., 141
Friling, R. S., 70–71
Fritz, L. C., 242
Frommer, W. B., 189–191
Frost, D. V., 67, 88
Frye, C. A., 206
Fryer, M. J., 187, 280
Fu, C., 116
Fuentealba, L. C., 73–74
Fuglsang, A. T., 189–190
Fujii, M., 53–54
Fujimoto, M., 249–253
Fujisawa, S., 241, 245, 247, 248–253
Fujiwara, M., 277–278
Fujiwara, T., 235
Fuleki, T., 217
Funato, Y., 52–53, 55–56, 67, 74, 77*f*, 79–80
Funk, R., 176
Furtmüller, P. G., 258–259
Furuta, S., 217–218

G

Gadalla, M. M., 123
Gadjev, I., 204–207
Gaillard, J., 248–249
Gakière, B., 186–187, 188–189
Galbraith, J. D., 141
Galvez-Valdivieso, G., 186–188
Ganther, H. E., 66, 91, 104, 105
Gao, Y., 54
Gapper, C., 223*t*
Garcia, F. J., 50–51
Garcia-Ruiz, C., 138
Gasnier, F., 118
Gates, K. S., 44–45, 45*t*
Gawlik, D., 104–105
Gechev, T. S., 204–207, 211, 214, 217
Gegg, M. E., 137–138
Geisler, M., 223*t*
Geldner, N., 281–282
Gelhaye, E., 43
Georgiou, G., 130
Gerard, M., 116
Gerats, T., 209–210
Gerber, C. A., 107
Gerlach, J. P., 74
Gerstmeier, S., 79–80
Gessner, H., 88–89
Gibson, B. E., 151–152
Gibson, B. W., 44–45
Gill, P. S., 44
Gill, V., 240
Gillingwater, T. H., 173–174
Gilon, P., 130–131
Gilroy, S., 189–190, 279, 280–281
Gjetting, K. S., 189–190
Gladwin, M. T., 46–47, 120–122, 123
Gladyshev, V. N., 68–69, 71, 88, 189–190
Glazebrook, J., 279
Glockshuber, R., 43
Glowinski, J., 130–131, 140–141
Godon, C., 190–191
Godzik, A., 26

Göken, E. M., 67, 68–69, 78–79
Gon, S., 130
González, M., 262–264
Gonzalez-Aller, C., 148, 149, 157, 157*np*, 159*np*
Gordan, J. D., 69–70
Gorokhovatsky, A. Y., 189–190, 194
Gough, D. R., 42
Grabher, C., 123–125, 190
Grant, C. M., 130
Greco, T. M., 45*t*
Green, E. W., 23–24, 115–116
Greenberg, J. T., 279
Gretes, M. C., 24–25
Griffith, O. W., 137
Gromadzinska, J., 108
Gromer, S., 92
Gross, A., 242
Gross, J. H., 92
Gruissem, W., 205, 217
Grundel, S., 6–8
Gruschus, J. M., 45*t*
Gu, D., 277–278
Gu, Z., 56, 116
Gualberto, J. M., 43
Gucciardo, S., 279
Guerois, R., 77*f*, 78
Guillon, B., 114–115, 116, 118*f*
Guinea, M., 258–259, 262–264
Gumper, I., 73–74
Günzler, W. A., 46*t*, 51–52, 66
Guo, S., 138
Gupta, V., 50–51
Gutscher, M., 55
Gutterer, J. M., 138
Gutteridge, J. M. C., 188–189
Gwosdow, A. R., 176

H

Hafeman, D. G., 66
Hager, J., 188–189
Hah, Y. S., 55–56
Hales, D. B., 177
Hales, K. H., 177
Halestrap, A. P., 190
Haley, B., 146–147
Hall, A., 22–23, 45*t*, 51–52, 114–115
Halliwell, B., 188–189
Hamprecht, B., 130–131, 137, 139–141
Hampton, M. B., 45, 51–52, 57–58, 114–115
Han, A., 74–75
Han, C. H., 159–160
Han, S. H., 23
Hancock, J. T., 223*t*
Handy, D. E., 141
Hannink, M., 69–70
Hannon, G. J., 153–154
Hansen, H. H., 116
Hantgan, R. R., 55–56
Hanukoglu, I., 130–131, 140–141, 171–172
Hara, T., 224–226
Harada, Y., 74
Harborth, J., 153–154
Hardingham, G. E., 173–174
Harper, J. F., 190–191
Harper, J. W., 69–70
Harris, J., 6–8
Hart, L. J., 265–266
Hashimoto, K., 189–190
Hatano, T., 104
Hatfield, D. L., 68–69, 88, 189–190
Havaux, M., 267
Hayashi, H., 187–188
Hayes, J. D., 69–70, 76–78, 77*f*
Haynes, A. C., 23
Hazlett, K. R., 33–34
He, S., 74–75
Heales, S. J., 137–138
Hebert, C., 104, 130, 137–138
Heinemeyer, G., 131, 133
Heinstein, P. F., 279
Heinzel, B., 140, 141
Held, J. M., 44–45, 50–51
Held, K., 189–190
Heller, M., 118
Hellström, E. K., 213
Hennigan, B. B., 148, 149
Herdman, M., 264–265
Herman, B., 242
Hernalsteens, J. P., 283–284
Hernandez, F., 78
Hervás, M., 262–264
Hesketh, J. E., 88–89
Hess, D. T., 56
Heupel, A. L., 278
Hewett, S. J., 114

Hibbs, J. B. Jr., 116–117
Hiester, A., 151–152, 155*f*, 156*f*
Higgins, L. A., 56–57
Higgins, L. G., 116, 123
Hildebrandt, A. G., 131, 133
Hill, K. E., 66, 88–89, 104–105, 106–107
Hille, J., 204–207, 217
Hillier, W., 187–188
Hilmert, H., 88–89
Himber, C., 190–191
Himeno, S., 70–71
Hincha, 232*t*
Hinks, J. A., 50
Hintze, K. J., 70–71
Hirabayashi, T., 241, 245, 247, 248–253
Hirai, M. Y., 235
Hirano, T., 243
Hirasawa, M., 43
Hirata, A., 241, 242, 253
Hirrlinger, J., 138
Hirt, H., 204–205, 278
Ho, S. M., 116
Hoeberichts, F. A., 186–187
Hoefig, C. S., 105–106
Hoekstra, W. G., 66
Hoffman, J. L., 104–105
Hoffman, N. E., 191–192
Hofte, H., 283–284
Hogg, N., 120–122, 123
Holmgren, A., 53–54, 104
Homann, A., 50–51
Hommel, B., 108
Homoyounpour, N., 153–154, 159–160
Horwich, A. L., 138
Howard, O. M., 79–80
Hu, H. C., 277–278
Hu, M. C., 23, 171, 172–174, 172*f*, 177–179, 178*f*
Hu, X., 54
Huang, B. W., 53–54
Huang, F., 74–75
Huang, H., 74
Huang, J., 146–147
Huang, X. D., 108
Huang, Y. L., 72–73
Hughes, S. M., 54
Hugo, M., 45*t*, 46*t*, 50*t*, 51–52
Hundertmark, 232*t*
Hundley, N. J., 141
Hunt, A. G., 208, 217
Hunt, M. D., 222
Hunter, K. J., 265–266
Huttenlocher, A., 123–125
Huycke, M. M., 68–69
Huynh, C., 123–125
Hwang, S. C., 22–23, 27, 55, 118, 170–171

I

Iadecola, C., 123–125
Iglesias-Baena, I., 267
Ihara-Ohori, Y., 249–253
Imai, H., 249–253
Imlay, J. A., 6*t*
Immenschuh, S., 114–115
Imura, N., 70–71
Inada, N., 249–253
Ince, G., 79–80
Infante, H. G., 89
Ings, J., 276
Innamorato, N. G., 77*f*
Innes, R. W., 206
Inzé, D., 205–207, 211, 214
Ip, C., 104
Irmak, M. B., 79–80
Isaac, J., 116
Ishii, T., 66–67
Ishikawa, T., 241, 245, 247, 248–253
Ishiwata, K., 105
Issakidis-Bourguet, E., 186–187
Ito, M., 240, 241, 247, 253
Itoh, K., 66–67, 69–70
Ivanov, B. N., 187–188
Iwata, H., 66

J

Jablonska, E., 108
Jabs, T., 240
Jacob, S., 258–259, 268
Jacobasch, G., 72
Jacobsen, M., 52
Jaeger, T., 45*t*
Jain, A. K., 76, 77*f*
Jain, M. K., 148, 149
Jain, M. R., 116
Jaiswal, A. K., 70*f*, 76, 77*f*
Jander, G., 209–210

Jang, H. H., 55–57, 114–115
Janicki, A., 115–116, 267
Janssen, L. J. J., 47–48
Jarvis, R. M., 54
Jat, P. S., 90
Jaworski, T., 77*f*
Jedlicka, A. E., 70–71
Jefcoate, C. R., 171–172, 177
Jeffery, E. H., 70–71
Jemmerson, R., 242
Jendrach, M., 240–241
Jeon, H. S., 56–57
Jeong, W., 23, 54, 55, 56–57, 115–116, 130–131, 132, 139, 140–141, 147–148, 170–171, 267
Jesaitis, A. J., 146–147
Jezierska-Drutel, A., 54
Jia, X., 67
Jiang, C. Z., 206
Jiang, H., 123
Jiang, K., 190
Jiang, Q., 74–75
Jin, D. Y., 8
Jin, J., 69–70
Jin, L. L., 44–45, 241, 242
Jo, H. Y., 163–165
Jo, Y., 176–177
John, C. F., 222
John, M., 140, 141
Johnson, F. A., 45*t*
Johnson, L. C., 50–51, 50*t*, 56
Johnston, R. B. Jr., 146
Jonas, J. C., 130–131
Jones, J. D. G., 204–205
Jones, M. A., 276
Jonsson, T. J., 51–52, 56, 115–116
Jordan, B., 222
Jordy, M. N., 43
Joseph, J. L., 90
Jun Jiang, Z., 76–78
Jung, B. G., 55–56, 114–115
Jung, H. W., 279
Jung, J. H., 56–57
Jürgensmeier, J. M., 242

K

Kajihara, Y., 44–45
Kajla, S., 74, 77*f*
Kalousek, F., 138
Kalwa, H., 123–125
Kanematsu, S., 278
Kang, D., 46–47, 48, 56–57, 130–131, 132, 139, 140–141, 187–188
Kang, M.-I., 69–70
Kang, S. W., 22–23, 27, 52, 55, 56–57, 118, 147–148, 150, 155–156, 157, 170–171, 267
Kangasjärvi, J., 205
Kanski, J., 138
Karai, N., 102–104
Karisch, R., 44–45
Karpinska, B., 186–187, 222, 240
Karpinski, S., 186–188, 222, 240
Karplus, P. A., 22–23, 24–26, 27, 33–34, 45*t*, 51–52, 55–56, 114–115, 259, 267
Karthaus, W. R., 74
Kato, G. J., 46–47
Kato, M., 74, 77*f*
Kawai, M., 241
Kawai-Yamada, M., 240–241, 242, 243, 244*f*, 245, 246–247, 249–253
Kawakita, K., 277–278
Kawazu, T., 229–230
Kay, S. A., 214
Kaya, A., 52
Kayo, T., 108
Keller, T., 277–278
Kelley, E. E., 141
Kelly, N., 141
Keng, Y. F., 45*t*, 46*t*
Kensler, T. W., 69–71, 75–76
Kettenhofen, N. J., 120–122, 123
Ketterer, B., 6–8
Khan, S. A., 176–177
Khoo, N. K., 141
Kiba, A., 240, 241, 247, 253
Kil, I. S., 23, 48, 114–115, 171, 172–174, 172*f*, 177–179, 178*f*
Kim, B. N., 116
Kim, C. N., 186–187, 242
Kim, G., 45*t*
Kim, H. J., 52, 54–55
Kim, H. K., 267
Kim, H. R., 116
Kim, J. M., 116
Kim, K. S., 22–23, 27, 55, 57, 118, 147–148, 155–156, 170–171

Kim, M. H., 163–165
Kim, S. Y., 56–57, 163–165, 189–190
Kim, T. S., 148, 149
Kim, W. Y., 55–56
Kim, Y. J., 54
Kim, Y. M., 23
Kimura, M., 226
King, S. R., 176–177
Kipp, A. P., 67, 68–69, 70–71, 72, 75, 77*f*, 78–80, 89, 91, 94, 102, 106–107, 108
Kirchsteiger, K., 258–259, 260–264, 270, 271
Kirsten, E., 9
Kistner-Griffin, E., 54
Klatt, P., 140, 141
Kleinberg, M. E., 146–147, 148, 149, 157, 157*np*, 159*np*
Klenell, M., 186–187
Klomsiri, C., 50–51, 50*t*
Klusener, B., 276
Kluth, D., 66–67, 70–71, 93
Knall, C., 151–152
Knaus, U. G., 159–160
Knight, A., 54, 55
Knight, J. A., 132
Knoops, B., 50–51
Kobayashi, A., 69–70
Kobayashi, H., 243
Kobayashi, M., 277–278
Kobayashi, T., 146–147
Kobayashi, Y., 105
Koc, A., 52
Köhrle, J., 105–106
Kolb-Bachofen, V., 6–8
Kollmus, H., 68, 88–89
Komeda, Y., 235
Koonin, E. V., 242
Koppenol, W. H., 43
Korsmeyer, S. J., 240–241, 242
Korswagen, H. C., 74, 76–78
Kotrebai, M., 89
Koussevitzky, S., 186–187
Koyama, H., 224–226, 229–230
Krajewski, S., 242
Kranich, O., 137
Kraus, R. J., 105
Krebs, M., 189–190
Krehl, S., 67, 70–71, 78, 91
Kreuzer, O. J., 72
Kriedemann, P. E., 212–213
Krieger-Liszkay, A., 187–188
Kristiansen, K. A., 47–48
Kryukov, G. V., 88
Kumar, C., 208, 217
Kumar, M. S., 176
Kumar, S., 104
Kun, E., 9
Kundu, J. K., 66
Kunimoto, M., 70–71
Kurasaki, K., 104–105
Kurihara, T., 104–105
Kurkchubasche, A., 151–152
Kurokawa, S., 104–105
Kussmaul, L., 130–131, 139–141
Kwak, J. M., 190–191
Kwon, J., 54
Kwon, K. S., 55–56
Kyriakopoulos, A., 79–80

L

Labarre, J., 23, 55, 66–67, 115–116, 170–171, 267
Labunskyy, V. M., 79–80
LaButti, J. N., 45*t*
Lacomme, C., 242
Lager, I., 190–191
Lai, A. G., 214
Lal, A. K., 133
Lalli, E., 174
Lally, E., 190
Laloi, C., 204–207, 211, 217
Lalonde, S., 190–191
Lam, E., 241, 249–253
Lam, Y. W., 116
Lamas, S., 116, 123
Lamb, C. J., 240, 245, 277–278
Lambert, A., 279
Lambeth, J. D., 42, 159–160
Lancelin, J. M., 50–51
Landry, J., 54
Langa, E., 78
Langerbartels, C., 205, 217, 231–232
Lardy, B., 42
Larrieu, A., 189–190
Larsson, A., 130, 137–138
Lassegue, B., 42

Last, R. L., 209–210
Latifi, A., 115–116, 267
Lavigne, M. C., 153–154, 159–160
Lawrence, R. A., 66
Lawson, T., 187
Laxa, M., 258–259, 268
Lázaro, J.-J., 267
Lázaro-Payo, A., 267
Le, T., 173–174
Leake, D. S., 66–67
Leaver, C. J., 248–249
Leavey, P. J., 148, 149, 151–152, 155*f*, 156*f*, 157, 157*np*, 159*np*
Lederer, W. J., 123–125
Ledgerwood, E. C., 54
Lee, B. J., 67
Lee, B. S., 116
Lee, C. H., 54–55
Lee, D. Y., 57
Lee, G. T., 52
Lee, H. E., 23, 116, 175
Lee, J., 171–172
Lee, J. I., 104–105
Lee, J. R., 55–56
Lee, J. S., 46–47
Lee, K. O., 55–56, 114–115
Lee, K.-P., 186–187
Lee, S., 277–278
Lee, S. K., 23, 116, 171, 172–174, 172*f*, 175, 177–179, 178*f*
Lee, S. M., 52, 56–57, 170–171
Lee, S. R., 54
Lee, Y. M., 56–57, 281–282
Leemans, J., 283–284
Leers-Sucheta, S., 174
Lehmann, R., 153–154
Lei, K., 123
Lei, X. G., 88–89
Leidal, K. G., 146
Leist, M., 91, 104–105
Lekstrom, K., 153–154, 159–160
Lendeckel, W., 153–154
Leonard, S. E., 50–51
Leslie, N. R., 54, 55
Leto, T. L., 148, 152, 153–154, 158*f*, 159–160, 161*f*, 163–165
Leung, W., 138
Levander, X. A., 66
Leveille, F., 116, 123
Levin, I. M., 209–210
Levine, A., 240–241, 248
Levine, R. L., 45*t*
Lewis, S. D., 45*t*
Ley, H. G., 150
Li, F., 74–75
Li, H.-Y., 222
Li, J. G., 108
Li, J. L., 108
Li, N., 67
Li, P., 106–107
Li, R. L., 277–278
Li, S., 108
Li, V. S., 74
Li, X. J., 146–147
Liang, H., 222
Liebler, D. C., 69–70, 76, 77*f*
Lim, J. C., 45*t*, 55–56, 57
Lin, C. S., 175
Linard, D., 50–51
Lindahl, M., 114–115, 258–259, 260–264, 265–266, 267, 268, 269, 270
Linsmaier, E. M., 246–247
Linton, G., 153–154, 159–160
Lippard, S. J., 123–125
Lippmann, D., 79–80
Lipton, S. A., 56, 116
Liu, M., 74–75
Liu, P., 277–278
Liu, T., 116
Liu, X., 242
Liu, Y., 54
Liu, Y. G., 209–210
Liu, Y. I., 74
Llopis, J., 190–191
Lo, E. H., 123–125
Lo, S.-C., 69–70
Locy, M. L., 78
Loewinger, M., 67, 70–71, 78, 91
Logan, C. Y., 67, 79–80
Logan, D. C., 190
Lohmann, 232*t*
Lok, J., 138
Longmate, J. A., 68–69
Look, A. T., 123–125, 190
Lopes, S. C. D. N., 6*t*
Lopez, M. G., 76

López-Jaramillo, F. J., 267
Loschen, G., 46t, 51–52
Lötzer, K., 91, 104–105
Low, F. M., 51–52, 114–115
Low, P. S., 279
Low, T. Y., 74
Lowther, W. T., 23, 50–52, 50t, 56, 115–116
Lu, C. C., 175
Lu, J., 53–54
Lubos, E., 141
Lucas, O., 78
Luche, S., 118
Lukyanov, K. A., 123–125, 130–131, 189–190, 194
Lukyanov, S., 189–190, 194
Luo, H., 74–75
Lv, D., 187–188
Lyons, P. R., 88–89

M

MacEwan, M. E., 104–105
Machen, T., 190
Machida, K., 123–125
MacIver, F. H., 130
Mackerness, A., 222
Mackinney, G., 265
Macri, F., 278
Maddipati, K. R., 133
Maghzal, G. J., 51–52, 114–115
Mahajan, N., 242
Mahmoudi, T., 74, 76–78
Mai, S., 240–241
Maiorino, M., 78–79, 130, 138
Malbon, C. C., 74
Malech, H. L., 151–152, 153–154, 159–160
Malik, K. A., 243
Malik, U. Z., 141
Malinouski, M., 52, 189–190
Mallonee, D. H., 108
Manabe, K., 221–238
Manabe, T., 226
Manevich, Y., 54, 123
Mankad, M. V., 52
Mankad, V. N., 52
Manna, P. R., 171–172, 177
Mannervik, B., 133
Manta, B., 44–45, 45t, 46t, 50t, 51–52, 78–79
Mari, M., 138
Marinho, H. S., 4–5, 6t, 16–17, 47–48
Marino, D., 279
Markvicheva, K. N., 189–190, 194
Marnett, L. J., 133
Marquez, R., 74, 76–78
Marra, E., 240
Marshall, O. J., 93–94
Martel, M. A., 116
Martinez-Ruiz, A., 116, 123
Maruta, T., 187–188, 223t, 232t
Masukawa, T., 66
Masumizu, T., 217–218
Mata-Cabana, A., 114–115, 258–259, 260–264, 265–266, 267, 268, 269, 270
Mateo, A., 186–187
Matias, A. C., 6t
Matsuda, Y., 247–248
Matsui, M., 226
Matsumura, H., 240, 241, 245, 247, 248–253
Matsumura, T., 44–45
Matsuno, K., 74, 77f
Matute, J. D., 146–147
Maurer, S., 91, 104–105
Mavis, R. D., 133
Mayer, B., 140, 141
McAlister-Henn, L., 8
McCafferty, D. G., 44–45
McCaskill, A. J., 191–192
McConnell, K. P., 104–105
McCord, J. M., 141
McCormick, F., 74, 76–78
McDonnell, J. M., 242
McMahon, M., 69–70, 76–78, 77f
McManus, E. J., 74, 76–78
McPhail, L. C., 146–147, 159–160
Medina, D., 67
Mehterov, N., 206–207, 217
Meister, A., 6–8, 137
Meller, J., 116
Mellman, I., 138
Melner, M. H., 174
Melvin, J. A., 44–45
Mendez-Sanchez, N., 123–125
Meng, T. C., 50

Menge, U., 45*t*
Méplan, C., 68–69, 71, 72, 75, 77*f*, 78–79, 89, 94, 102, 106–107, 108
Merrill, G. F., 78
Meskauskiene, R., 186–187
Messerli, M. A., 189–190, 279, 280–281
Metcalfe, C., 72
Metodiewa, D., 45*t*, 46*t*
Meyer, A. J., 55, 189–190
Meyer, J., 26, 38
Meyerhof, W., 72
Miao, C., 187–188
Miao, Y. C., 187–188
Michel, T., 123–125
Michiue, T., 67, 74, 77*f*
Middleton, L. M., 107
Miedema, H., 276, 279
Mieyal, J. J., 56
Mihara, H., 104–105
Miki, H., 52–53, 55–56, 67, 74, 77*f*, 79–80
Miles, P. R., 140–141
Millar, A. H., 248–249
Miller, A. G., 265–266
Miller, G., 186–188, 278, 279
Miller, W. L., 171–172
Milliman, C. L., 241
Mills, G. C., 133
Min Hua, T., 76–78
Minamino, T., 76–78
Minard, K. I., 8
Miniard, A. C., 107
Minkov, I. N., 205–207, 214
Mirabet, V., 189–190
Miranda, K. M., 114
Mishina, N. M., 189–190, 194
Mitchison, T. J., 123–125, 190
Mitsuhara, I., 243
Mitsukawa, N., 209–210
Mittler, R., 186–188, 222, 276
Miura, M., 243
Miura, Y., 279
Miyawaki, A., 189–190
Mizel, D., 140–141
Moffatt, B. A., 208, 217
Mohn, C. E., 176
Moinova, H. R., 66–67
Moissiard, G., 190–191
Moles, J. P., 76–78
Molina, A., 222
Moller, A. L., 47–48
Møller, I. M., 47–48, 242
Mondol, A. S., 74, 77*f*
Monshausen, G. B., 189–190, 279, 280–281
Monteiro, G., 45*t*, 50*t*, 51–52
Montiel, J., 279
Moon, J. C., 55–56
Moore, R. B., 52
Morales, A., 138
Morel, J. B., 190–191
Moreno, J., 258–259, 260–264, 270, 271
Morgan, B. A., 42, 55, 123–125
Mori, I. C., 276–277
Morinaka, A., 52–53, 55–56
Morsa, S., 205
Moseley, M. A., 44–45, 116
Moskowitz, M. A., 123–125
Mostert, V., 66
Motley, A. K., 66, 104–105, 106–107
Motohashi, H., 69–70
Mourrain, P., 190–191
Moustafa, M. E., 71
Mubarakshina, M. M., 187–188
Mueller-Roeber, B., 206–207, 214, 217
Mühlenbock, P., 186–187
Mukhopadhyay, P., 189–190
Mulcahy, R. T., 66–67
Mullen, M. L., 146–147
Müller, C., 91
Müller, M. F., 67, 68–69, 71, 78–80
Mullineaux, P. M., 186–188, 222, 240, 280
Murad, E., 42
Murata, Y., 276
Murphy, C. F., 44–45
Murphy, M. P., 190
Murray, L. M., 173–174
Murray, M. S., 56
Mustafa, A. K., 123
Myers, M. P., 50
Mylona, P., 276, 279

N

Na, H. K., 66
Nadeau, P. J., 54
Nagahara, N., 44–45
Nagai, T., 76–78, 189–190
Nagano, M., 241, 249–253

Nagasawa, A., 74, 77*f*
Nagata, T., 217–218
Naito, A. T., 76–78
Naito, S., 235
Nakajima, H., 245
Nakakita, J., 146–147
Nakamura, T., 56, 102–104, 116, 146–147
Nakayama, A., 66
Nakayashiki, H., 240–241
Nam, H. W., 55–56, 57
Naranjo-Suarez, S., 67
Narayanan, B. A., 74–75
Narayanan, N. K., 74–75
Narvaez-Vasquez, J., 279
Nauseef, W. M., 146
Navarro, J. A., 262–264
Naydov, I. A., 187–188
Nayeem, A., 26
Neill, S. J., 223*t*
Nelson, K. J., 24–25, 26, 33–34, 45*t*, 50–51, 50*t*
Netto, L. E. S., 45*t*, 50*t*, 51–52, 155–156
Neubert, H., 50–51
Neuburger, M., 248–249
Neuenschwander, U. H., 222
New, L. S., 137–138
Ng, S. S., 74, 76–78
Nicholls, P., 4–5
Niehaus, K., 277–278
Niethammer, P., 123–125, 190
Nirasawa, S., 240, 241, 247, 253
Nishimoto, M., 70–71
Nishimura, T., 66
Nishitoh, H., 53–54
Nishizawa, A., 223*t*, 232*t*
Nishizawa-Yokoi, A., 187–188
Niwa, Y., 240–241, 242, 243, 244*f*, 245, 253
Njalsson, R., 130, 137–138
Noble, M. D., 90
Noctor, G., 188–189
Nogueira, J. M., 6*t*
Nogueira, L., 116
Noh, Y. H., 23
Nordeng, T. W., 258–259
Norgren, S., 130, 137–138
Noritake, T., 279
Noronha-Dutra, A. A., 114
Norris, S. R., 209–210
Norsworthy, B. K., 104–105
Nosaka, R., 187–188
Novoselov, S. V., 71, 79–80
Nugent, J. H., 146–147
Nunes de Miranda, S. M., 258–259, 268
Nusse, R., 67, 79–80

O

Obata, T., 214
Obinger, O., 258–259
Oelze, M.-L., 258–259, 268
Ogawa, K., 278
Ogawa, S., 104
Oger, E., 279
Ogra, Y., 104, 105
Ogusucu, R., 45*t*, 51–52
Oh, G. S., 116
Ohashi, Y., 243
Ohmichi, M., 104
Ohori, Y., 241, 243, 246–247, 249–253
Ohta, Y., 105
Ohura, I., 277–278
Oka, S., 116
Okamoto, R., 44–45
O'Kane, D., 240
Okuda, T., 247–248
Okumoto, S., 189–191
O'Leary, S. J. B., 265–266
Oliva, M., 189–190
Oliveira-Marques, V., 6*t*
Oliver, D. J., 211
Olm, E., 104
Olmedo, M., 23–24, 115–116
Oltersdorf, T., 242
Oltvai, Z. N., 241
O'Neill, J. S., 115–116
Onodera, K., 69–70
Oosumi, T., 209–210
Orozco-Cardenas, M. L., 222, 279
Ortiz, C., 45*t*, 46*t*, 51–52
Osna, N. A., 123–125
Oxborough, K., 187
Ozturk, M., 79–80

P

Paddison, P. J., 153–154
Pae, H. O., 116
Palais, G., 116

Pallast, S., 138
Pallauf, J., 108
Pan, L., 241
Panepinto, J., 151–152
Papadia, S., 116, 123
Park, E. S., 189–190
Park, H. J., 279
Park, J. H., 55–56, 114–115
Park, J. W., 56
Park, K. J., 55
Park, P., 240–241
Park, S., 170–171
Park, S. J., 55, 56–57
Park, S. K., 55–57, 114–115
Park, Y. S., 55–56
Parkhill, L. K., 66
Parkos, C. A., 146
Parmigiani, R. B., 57
Parniske, M., 189–190
Parsonage, D., 26, 33–34, 45*t*, 51–52
Parsons, Z. D., 44–45
Pascual, M. B., 114–115, 258–259, 260–264, 265–266, 267, 268, 269, 270
Passarella, S., 240
Patino, P. J., 146–147
Paton, L. N., 51–52, 114–115
Paul, B. D., 123
Paulsen, C. E., 50–51
Pearson, A. G., 51–52
Pedroso, N., 6*t*
Pei, Z. M., 276
Pellinen, R., 205
Peng, C. L., 217–218
Peng, X. X., 217–218
Peragine, A., 190–191
Pérez-Pérez, M. E., 258–259, 260–262
Pérez-Ruiz, J. M., 258–259, 260–264, 270, 271
Perier, C., 45*t*
Perkins, A., 24–25
Peschek, G. A., 258–259
Peskin, A. V., 51–52, 114–115
Peters, J. L., 209–210
Peterson, G. L., 14
Peterson, V. M., 151–152
Pezzi, V., 174
Pflieger, D., 52, 54–55
Pham, J. A., 245
Pick, E., 140–141, 146–147
Pickett, C. B., 70–71
Pinto, J. T., 104–105
Piper, W. N., 9
Piszczek, G., 56, 57
Pittman, B., 74–75
Pizzo, P., 186–188, 190, 191, 281
Planson, A. G., 116, 124*f*
Plaszczyca, M., 186–187
Plishker, G. A., 52
Plouhinec, J. L., 73–74
Pluta, R. M., 114
Pluth, M. D., 123–125
Poethig, R. S., 190–191
Pogson, B. J., 186–187, 223*t*
Pohl, S., 240–241
Polavarapu, R., 42
Pons, J., 50–51
Poole, L. B., 22–23, 24–26, 27, 31*f*, 33–34, 38, 45*t*, 46–47, 51–52, 55–56, 114–115, 147–148, 170–171, 259, 267
Pope, A. L., 66
Porra, R. J., 212–213
Porse, B., 116–117
Potocky, M., 276
Power, R. F., 108
Powrie, F., 6–8
Premont, J., 130–131, 140–141
Prestifilippo, J. P., 176
Price, M. O., 159–160
Prigge, J. R., 78
Prosser, B. L., 123–125
Puckett, R. L., 50–51
Puerto-Galán, L., 262–264
Pugh, T. D., 108
Pulido, P., 258–259, 262–264
Pullar, J. M., 51–52
Puri, R. N., 6–8
Puype, M., 56

Q

Qin, X., 23–24, 115–116
Quan, L. J., 222
Queval, G., 186–187, 188–189
Quinn, M. T., 146–147
Quintana-Cabrera, R., 130, 131–132, 133, 137–138

Quinto, C., 276, 279, 280–281
Qureshi, K., 206–207, 217

R

Rabilloud, T., 118
Rachakonda, G., 69–70, 76, 77*f*
Rada, P., 76–78, 77*f*
Radi, R., 44–45, 51–52, 78–79, 114–115
Ragan, C. I., 245
Raines, A. M., 88–89, 108
Rainey, W. E., 174
Rajashekar, K., 245
Ramadori, G., 114–115
Rand, J. D., 55
Rao, C. V., 74–75
Rao, M. V., 71, 88, 240, 245
Rao, S., 208, 217
Rapoport, R., 130–131, 140–141, 171–172
Rasband, 195
Ray, P. D., 53–54
Rayman, M. P., 88, 89
Reddy, B. S., 74–75
Reddy, S. P., 70–71
Reed, J. C., 241, 242, 249–253
Reeves, S. A., 26
Reilly, T. J., 45*t*
Reiter, R., 66
Remington, S. J., 130–131
Rempinski, D. R. Jr., 74–75
Renko, K., 105–106
Reszka, E., 108
Rey, C., 118
Rey, P., 267
Reynaert, A., 283–284
Reynolds, C. M., 26, 38
Reynolds, H., 186–187, 222
Rhee, S. G., 22–23, 46–47, 48, 52, 54, 56–57, 114–116, 118, 130–131, 132, 139, 140–141, 147–148, 150, 155–156, 157, 170–171, 187–188, 267
Richard, O., 223*t*
Ridgway, D., 47–48
Rihua, W., 108
Rinckel, L., 148, 149, 157, 157*np*, 159*np*
Rippka, R., 264–265
Ristoff, E., 130, 137–138
Ritskes-Hoitinga, M., 68–69
Ritzenthaler, C., 190–191
Rizhsky, L., 222
Robin Harris, J., 46–47, 170–171
Robison, K., 147–148
Rochat, A., 76–78
Rodermel, S. R., 282
Rodriguez, J. G., 74–75
Rogers, L. K., 78
Rojo, A. I., 76–78, 77*f*
Rollins, M. F., 78
Roma, L. P., 130–131
Rombauts, S., 205, 217, 231–232
Ronaldson, L., 74, 76–78
Roos, D., 146, 151–152
Roots, I., 131, 133
Rooyackers, O., 130, 137–138
Roppolo, D., 281–282
Rosa, A. O., 76
Rosenberg, L. E., 138
Rosenthal, J., 151–152
Roth, S., 78–79
Rotruck, J. T., 66
Rouhier, N., 43
Rounsley, S. D., 209–210
Rubio, M. C., 281–282
Rudolph, J., 50*t*
Ruiz, E., 66–67
Rumeau, D., 267
Rushmore, T. H., 70–71
Rustum, Y. M., 74–75
Ruzin, S., 190
Ryals, J. A., 222
Ryan, C. A., 222, 279
Ryu, K. W., 23, 171, 172–174, 172*f*, 177–179, 178*f*

S

Sagi, M., 278
Sagisaka, S., 247–248
Sahrawy, M., 258–259, 260–264, 270, 271
Saifo, M. S., 74–75
Saitoh, H., 240, 241, 247, 253
Saitoh, M., 53–54
Sakamoto, K., 74, 76–78
Sakamoto, M., 240–241
Sakurai, A., 70–71
Salazar, M., 76
Salmeen, A., 50
Salmona, M., 56

Samstag, Y., 55
Sanchez, L. M., 279
Sánchez-Riego, A. M., 258–259, 260–262
Sanchez-Valle, V., 123–125
Sandalio, M. L., 262–264
Saraste, M., 130
Sargent, M., 89
Sarma, G. N., 33–34
Sartoretto, J. L., 123–125
Sassone-Corsi, P., 174
Sastre, J., 130, 131–132, 133, 137–138
Satomi, S., 69–70
Saurin, A. T., 50–51
Sawada, Y., 53–54
Sawano, A., 189–190
Scheid, S., 88–89
Schimmer, B. P., 173–174
Schippers, J. H., 214
Schjoerring, J. K., 47–48
Schlauch, K., 278, 279
Schmehl, K., 72
Schmidt, E. E., 78
Schmitmeier, S., 91
Schmitt, J. D., 26
Schneider, C. J., 88
Schneider, M., 78–79
Schock, H. H., 66
Scholz, R., 150
Schomburg, L., 68–69, 71, 72, 75, 77*f*, 89, 94, 102, 105–106, 108
Schopfer, P., 278
Schrauzer, G. N., 88
Schröder, E., 46–47, 170–171
Schroder, K., 42
Schroeder, J. I., 186–188, 190, 191, 276–277, 281
Schulte, J., 54, 55
Schultz, M., 91, 104–105
Schulz, A., 47–48, 189–190
Schwarzer, C., 190
Schwarzländer, M., 190
Schweizer, U., 71
Seaver, L. C., 6*t*
Segal, A. W., 146–147
Seguin, C., 116, 124*f*
Seidlitz, H. K., 205, 231–232
Seiler, A., 78–79
Sekhar, K. R., 69–70, 76, 77*f*
Sekine, Y., 221–238
Sellins, K. S., 151–152, 155*f*, 156*f*
Sen, N., 123
Sen, T., 123
Sengupta, A., 71
Seo, J. H., 57
Seo, M. S., 155–156
Serrato, A. J., 262–264
Sevilla, F., 267
Shafer, J. A., 45*t*
Shakhbazov, K. S., 123–125, 130–131, 189–190, 194
Shamberger, R. J., 67, 88
Sharp, P. A., 153–154
Shen, W. J., 174
Shepherd, F., 67
Shi, C., 279, 280–281
Shi, J., 42
Shi, K., 74–75
Shi, W.-W., 222
Shibata, T., 69–70
Shigeoka, S., 223*t*, 232*t*
Shim, J. H., 163–165
Shimada, Y., 221–238
Shimamoto, K., 277–278
Shimizu, M., 243
Shimizu, S., 242
Shin, D. H., 46–47, 48, 56–57, 187–188
Shin, S., 75–76
Shinshi, H., 222
Shirasu, K., 245
Shirley, P. S., 146–147
Shpiro, N., 74, 76–78
Shrimali, R., 71
Shriver, S. K., 52
Shulaev, V., 211, 278, 279
Sies, H., 4–5, 6–8
Silliman, C. C., 151–152, 155*f*, 156*f*
Simi, B., 74–75
Singh, M., 45*t*
Sinha, R., 104–105
Sirichandra, C., 277–278
Siu, K. L., 8
Sklan, D., 130–131, 140–141, 171–172
Skoko, J. J., 75–76
Skoog, F., 246–247
Slate, E. H., 88
Slattery, K., 187

Slocum, S. L., 75–76
Small, I. D., 186–187
Smeets, A., 45*t*, 46*t*, 50*t*, 51–52
Smirnoff, N., 187, 276
Snyderman, R., 146–147
Snyrychova, I., 188–189
Soboll, S., 6–8
Sobotta, M. C., 55, 123–125
Soda, K., 102–104
Sohn, J., 50*t*
Soito, L., 50–51, 50*t*
Sokolic, R. A., 153–154, 159–160
Somekawa, L., 104
Son, J., 146–147
Sorescu, D., 42
Soriano, F. X., 116, 123, 173–174
Sousa-Lopes, A., 4–5, 6*t*
Souza, J. M., 50*t*
Spallholz, J. E., 67, 104
Spínola, M. C., 258–259, 260–264, 270, 271
Sporn, M. B., 173–174
Srivastava, S. K., 133
Stacy, R. A., 258–259
Stadler, P., 208, 217
Stadtman, E. R., 54, 155–156
Stadtman, T. C., 66–67
Stamer, S. L., 69–70, 76, 77*f*
Stamler, J. S., 56, 116
Stanbury, D. M., 43
Stanier, R. Y., 264–265
Starnes, T. W., 123–125
Staroverov, D. B., 123–125, 130–131, 189–190, 194
Steiner, H., 222
Stellwagen, E., 133
St-Germain, J. R., 44–45
Stocco, D. M., 171–172, 174, 176–177
Stokman, P. M., 146, 151–152
Stone, J. M., 204–206, 211
Stone, J. R., 45*t*
Stoppani, A. O., 245
Stork, L. C., 151–152
Storz, G., 38, 147–148, 189–190
Straube, R., 47–48
Stults, F. H., 6–8, 14
Sturm, M., 108
Su, L. J., 217–218
Suda, I., 217–218
Suh, Y. A., 42
Sujeeth, N., 214
Sukrong, S., 208, 217
Sun, C. Q., 42
Sun, Y., 33–34
Sunde, R. A., 68, 88–89, 106–107, 108
Sung, S. H., 23, 116, 175
Surh, Y. J., 66
Sutherland, C., 76, 77*f*
Suvorova, E. S., 78
Suzuki, K., 222
Suzuki, K. T., 104–105
Suzuki, N., 104–105, 186–187
Suzuki, T., 69–70
Suzuki, Y., 229–230
Swanson, A. B., 66
Swanson, S. J., 190, 280–281
Szabo, G., 123–125
Szechynska-Hebda, M., 186–187

T

Tacken, R. A., 47–48
Tada, Y., 240–241
Taelman, V. F., 73–74
Taguchi, K., 69–70
Tainaka, H., 187–188
Takac, I., 42
Takahara, K., 241, 245, 247, 248–253
Takahashi, H. K., 130–131
Takano, H., 76–78
Takaya, K., 69–70
Takayama, H., 105
Takeda, K., 53–54
Takehashi, M., 104–105
Takemori, H., 171–172
Tam, R., 278, 279
Tamai, K., 74
Tamoi, M., 187–188
Tamura, T., 66–67
Tan, M., 114–115
Tanaka, H., 102–105
Tanaka, S., 104–105
Tanner, J. J., 44–45
Tappel, A. L., 6–8, 14
Tarin, C., 123
Tarpey, M. M., 141
Taulavuori, E., 213
Taulavuori, K., 213

Taylor, P., 44–45
Taylor, S. R., 55
Tello, D., 123
Templeton, D. J., 69–70
Tempst, P., 57
Tenhaken, R., 240
Teodoro, N., 6*t*
Terauchi, R., 241, 245, 247, 248–253
Terskikh, A. V., 123–125, 130–131, 189–190, 194
Tew, K. D., 123
Thévenaz, 195
Thomas, B., 45*t*, 222
Thomine, S., 276
Thompson, H. J., 104
Thompson, J. W., 44–45, 116
Thompson, K. M., 88–89
Thompson, W. A., 212–213
Thomson, L., 114–115
Thrasher, A. J., 146
Thurman, G., 148, 149, 151–152, 155*f*, 156*f*, 157, 157*np*, 159*np*
Thurman, G. W., 148, 149, 151–152, 153, 154, 158*f*, 161*f*, 163–165, 163*f*, 164*f*
Thurman, R. G., 150
Tichauer, Y., 70–71
Tichy, M., 266–267
Tipple, T. E., 78
Tiwari, B. S., 240–241, 248
Tjandra, N., 45*t*
Tjoe, M., 131, 133
Tobe, R., 67
Tobiume, K., 53–54
Toda, T., 224–226
Todoriki, S., 217–218
Tognetti, V., 258–259, 268
Tohge, T., 214
Toledano, M. B., 23, 52, 54–55, 66–67, 77*f*, 78, 115–116, 130, 170–171, 267
Toledo, J. C. Jr., 45*t*, 51–52
Tomasello, F., 114–115, 116, 118*f*
Toneva, V., 205–207, 211, 214, 217
Tong, K. I., 69–70
Tonks, N. K., 50
Toppo, S., 130, 138
Torres, M. A., 276, 278, 279
Tosa, Y., 240–241
Townsend, D. M., 123
Troein, C., 115–116
Trujillo, M., 44–45, 45*t*, 46*t*, 50*t*, 51–52, 78–79, 114–115
Truman, W., 187
Truong, T. H., 50–51
Tsai, S. C., 175
Tschaplinski, T. J., 279
Tsien, R. Y., 190–191
Tsoras, M., 138
Tsuji, P. A., 67
Tsuji, Y., 53–54
Tsujimoto, M., 70–71
Tsujimoto, Y., 242
Tsukagoshi, H., 276
Tsutsumi, N., 240–241, 242, 244*f*, 245, 249–253
Turanov, A. A., 88
Turell, L., 50*t*
Turnbull, B. W., 88
Turner-Ivey, B., 54
Tuschl, T., 153–154
Twell, 282
Tyson, J. F., 89

U

Uchida, K., 69–70
Uchimiya, H., 240–241, 242, 243, 244*f*, 245, 246–247, 249–253
Uden, P. C., 89
Uesugi, K., 52–53, 55–56
Ueyama, T., 146–147
Uhm, T. B., 147
Underwood, G. J. C., 186–187
Unoki, H., 66–67
Urasaki, N., 240, 241, 247, 253
Uribe, M., 123–125
Ursini, F., 78–79, 130, 138
Urusova, D. V., 56–57
Ushio-Fukai, M., 187–188, 281–282

V

Vacca, R. A., 240
Vakorina, T. I., 56–57
Valenti, D., 240
Valvekens, D., 235–236
Van Breusegem, F., 204–207, 211
Van Camp, W., 211, 214
Van De Cotte, B., 205

Van Der Kelen, K., 205
van Deurs, B., 277–278
van Es, J. H., 72
van, L. K., 138
van Lijsebettens, M., 235–236
van Meir, E. G., 42
Van Montagu, M., 211, 214, 283–284
van Montagu, M., 235–236
van Ooijen, G., 23–24, 115–116
van Schothorst, E. M., 68–69, 71, 72, 75, 77*f*, 78–79, 89, 94, 102, 106–107, 108
Van Scott, M. R., 140–141
van Stroe-Blezen, S. A. M., 47–48
Vanacker, H., 186–187
Vandenabeele, S., 205, 217, 231–232
Vanderauwera, S., 205, 217, 231–232
Vandorpe, M., 186–187
Vandromme, M., 76–78
VanEeden, P. E., 90
Vanhaecke, F., 105
Vanin, S., 130, 138
Vann, J. A., 108
Varadarajan, S., 138
Varley, J., 116, 123
Vasquez, O. L., 137
Veal, E. A., 42, 55
Veenstra, T. D., 123
Veeravalli, K., 130
Velasco, D., 76
Venta-Perez, G., 57
Verhoeven, A. J., 146, 151–152
Verma, S., 171–172
Vermaas, W., 266–267
Verniquet, F., 248–249
Vianello, A., 278
Vilhardt, F., 277–278
Villanueva, N., 78
Vinh, J., 52, 54–55
Vinokurov, L. M., 189–190, 194
Virtanen, C., 44–45
Voinnet, O., 190–191
Volpp, B. D., 146
Vorwald, P. P., 73–74
Vöth, M., 240–241
Vuylsteke, M., 205, 231–232

W

Wabnitz, G. H., 55
Wakabayashi, N., 75–76
Wald, K. A., 70–71
Wallenberg, M., 104
Wands, J. R., 123–125
Wang, H. Y., 74
Wang, K. N., 108
Wang, L., 279
Wang, P. C., 187–188
Wang, P. S., 175
Wang, Q. L., 277–278
Wang, X., 120–122, 123, 242
Wang, X. C., 187–188
Ward, C. W., 123–125
Waring, J., 186–187
Wasowicz, W., 108
Watai, Y., 69–70
Watanabe, N., 241, 249–253
Waterbury, J. B., 264–265
Waterhouse, C. C., 146–147
Weaver, J. A., 71
Weber, K., 153–154
Wei, Q., 123
Weigel, D., 207
Weiner, L., 171–172
Weinman, S. A., 123–125
Weischenfeldt, J., 116–117
Weisend, C. M., 78
Weisenseel, M. H., 279
Welch, W. J., 44
Wells, D. M., 189–190
Wendel, A., 66
Weng, Z., 54
Werner, D., 277–278
Wessjohann, L. A., 67, 70–71, 78
West, I., 146–147
Whalen, A. M., 42
Wheeler, A. D., 106–107
White, D. A., 88
Whitehead, R. H., 90
Whiteman, M., 188–189
Whittier, R. F., 209–210
Wiederkehr, A., 130–131
Wientjes, F., 146–147
Wilczynski, S., 68–69
Wild, A. C., 66–67
Wilkins, K. A., 276
Willekens, H., 211, 214
Willits, M. G., 222
Wilson, G. M., 242
Wingler, K., 68, 72, 88–89, 91

Wingsle, G., 186–187, 222
Wink, D. A., 114
Winterbourn, C. C., 22–23, 45, 45*t*, 46*t*, 51–52, 57–58, 114–115
Wirth, E. K., 78–79
Wolf, N. M., 108
Wolff, S. P., 260, 262
Wong, C. M., 8
Woo, H. A., 22–23, 27, 46–47, 48, 52, 55–57, 114–115, 116, 118, 170–171, 172–174, 172*f*, 175, 177–179, 178*f*, 187–188, 267
Woo, J., 189–190
Woo, N. S., 186–187
Wood, Z. A., 22–23, 24–26, 27, 33–34, 46–47, 55–56, 114–115, 170–171, 259, 267
Woolf, N., 114
Wright, N. A., 146–147
Wrobel, I., 146–147
Wu, C., 116
Wu, G., 190–191
Wu, J. T., 132
Wu, L. H., 132
Wunderlich, M., 43

X

Xia, X. J., 108
Xia, Y., 106–107
Xiao, Z., 123
Xie, Z., 242
Xing, L., 76–78
Xiong, B., 74–75
Xiong, Y., 69–70, 76, 77*f*
Xu, J., 106–107
Xu, Q., 241, 249–253
Xu, R., 123, 208, 217
Xu, W. S., 57
Xu, X. M., 42, 68–69, 88

Y

Yabuta, Y., 223*t*, 232*t*
Yaffe, M. P., 242
Yalcin, A., 153–154
Yamamoto, M., 66–67, 69–71
Yamamoto, Y. Y., 223*t*, 226, 231–232
Yamanaka, A., 247–248
Yan, P., 76–78
Yan, S., 74–75
Yaneva, M., 57
Yang, C. S., 67
Yang, H., 67
Yang, J., 242
Yang, J. S., 55
Yang, K.-S., 22–23, 27, 54, 55, 118, 170–171, 267
Yang, S. H., 46–47, 74–75
Yang, W., 74–75
Yang, Y., 74–75, 141
Yano, A., 222
Yao, N., 240–241
Yim, M. B., 45*t*, 46*t*
Yim, S. H., 46–47, 48, 56–57, 187–188
Yokota, C., 74
Yokota, N., 277–278
Yoo, M. H., 67, 88
Yoo, S. K., 123–125
Yoshida, E., 223*t*, 232*t*
Yoshikawa, M., 190–191
Yoshimoto, K., 243
Yoshimura, K., 223*t*, 232*t*
Yoshinaga, K., 240–241, 242, 244*f*, 245, 253
Yoshioka, H., 279
Yoshizumi, T., 226
Young, M. R., 123
Youngblood, D. S., 33–34
Ytting, C. K., 189–190
Yu, D.-Y., 46–47, 48, 56–57, 187–188
Yu, S., 56–57
Yuan, Y., 116
Yun, D. J., 242, 253
Yun, K. Y., 208, 217

Z

Zagozdzon, A., 54, 55
Zaidi, S. M., 174
Zamboni, A., 223*t*
Zamocky, M., 258–259
Zamore, P. D., 153–154
Zangar, R. C., 171–172
Zarsky, V., 276
Zayed, A. M., 191–192
Zeng, H., 70–71
Zeng, X., 74
Zeng, X. Q., 217–218
Zha, H., 242
Zhang, B., 222
Zhang, C.-C., 115–116, 267

Zhang, D. D., 69–70
Zhang, G., 74–75
Zhang, H., 174
Zhang, J. L., 56–57, 108, 208, 217
Zhang, L. Y., 42, 45*t*, 54, 70–71, 277–278
Zhang, S. B., 186–187, 190
Zhang, W., 74–75
Zhang, X., 74
Zhang, Y., 74, 76, 77*f*
Zhang, Y. J., 108
Zhang, Y. Y., 141
Zhang, Z. Y., 45*t*, 46*t*
Zhao, D., 177
Zhao, H., 108
Zhao, S., 138
Zhao, Y., 23–24, 115–116
Zheng, M., 38
Zhou, D., 106–107
Zhou, H., 44–45
Zhou, J. C., 108
Zhou, Y., 189–190
Zhou, Z., 66–67, 70–71, 93
Zhu, F., 56–57
Zhu, Z., 104
Zimmermann, P., 205, 217
Zimmermann, T., 72–73
Zottini, M., 186–188, 190, 191, 281
Zykova, T. A., 56–57

SUBJECT INDEX

Note: Page numbers followed by "*f*" indicate figures, and "*t*" indicate tables.

A

Abiotic stress, *Arabidopsis*
  cold treatment, 228
  equipments, 226–227
  HL treatment, 227
  preculture, 227
  reagents, 226
  UV-B treatment, 228
Aconitase activity, 248–253, 251*f*
ACTH. *See* Adrenocorticotropic hormone (ACTH)
Adenomatous polyposis coli (APC), 72
Adrenocorticotropic hormone (ACTH)
  clonal cells, 174, 175*f*
  corticosterone and PrxIII–SO_2, 172–173
  CRH and AVP, 171–172
  expression, antioxidant proteins, 173–174
  MAPK, 172–173
  Nrf2 KO mice, 173–174
  p38 inhibitor SB202190, 177, 178*f*
  plasma corticosterone, 177–179
Aminotriazole (AT)
  anthocyanins, 216, 216*f*
  oxr1, evaluation, 211, 212*f*
  screening, EMS mutagenesis
    caution, 208–209
    germination, 208
    immersion, seeds, 207
    pots, 207
    sterilization, 208
Anthocyanins, 216–217, 216*f*
APC. *See* Adenomatous polyposis coli (APC)
Apoptosis signaling kinase 1 (ASK1)
  6-hydroxydopamine, 54
  oxidation, 54
  Trx activity, 54
Arabidopsis plants
  abiotic stress, 226–228
  H_2O_2 treatment
    equipments, 224, 225*f*
    method, hydroponic culture, 224–225, 225*f*
    reagents, 224
Arabidopsis thaliana, H_2O_2-induced oxidative stress
  anthocyanins, 216–217, 216*f*
  CAT activity (*see* Catalase (CAT) activity)
  chlorophyll, 212–213
  description, 205–206
  EMS and T-DNA mutagenesis (*see* Ethyl methanesulfonate (EMS) mutagenesis)
  evaluation, *Oxr1*
    fresh weight and chlorophyll, 211, 212*f*
    germination, 211
    herbicide, 211
    parameters, 211
  MDA determination, 213
  ROS (*see* Reactive oxygen species (ROS))
  TAIL-PCR reaction, 209–211
Arginine-vasopressin (AVP), 171–172
ASK1. *See* Apoptosis signaling kinase 1 (ASK1)
AT. *See* Aminotriazole (AT)
AVP. *See* Arginine-vasopressin (AVP)

B

Bax inhibitor-1 (BI-1)
  absorbance and inhibitory effects, 247–248
  aconitase activity, 248–253
  AtBI-1 overexpression, 249–253, 251*f*
  catabolic activity, 249–253
  cultured cells, treatment, 247
  description, 253
  formation, *cis*-aconitate, 248–249
  in vivo antioxidant activity, 248
  lethality, mouse Bax expression, 241
  menadione, 247–248
  rice *(Oryza sativa)*, 247
  sphingolipids, 249–253

Bax inhibitor-1 (BI-1) (*Continued*)
suspension cultures, tobacco, 246–247
tobacco BY-2 cells, 246, 249–253, 250*f*
transcription, 241
treatment, rice suspension cells, 249–253
BI-1. *See* Bax inhibitor-1 (BI-1)
Biomarkers, selenium status
GPx activity, 91–92
quantitative real-time PCR, 93–94
SDS-PAGE and Western blotting, 93
TrxR activity, 92–93

C

Casparian strip, 281–282
CAT activity. *See* Catalase (CAT) activity
Catalase (CAT) activity
anthocyanins, 217–218
atr1 and *atr7* mutants, 206–207, 217–218
measurements
in-gel assays, 214
native PAGE, 214–215
wild type and *oxr1* plants, 214, 214*f*
oxt1 mutation, 217
paraquat, 206
photometric determination, 215
Cell-free system, Nox2 activity
changes, velocity, 158*f*, 159
description, 157, 157*t*
kinetic parameters, 157–158, 159*t*
recombinant oxidase proteins, 157, 158*f*
Cellular thiols, H_2O_2 reaction
biological thiols, 44–45
vs. diffusion
cellular compartment, 48
peroxidases, 48
water, 47–48
erythrocytes, 46–47
floodgate hypothesis, 47, 47*f*
mass balance, 43–44
protein cysteines, 45, 45*t*
reaction kinetics and diffusion distances, 45–46, 46*t*
redox signaling pathways, 43
reduction potentials, 43
thermodynamics, 43
thioredoxin, 44
Chlorophyll, 212–213
CHP. *See* Cumene hydroperoxide (CHP)
cHyPer. *See* Cytosolic HyPer (cHyPer)
Circularly permutated YFP (cpYFP), 281
Confocal microscopy, 194–195
Corticotropin-releasing hormone (CRH), 171–172
cpYFP. *See* Circularly permutated YFP (cpYFP)
CRH. *See* Corticotropin-releasing hormone (CRH)
Cumene hydroperoxide (CHP)
vs. H_2O_2, 31–32
protein, 35
Tpx, 27–29
Cyanobacteria
description, 258–259
overoxidation, 267–271
peroxide decomposition (*see* Peroxide decomposition)
photosynthetic prokaryotes, 258–259
Prxs (*see* Peroxiredoxins (Prxs))
ROS (*see* Reactive oxygen species (ROS))
2-Cys-Prx overoxidation
cell culture, 116–117
electrophoresis analysis, 118–119
glucose oxidase, 117–118
immunoblotting, 117–118
Cytosolic HyPer (cHyPer)
cotyledon epidermal cells, 192–193, 193*f*
expression, *Arabidopsis* seedlings, 191–192, 192*f*
fluorescence, cotyledon, 196–198, 197*f*
pH measurements, 190

D

DAB. *See* 3,3'-Diaminobenzidine (DAB)
3,3'-Diaminobenzidine (DAB)
and NBT, 280
polymerization, 280
rapid *in situ* detection, H_2O_2 production, 280
Dihydrodichlorofluorescein diacetate (H2DCF-DA), 280–281
Dithiothreitol (DTT)
Cys-SOH, 148
ferric chloride, 150
H_2O_2 degradation, 156
and peroxide levels, 34–35

Prdx6 and 1-cys Prdxs, 155–156
and Tpx, 33
DTT. *See* Dithiothreitol (DTT)

E

eIF4a3. *See* Eukaryotic initiation factor 4a3 (eIF4a3)
Environmental stress, plants
Arabidopsis (*see Arabidopsis* plants)
description, 222
identification, marker genes, 222
signal transduction pathways, 222
total RNA extraction, 228–229
Ethyl methanesulfonate (EMS) mutagenesis
advantages, 206
CAT activity, 206–207
caution, 208–209
isolation, *oxr1* mutant, 207, 208*f*
8600 M3 T-DNA activation, 207, 208*f*
paraquat, 206
and AT screening, 207–209
Eukaryotic initiation factor 4a3 (eIF4a3), 107

F

Ferrous ion oxidation (FOX) assay, 260
Floral dip method, 282–283, 285*f*

G

Germination medium (GM), 226, 235–236
GFP-based sensors. *See* Green fluorescent protein (GFP)-based sensors
γGC. *See* γ-Glutamylcysteine (γGC)
Glutamine synthetase protection, Prdx6, 150, 155–156, 155*f*
γ-Glutamylcysteine (γGC)
acivicin, tissue lysates, 138
formation, ferric thiocyanate, 135–136, 135*f*
GPx1 activity, 138
in vitro quantification, 141
mitochondria, 138
oxidation, 132–133
peroxide levels, H2O2 concentration, 136–137, 136*f*
preparation, reactants, 134–135, 134*t*
scrape and resuspend cells, PBS, 138–139
silencing, GSH metabolism, 137
tissue levels, 130
Glutathione (GSH)
depletion, 138
GPx, 132–133
NADPH(H^+), 132
peroxide levels, 136–137, 136*f*
silencing, 137
tissue homogenates, 138–139
Glutathione peroxidase-1 (GPx1)
antioxidants, 136–137, 136*f*
bovine erythrocytes, 131–132, 133
mitochondria, γGC, 138
pH and temperature, 133
reactions, GSR, 134
Glutathione peroxidase (GPx)
catalase, 6–8
titration, 6–8
Glutathione-S-transferase (GST), 66
GM. *See* Germination medium (GM)
GPx. *See* Glutathione peroxidase (GPx)
Green fluorescent protein (GFP)-based sensors
cHyPer, 190
cytoplasm and peroxisomes, 190
Escherichia coli (E. coli), 189–190
fluorescence emission, 189–190
pH sensitivity, 190
redox-active cysteine residues, OxyR, 189–190
GSH. *See* Glutathione (GSH)
GST. *See* Glutathione-S-transferase (GST)

H

H2DCF-DA). *See* Dihydrodichlorofluorescein diacetate (H2DCF-DA
High light (HL)
Arabidopsis thaliana, 187
chlorophyll fluorescence imaging, 196, 197*f*
cotyledon epidermal cells, 197*f*, 198
experimental setup, 196
pilot experiment, 196–198, 197*f*
HL. *See* High light (HL)
H_2O_2. *See* Hydrogen peroxide (H_2O_2)
H_2O_2 detoxification, GSH and γGC
antioxidant activity, 137–139
buffer conditions, 132
description, 131

H_2O_2 detoxification, GSH and γGC (*Continued*)
 determination and samples collection, 139–140
 enzymes and substrates purification, procedure, 133–137
 GPx1 (*see* Glutathione peroxidase-1 (GPx1))
 in vitro quantification, 140–141
 materials, 131–132
 NOS, 140
 oxidative phosphorylation, 130
 substrates and enzymes, 132–133
 techniques, *in vitro* and *in vivo*, 130–131
Hydrogen peroxide (H_2O_2)
 Arabidopsis thaliana, 187
 assay, catalase activity, 15–16
 calibration curve, 10
 catalase, 12
 cellular capability, 4–5
 cellular thiols (*see* Cellular thiols, H_2O_2 reaction)
 chloroplasts and protoplast, 187–188
 consumption, intact and yeast cells, 11–12
 data handling/processing
  cell lines, 16–17
  yeast cells, 16
 determination, intact cells, 13–14
 gene expression, 52
 glutathione peroxidase activity, 14–15
 HyPer (*see* HyPer)
 kinetic reasons, 4, 5*f*
 in living plant cells
  biochemical assays and fluorescent markers, 280
  biotic and abiotic responses, 280
  DAB, 280
  H2DCF-DA, 280–281
  Hyper, 281–282
 localization, methods, 188–189
 mammalian cell lines, 6–8
 NADPH oxidase activity, 187
 oxygen electrode, 10–11, 11*f*
 photosynthesis, 186–187
 plant and algal cells, 186–187
 plasma membrane, 4–5, 6*t*
 preparation, postnuclear protein extract, 14
 redox signaling pathways, 42–43
 signal transduction, 187–188
 subcellular level, determination, 188
 yeast cells, 8
HyPer
 Casparian strip, 281–282
 cHyPer, 191–192
 confocal microscopy, 194–195
 cpYFP, 281
 epidermal cells, 193*f*, 194
 expression, silencing, 190–191
 GFP-based sensors, 189–190
 H_2O_2 levels, 281
 image processing and analysis, 194*f*, 195–196
 mounting live seedlings, 192–193
 OxyR domain, 281
Hyperoxidation, PrxIII
 adrenal gland, endocrine tissues, 171–172
 corticosterone, ACTH, 172–173, 172*f*
 CRH and AVP, 171–172
 CYP11A1 and CYP11B1, 171–172
 GPx1 and PrxV, 171
 MAPK, 172–173

I

Immunoblotting
 macrophages, 120–122, 121*f*
 proteins, 120–122
 Prxs, 120
In vitro activity assays, Prxs
 NTRC/2-Cys Prx assay, 262–264
 peroxide decomposition, 260–262

L

Leaf disc transformation method, 283–284
Living plant cells
 external and internal stimuli, 279
 growth conditions, 285–286
 guard cells, 276
 H_2O_2, 280–281
 monochromatic UV and LED illumination sources, 286–288
 NOXs (*see* Nicotinamide adenine dinucleotide phosphate (NADPH), oxidases (NOXs))
 Petri dishes, preparation and sterilization, 284, 287*f*

ratio imaging, *Arabidopsis* roots and tobacco pollen tubes, 286–288, 288*f*
ROS (*see* Reactive oxygen species (ROS))
scavenging system, 276–277
seeds sterilization and stratification, 284–285
vector description and plant transformation, 282–283

M

Malondialdehyde (MDA), 213
Mammalian cell lines, H_2O_2
GPx, 6–8
plasma membrane, 6–8
Marker gene
abiotic stress response, 232–234, 234*f*
description, 222, 223*t*
expression, H_2O_2 treatment, 231–232, 233*f*
selection, primers, 231, 232*t*
Mass spectrometry (MS) analysis
CHP, 35
E. coli Tpx, 35, 36*f*
hyperoxidation, 34–35
MALDI, 119–120
peptide preparation, 119
MDA. *See* Malondialdehyde (MDA)
MetaMorph/MetaFluor software, 286–288
Mitochondria, ROS stress
bacilliform shape, 243–245, 244*f*
Bcl-2 family, 240–241, 242
Cyclosporin A, 240–241
ectopic expression, mammalian Bax, 242
fluorescent proteins, 253
ion leakage measurement, 243, 244*f*
myosin ATPase, 246
paraquat and menadione, 242, 245
plant materials and treatment, 243
MS. *See* Mass spectrometry (MS) analysis
Multivesicular bodies (MVBs)
axin, 73–74
GSK3β complex, 76
MVBs. *See* Multivesicular bodies (MVBs)

N

NADPH. *See* Nicotinamide adenine dinucleotide phosphate (NADPH)
NADPH-dependent thioredoxin reductase C (NTRC)/2-Cys Prx assay
absorbance, 264
description, 262–264
electron transfer, 262–264
in vivo and *in vitro*, 262–264
purification, 262–264
rice and cyanobacterium *Anabaena* sp. PCC 7120, 262–264
NADPH oxidase (Nox2)
diphenylene iodonium, 240
and Prdx6
cell-free system (*see* Cell-free system, Nox2 activity)
LUMIstar Optima luminometer, 153
magnitude, luminescence, 153
neutrophils, KRPD, 152
phox proteins and Rac1, 152
shRNA suppression, PLB-985 cells, 160–163
siRNA suppression, transgenic K562 cells, 159–160
SOD-inhibitable cytochrome *c* reduction, 152
Nicotinamide adenine dinucleotide phosphate (NADPH)
micromolar, 29
oxidases (NOXs)
animal cells, 277–278
Arabidopsis, 278
capacity to transport electrons, 277–278
H_2O_2, 278
and RBOHs, 277–278
Nitric oxide (NO)
GOX and PMA, 117–118
and ROS, 114
Srx, 116
Nitric oxide synthase (NOS) activity, 140
NO. *See* Nitric oxide (NO)
Nonreducing SDS-PAGE and Western blot analysis
Anabaena sp. PCC 7120 2-Cys Prx, 269
isolation, 269–270
one-dimensional, 270, 271*f*
thiol-/sulfenic acid forms, 268–269
treatment, cyanobacterial cultures, 269
two-dimensional isoelectric, 270–271, 271*f*

NOS activity. *See* Nitric oxide synthase (NOS) activity
NOXs. *See* Nicotinamide adenine dinucleotide phosphate (NADPH)

O

Overoxidation, 2-Cys Prxs
catalytic cysteines, 267
in vitro immunological detection
antigenic peptide, 267
polyclonal antibodies, 267
rice, 268
Western blot, 267
nonreducing SDS-PAGE and Western blot analysis, 268–271
Oxidative burst, 279

P

PBS. *See* Phosphate buffered saline (PBS)
Peroxidase
activity, 261–262
cyanobacterium, 264
thiol-dependent, 258–259
Peroxide decomposition
cyanobacteria *in vivo*
chlorophyll measurement, 265
growth conditions, 264–265
quantification, 265–267, 266*f*
Prxs *in vitro* activity assays
dilute acid hydroperoxides oxidize, 260
DTT, 261–262
electron donors, 261–262
FOX assay, 260
measurement, 260–261
Trx concentration functions, 262, 263*f*
xylenol orange, 260
Peroxiredoxin 6 (Prdx6)
cell isolation and preparation, subcellular fractions, 151–152
Cys-SOH, 148
cysteine residues, 147–148
description, 147, 163–165
direct oxidation, H_2O_2, 150, 156, 156*f*
DTT, 147
glutamine synthetase protection, 150, 155–156, 155*f*
29-kDa protein, neutrophils, 148
neutrophils and monocytes, 146
Nox2 (*see* NADPH oxidase (Nox2), and Prdx6)
$p47^{phox}$, $p67^{phox}$, $gp91^{phox}$ and $p22^{phox}$, 146–147
production, oxygen metabolites, 146
protein purification, 149
translocation, phox-competent K562 cells, 148
Peroxiredoxins (Prxs)
activity, Prdx6
direct oxidation, H_2O_2, 150, 156, 156*f*
glutamine synthetase protection, 150, 155–156, 155*f*
catalytic cycle, 24–25
C. pasteurianum Cp20, 35–38, 37*f*
Cys–SOH, 170–171
dithiothreitol, 24–25
gel approach, 38
glutaredoxin, 38
H_2O_2, 42
hyperoxidation, PrxIII, 171–173
in vitro activity assays
NTRC/2-Cys Prx assay, 262–264
peroxide decomposition, 260–262
inactivation pathways and catalytic cycle, 22–23, 22*f*
intracellular signaling molecule, 42–43
isoforms, mammalian cells, 170–171
materials
proteins, 26
solutions, 26
NADPH, 27–29
plasma membrane, 42
protein interactions, 52–55
proteins, 27
PTMs, 55–57
"sensitive" and "robust", 23
sensors and transducers, H_2O_2, 52
sulfenic acids, 24–25, 49–51
and sulfiredoxin
2-Cys-Prx overoxidation, 116–122
Cys-Prxs, 115–116
electrophiles, 116
isoforms, 114–115
macrophages, 116
NO, 114
protein expression, 123, 124*f*
SDS-PAGE, 122–123

thiol, 114–115
thiol peroxidases, 51–52
and Trx
cloning and expression, 259
purification, 260
Petri dishes, 284, 287*f*
Phosphatase and tensin homolog (PTEN), 54
Phosphate buffered saline (PBS)
NADPH(H^+), 134
scrape and resuspend cells, 138–139
Posttranslational modifications (PTMs)
cysteine, 56
N-acetylation, 57
noncatalytic residues, 56–57
protein glutathionylation, 56
protein interactions and signaling, 57
Prxs, 55
S-nitrosylation, 56
sulfiredoxin, 56
Prdx6. *See* Peroxiredoxin 6 (Prdx6)
Protein interactions, Prxs
ASK1, 53–54
kinases/phosphatases, 52
MST1, 52–53
oxidation signal, 55
PTEN, 54
PTPs, 54–55
redox signaling, 52, 53*f*
Protein purification, Prdx6, 149
Prxs. *See* Peroxiredoxins (Prxs)
PTEN. *See* Phosphatase and tensin homolog (PTEN)
PTMs. *See* Posttranslational modifications (PTMs)

Q

Quantitative real-time RT-PCR
equipments, 230
methods, 230
reagents, 230

R

Reactive oxygen species (ROS)
accumulation, 276
aerobic metabolism, 258–259
BI-1 (*see* Bax inhibitor-1 (BI-1))
CAT activity, 205
cell wall modifications, 281–282
cytoplasm/organelle, 280
deleterious effect, plant cells, 276
developmental processes, 204–205
ion channels, 276–277
irreversibly fluorescent, 280–281
mitochondria, 240–241, 242–246
NADPH oxidase, 240
oscillation in extracellular ROS levels, 281
oxidative stress, 204–205
polar growth, 276
production and removal, 276–277
signaling, biological effects, 205
superoxide anion and hydrogen peroxide, 277–278
Redox signaling pathways
H_2O_2, 42–43
mechanism, 46–47
Prx–protein interactions, 55
thiols, 44–45
Root hair cells, 279
ROS. *See* Reactive oxygen species (ROS)

S

SAA. *See* Systemic acquired acclimation (SAA)
SAR. *See* Systemic acquired resistance (SAR)
Selenium
cancer development, 67
compounds
description, 104
GPx activity, 104–105
metabolism, 104–105
mRNA stabilization, 106–108
selenoprotein level and activity, 105–106
SeMeSeCys, 105
Western blotting, 106
GPx, 66
GPx2 expression, 70–71
GST, 66
hydroperoxides, 78–79
Keap1/Nrf2 system, 69–72
microarray analyses, 71
Nrf2 pathway, 66–67
oxidant, 79–80
phosphorylation, Fyn, 79

Selenium (*Continued*)
selenoproteins, 72
and selenoprotein synthesis, 68–69
TrxR/Trx, 78
Wnt pathway (*see* Wnt pathway)
Selenomethionine, 89
Selenoprotein
biomarkers
GPx activity, 96
TrxR activity, 96–99
compounds, 89–90
description, 88
glutathione peroxidase, 88–89
GPx activity, 91–92
HepG2, 90
metabolism, 104–105
mRNA levels, 99–102
protein levels, 99
quantitative real-time PCR, 93–94
sample preparation, 91
SDS-PAGE and western blotting, 93
selenomethionine, 89
SeMeSeCys, 105
SeMet, 106
structures, 89–90
toxicity, 94–95
treatment, 91
TrxR activity, 92–93
YAMC, 89–90
shRNA suppression of Prdx6, PLB-985 cells
DMSO-mediated terminal differentiation, 162
knockdown, myeloid cells, 160–162, 163*f*
penicillin and streptomycin, 154
phox proteins, 162
plasmid TI340760, 154
PMA and fMLP, 162–163
WT and KD cells, 162–163, 164*f*
siRNA suppression of Prdx6, transgenic K562 cells
description, 153–154
growth and resuspension, 154
nonsilencing siRNA (Nsi), 159–160
phox proteins and actin, 160, 161*f*
production, 159–160
Srx. *See* Sulfiredoxin (Srx)
Srx–PrxIII regulatory system
clonal cells, 174, 175*f*
in vitro model, adrenal glands
description, 176
incubation, 176–177
p38 inhibitor SB202190, 177, 178*f*
plasma corticosterone levels, 177–179
rabbit antibodies, 176–177
regulation, StAR function, 177
primary adrenocortical cells, 175, 176*f*
Sulfenic acids
description, 49
rate constants, 50, 50*t*
redox signaling, 51
thiols, 51
Sulfiredoxin (Srx)
ACTH effect, 172*f*, 173–174
adrenal cortex, 179
mitochondria, 174
and PrxIII (*see* Srx–PrxIII regulatory system)
Systemic acquired acclimation (SAA), 279
Systemic acquired resistance (SAR), 279

T

TAIL-PCR reaction, 209–211
Thiol peroxidase (Tpx)
CHP, 27–29, 28*f*, 31, 31*f*
decay rate, 29–31
human Prx I, 31
NADPH, 27–29
protein, 27
Prx reaction rate, 29
S. typhimurium AhpC, 33–34
Trx1 *vs.* Trx2, 33
Thioredoxin (Trx)
concentrations, 261–262, 263*f*
and Prxs
cloning and expression, 259
purification, 260
Synechocystis, 258–259
Thioredoxin reductase (TrxR), 89–90
Tobacco pollen tubes, 286–288, 288*f*
Total RNA extraction
equipments, 228
methods, 229
reagents, 228
Tpx. *See* Thiol peroxidase (Tpx)
Trx. *See* Thioredoxin (Trx)
TrxR. *See* Thioredoxin reductase (TrxR)

V

Vector description and plant transformation
CaMV 35s:Hyper construct, 282, 282*f*
floral dip method, 282–283, 285*f*
leaf disc transformation method, 283–284
YFP positive seedlings, 283

W

Western blot analysis
protocol, 267
and SDS-PAGE (*see* Nonreducing SDS-PAGE and Western blot analysis)
Wnt pathway
APC, 72
axin, 72–73
GPx2, 72
GSK3β, 73–74
MVBs, 74
and Nrf2
GSK3β, 76
protein kinases, 75–76
putative interplay, 76, 77*f*
redox-sensitive phosphatases, 78
proteins, 72
redox regulation, 74

Y

YAMC. *See* Young adult mouse colon (YAMC)
Young adult mouse colon (YAMC), 89–90

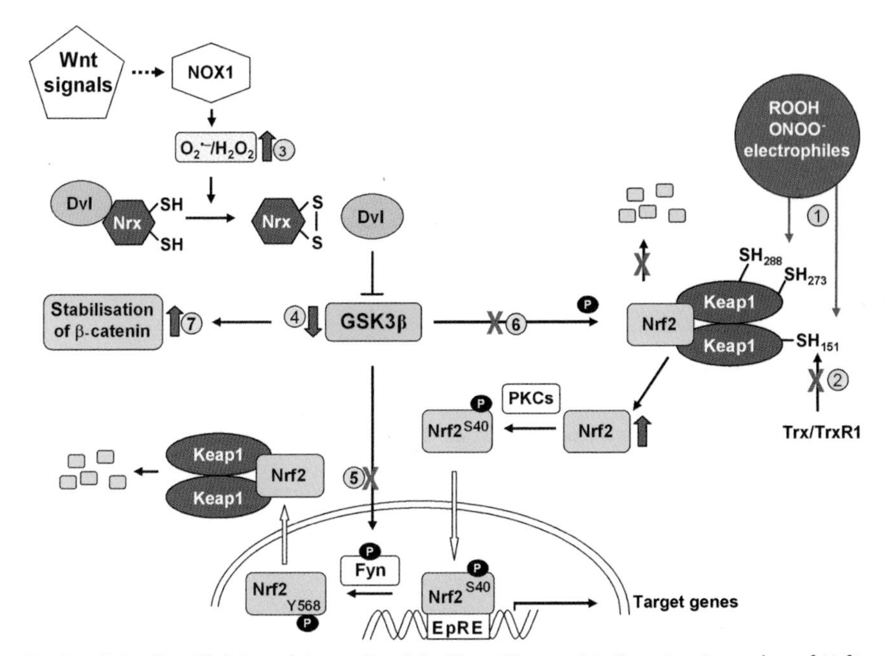

Regina Brigelius-Flohé and Anna Patricia Kipp, Figure 4.3 Putative interplay of Nrf2 and Wnt signaling and role of a low-selenium state. Processes influenced by selenium deficiency are indicated by 1–7 and explained in detail in the text. Blocks in processes or up/downregulation by a low-selenium state are indicated by brown crosses or arrows, respectively. Abbreviations (not explained in Figs. 4.1 and 4.2): TrxR1, thioredoxin reductase-1; NOX1, NADPH oxidase-1; Fyn, member of the nonreceptor protein-tyrosine kinase subfamily Src-A. (1) Lack of hydroperoxide removal by, for example, GPxs leads to an enhanced stabilization of Nrf2 in the cytosol and nuclear translocation. (2) Inhibition of rereduction of Keap1 disulfides by a low activity of the Trx/TrxR1 system (Fourquet, Guerois, Biard, & Toledano, 2010) or by peroxiredoxins and their subsequent reduction by Trx/TrxRs, prevents Keap1 restoration. Newly synthesized Nrf2 can enter the nucleus. (3) Enhanced production of H_2O_2 via NOX1 upon stimulation with Wnt signals (Kajla et al., 2012) by lack of GPxs or TrxR1-mediated reduction of Prxs leads to persistent oxidation of Nrx and release of Dvl (Funato et al., 2006). (4) Inhibition of GSK3β activity by deliberation of Dvl (Funato et al., 2006) or decreased expression (Kipp et al., 2009) prevents steps 5–7. (5) Suppressed GSK3β activity prevents phosphorylation of Fyn and thus the export of Nrf2 out of the nucleus (Jain & Jaiswal, 2007). (6) Inhibition of phosphorylation of Nrf2 by GSK3β prevents Keap1-independent Nrf2 degradation (Chowdhry et al., 2012; Rada et al., 2011, 2012). (7) Putative stabilization of β-catenin due to decreased action of GSK3β (needs to be validated).

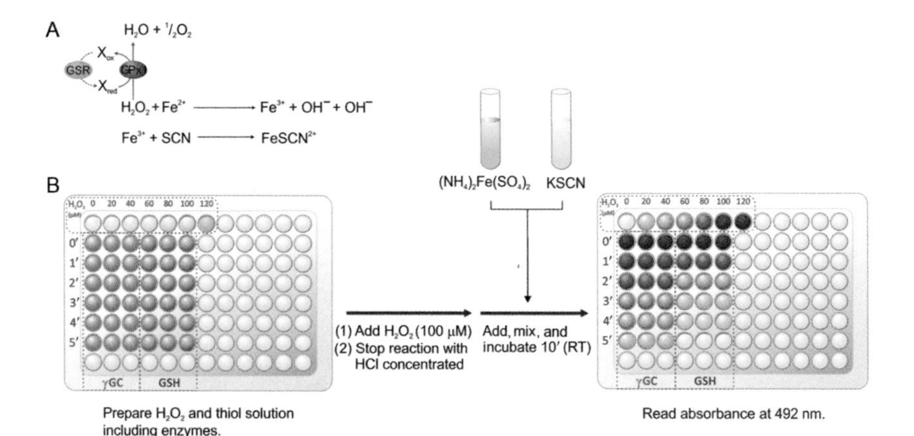

Ruben Quintana-Cabrera and Juan P. Bolaños, Figure 7.1 (A) H_2O_2 quantification is based on the formation of ferric thiocyanate ($FeSCN^{2+}$) in the presence of KSCN after Fe^{2+} oxidation by the peroxide. Gray lines represent the reduction of H_2O_2 by GPx1 at the expense of the thiolic compounds; X, GSH, or γGC. GSR only regenerates GSH (dotted line). (B) Schematic schedule of the assay to analyze the reduction of 100 μ*M* H_2O_2 by a fixed concentration of γGC or GSH. Triplicates for each time point ranging from 1 to 5 min are represented.

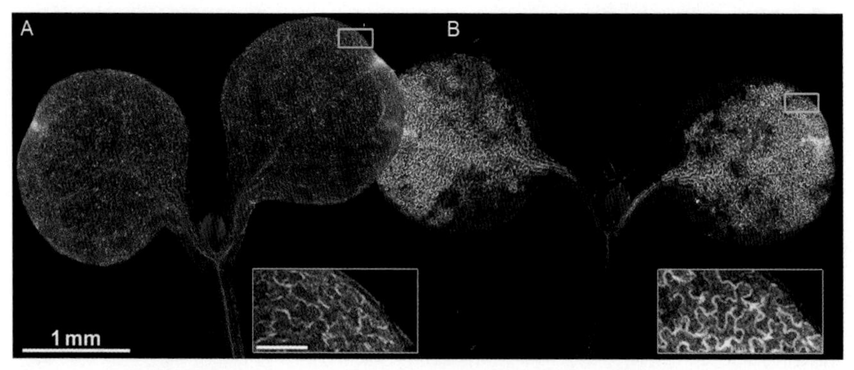

Marino Exposito-Rodriguez *et al.*, Figure 10.1 Expression of cytosolic HyPer (cHyPer) in 8-day-old *Arabidopsis* seedlings. (A) cHyPer in wild type. (B) cHyPer in *sgs3–11* background. The seedlings were placed next to each other on the same slide, ensuring image acquisition conditions were identical. Inset cyan rectangles show epidermal cells at higher magnification (scale bar: 100 μm).

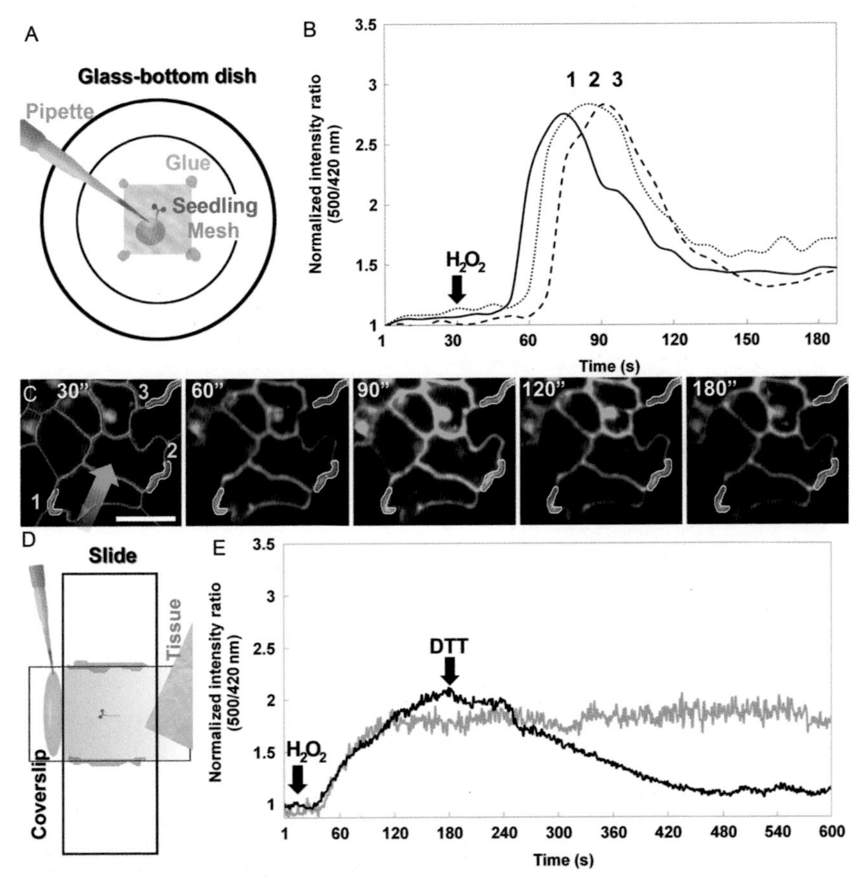

Marino Exposito-Rodriguez *et al.*, Figure 10.2 Cytosolic HyPer (cHyPer) response to added H_2O_2 in cotyledon epidermal cells. (A) Experimental setup using 35 mm glass-bottom petri dishes. (B) Time course of HyPer response to H_2O_2 addition (arrow). 1, 2, and 3 indicate the maxima in regions of interest, indicated in (C). (C) Image sequence of cotyledon epidermal cells expressing cHyPer. Only the green channel (excitation at 500 nm) is shown. Cells were exposed to 1 m*M* H_2O_2. A blue arrow shows the general direction of H_2O_2 flux. Time is indicated in the bottom right. Numbers 1, 2, 3 indicate the positions of the analyzed regions of interest. Cell borders are overlaid in red. Scale bar: 10 μm. (D) Slide-based experimental setup using custom-built perfusion chamber. (E) Time course of redox changes in cHyPer after the addition of H_2O_2 (1 m*M*) and the reducing agent dithiothreitol (DTT; 1 m*M*) using the experimental setup indicated in (D). All ratio values were normalized to a minimum value of 1.0.

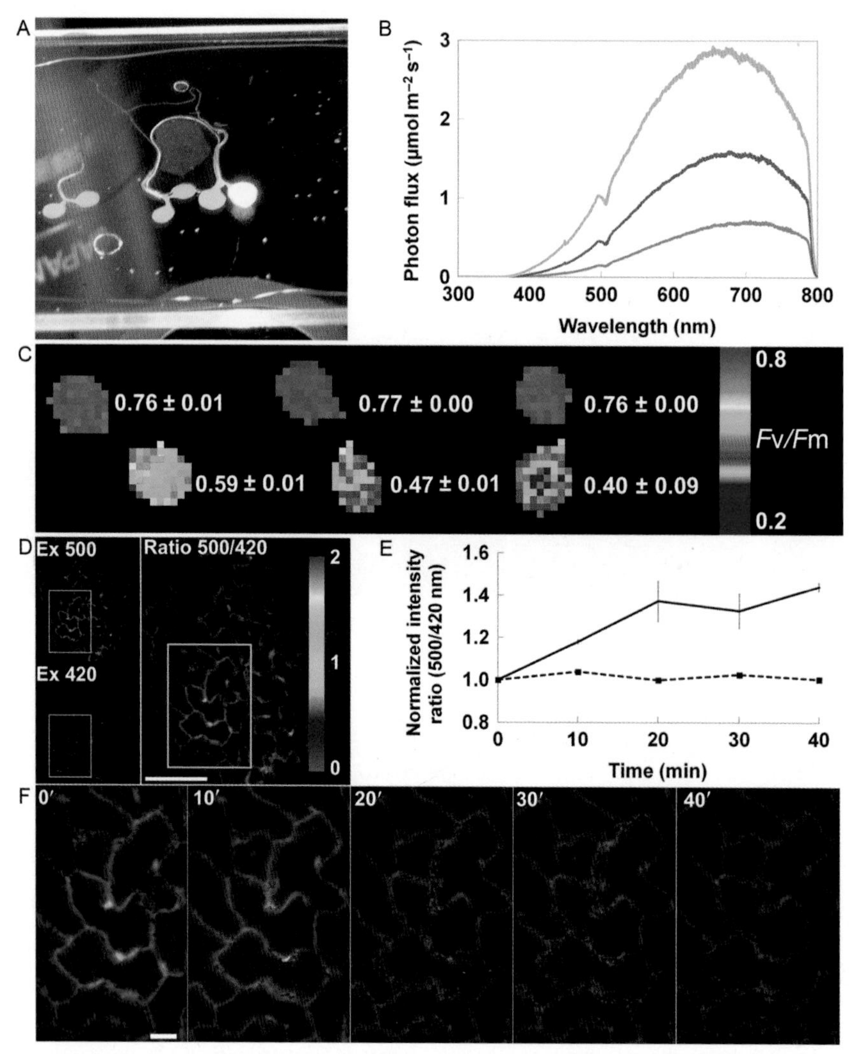

Marino Exposito-Rodriguez *et al.*, Figure 10.4 Pilot experiment inducing high light stress on the confocal microscope. (A) A precise, small area of light was applied to only one cotyledon. (B) Spectral measurement of the light used in this study. (C) Fluorescence images of cotyledons showing F_v/F_m values after 20 min exposures at 1200, 1600, and 2000 μmol m^{-2} s^{-1} (left to right, respectively). (D) Combining the two channels to produce a ratiometric image. The inset cyan rectangles demarcate the enlarged area shown in (F). (E) cHyPer response of cotyledon after high light treatments in 10 min intervals. Time point zero is a low light (LL) value taken immediately prior to exposure of the cotyledon to HL. Solid line and dashed line are values from cotyledons exposed to HL or LL parallel control, respectively. Values are the means (±SD; $n=3$). All ratios were normalized to a minimum value of 1.0. (F) False-color ratio images corresponding to (E).

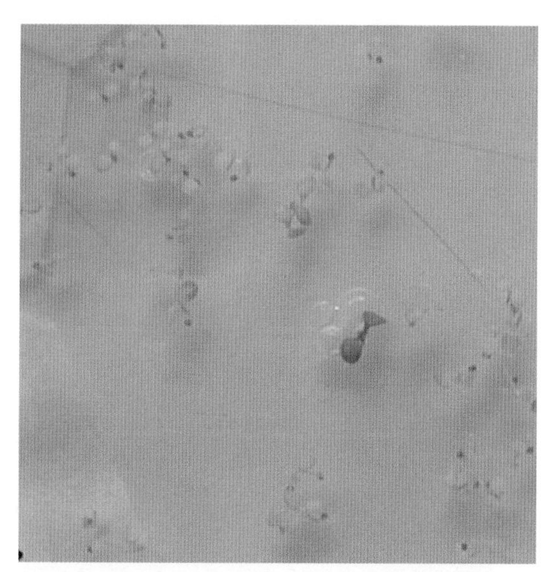

Tsanko Gechev *et al.*, Figure 11.1 Isolation of the *oxr1* mutant with enhanced tolerance toward oxidative stress. Seeds from 8600 M3 T-DNA activation tagged mutant lines were plated on MS plant growth media supplemented with 9 μM AT. Plants were grown in a climate room under 16-h light/8-h dark photoperiod and light intensity 60 $\mu mol\ m^{-2}\ s^{-1}$, 22 °C. Results were scored 10 days after germination.

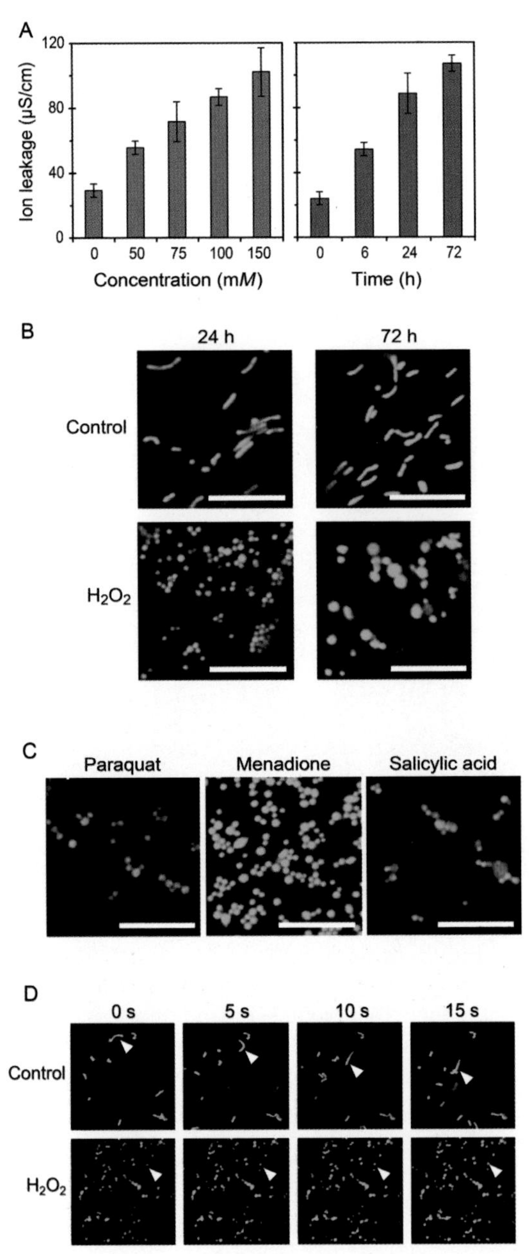

Toshiki Ishikawa *et al.*, Figure 13.1 H_2O_2-induced cell death and morphological change of mitochondria in Arabidopsis leaves. (A) Leaf discs prepared from 3-week-old Arabidopsis plants were floated on distilled water and treated with H_2O_2 for 3 days at 23 °C under continuous light. Ion leakage is determined using various concentrations of H_2O_2 (0–150 m*M*) and treatment times (0–72 h at 100 m*M* H_2O_2). Data are means $\pm$ SD ($n = 3$). (B) Morphology of mitochondria in leaves incubated with 100 m*M* H_2O_2 for 24 and 72 h. (C) Morphology of

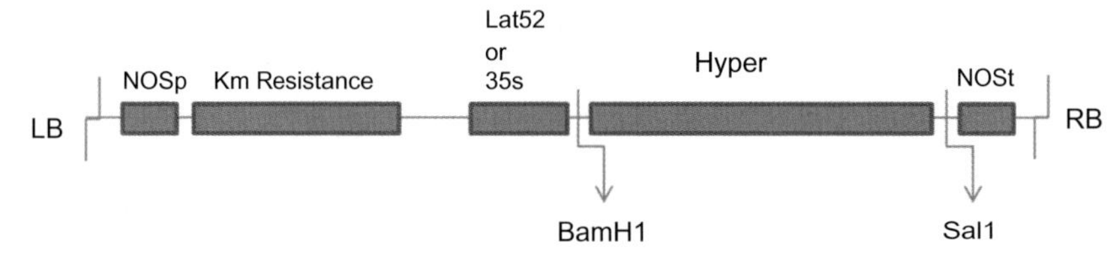

Alejandra Hernández-Barrera *et al.*, Figure 15.1 A restriction map for the CaMV 35s: Hyper construct.

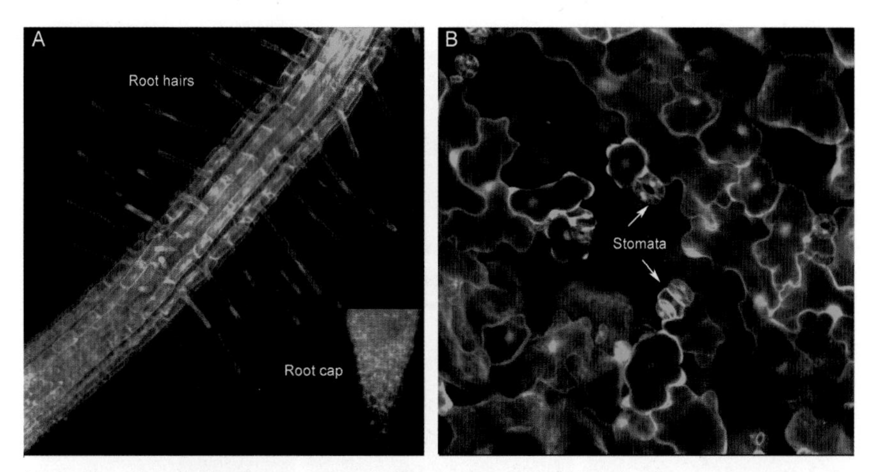

Alejandra Hernández-Barrera *et al.*, Figure 15.3 *Arabidopsis* plants expressing the Hyper molecular probe in transgenic stable lines. (A) Images from *Arabidopsis* roots under the confocal microscope, note that epidermal cell and root hairs present a good expression of hyper. (B) A region of the leave showing the pavement cells and stomata showing Hyper expression. Inset, shows the cap of the root tip with hyper expression of the same root.

---

mitochondria incubated with 0.3 μ*M* paraquat (24 h), 60 μ*M* menadione (72 h), or 300 μ*M* salicylic acid (24 h). Epidermal cells of leaf discs obtained from transgenic mtGFP Arabidopsis were examined using a confocal laser scanning microscope at 488 nm excitation wavelength to detect mtGFP. Controls were treated with water (0 m*M* H_2O_2). (D) Arrest of mitochondrial movement by exogenous treatment with H_2O_2. Movement of mitochondria in epidermal cells of Arabidopsis leaves expressing mtGFP was observed after treatment with 100 m*M* H_2O_2 for 72 h. Images were taken at 5-s intervals. Arrowheads indicate traces of one mitochondrial particle, actively moving in control cells, but remaining motionless in H_2O_2-treated cells. Scale bars = 10 μm. *Modified from Yoshinaga, Arimura, Niwa, et al. (2005).*

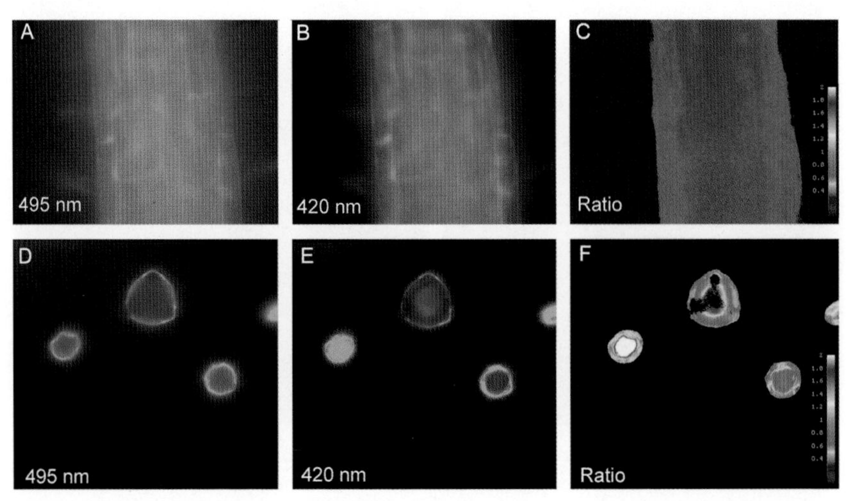

Alejandra Hernández-Barrera *et al.*, Figure 15.4 Ratio imaging of *Arabidopsis* roots and tobacco pollen tubes expressing Hyper. Panels A, B, and C represent the same region excited at 495 and 420 nm and the corresponding ratio image, respectively. Emission was collected at 530 nm with a band pass of 30 nm. Panels D, E, and F represent another ratio image experiment using the transgenic tobacco pollen tube. Note that the red color indicates a higher ROS concentration and blue a low level.